Food Science and Engineering for Sustainability

Food Science and Engineering for Sustainability

Guest Editors

Long Yu
Ping Dong

MDPI

Basel • Beijing • Wuhan • Barcelona • Belgrade • Novi Sad • Cluj • Manchester

Guest Editors

Long Yu
College of Medicine and Public Health
Flinders University
Adelaide
Australia

Ping Dong
Haide College
Ocean University of China
Qingdao
China

Editorial Office
MDPI AG
Grosspeteranlage 5
4052 Basel, Switzerland

This is a reprint of the Special Issue, published open access by the journal *Sustainability* (ISSN 2071-1050), freely accessible at: https://www.mdpi.com/journal/sustainability/special_issues/85TBC45096.

For citation purposes, cite each article independently as indicated on the article page online and as indicated below:

Lastname, A.A.; Lastname, B.B. Article Title. *Journal Name* **Year**, *Volume Number*, Page Range.

ISBN 978-3-7258-5387-8 (Hbk)
ISBN 978-3-7258-5388-5 (PDF)
https://doi.org/10.3390/books978-3-7258-5388-5

Contents

About the Editors

Long Yu

Long Yu is a research fellow at Flinders University, Australia. He obtained his PhD from the University of Queensland and completed postdoctoral research at the University of Adelaide. His research focuses on carbohydrate analysis, the mechanical characterization of hydrocolloids and gels, and the discovery of novel bioactive molecules from diverse biomass using advanced mass spectrometry techniques. His work lies at the interface of carbohydrate polymers, bioactive compounds, green extraction technologies, and hydrocolloid science, aiming to develop sustainable and innovative approaches for food, health, and material applications. Since 2013, he has published 33 high-quality refereed papers that have attracted 1,376 citations, with approximately 72% of these citations recorded since 2020, reflecting the growing impact of his research. Long Yu has received numerous awards and recognitions, highlighting his international and national achievements. These include the 2019 Plant Biology travel grant from the American Society of Plant Biologists, where only 80 awardees were selected worldwide, and his selection as the sole Australian Early Career Researcher representative to the Early Career Colloid Conference (UK) in 2018, supported by the Australasian Colloids and Interface Society.

Ping Dong

Ping Dong is a professor and Deputy Dean of Haide college at Ocean University of China and completed postdoctoral research at the University of Massachusetts, USA. Her research focuses on the nutrition and sustainable utilization of marine biological resources. She has led or participated as a key member in over ten research projects, including a sub-project of the National Key R&D Program, two National Natural Science Foundation projects, the Shandong Key R&D Program, and the Shandong Natural Science Foundation. She has published more than 60 SCI papers and holds over ten patents. As an initiator, she secured the first FDA-authorized BPCS training qualification in the Asia-Pacific region. She was selected as a member of the inaugural Youth Committee of the Agricultural Branch of the Chinese Society of Biochemistry and Molecular Biology. Her work has been recognized with multiple awards, including the First Prize of the Shen Nong China Agricultural Science and Technology Award, the Second Prize of Marine Engineering Science and Technology, the First Prize of the Dongsheng Course Teaching Excellence Award, and the May Fourth Youth Award of Ocean University of China. Through her research and leadership, Ping Dong has made significant contributions to advancing sustainable marine biotechnology, promoting the development and application of marine bioresources, and fostering innovation in both science and education.

sustainability

MDPI

Article

Beyond the Hype: Stakeholder Perceptions of Nanotechnology and Genetic Engineering for Sustainable Food Production

Madison D. Horgan [1,*], Christopher L. Cummings [2], Jennifer Kuzma [2,3], Michael Dahlstrom [4], Ilaria Cimadori [5], Maude Cuchiara [6,7], Colin Larter [8], Nick Loschin [1,2,6] and Khara D. Grieger [1,2,6]

1 Department of Applied Ecology, North Carolina State University, Raleigh, NC 27695, USA
2 Genetic Engineering and Society Center, North Carolina State University, Raleigh, NC 27695, USA
3 School of Public and International Affairs, North Carolina State University, Raleigh, NC 27695, USA
4 Greenlee School of Journalism and Communication, Iowa State University of Science and Technology, Ames, IA 50011, USA
5 Yale School of the Environment, Yale University, New Haven, CT 06511, USA
6 North Carolina Plant Science Initiative, North Carolina State University, Raleigh, NC 27606, USA
7 Department of Materials Science and Engineering, North Carolina State University, Raleigh, NC 27695, USA
8 Department of Sociology and Criminal Justice, Iowa State University of Science and Technology, Ames, IA 50011, USA
* Correspondence: mdhorgan@ncsu.edu

Abstract: Ensuring sustainable food systems is an urgent global priority as populations grow and environmental pressures mount. Technological innovations such as genetic engineering (GE) and nanotechnology (nano) have been promoted as promising pathways for achieving greater sustainability in agriculture and food production. Yet, the sustainability of these technologies is not defined by technical performance alone; it hinges on how they are perceived by key stakeholders and how well they align with broader societal values. This study addresses the critical question of how expert stakeholders evaluate the sustainability of GE and nano-based food and agriculture (agrifood) products. Using a multi-method online platform, we engaged 42 experts across academia, government, industry, and NGOs in the United States to assess six real-world case studies—three using GE and three using nano—across ten different dimensions of sustainability. We show that nano-based products were consistently rated more favorably than their GE counterparts in terms of environmental, economic, and social sustainability, as well as across ethical and societal dimensions. Like prior studies, our results reveal that stakeholders see meaningful distinctions between nanotechnology and biotechnology, likely due to underlying value-based concerns about animal welfare, perceived naturalness, or corporate control of agrifood systems. The fruit coating and flu vaccine—both nano-enabled—received the most positive ratings, while GE mustard greens and salmon were the most polarizing. These results underscore the importance of incorporating stakeholder perspectives in technology assessment and innovation governance. These results also suggest that responsible innovation efforts in agrifood systems should prioritize communication, addressing meaningful societal needs, and the contextual understanding of societal values to build trust and legitimacy.

Keywords: agrifood technologies; genetic engineering; nanotechnology; perceptions of sustainability; stakeholder engagement; responsible innovation; technology governance

1. Introduction

Food and agriculture (agrifood) systems are under growing pressures to deliver safe, plentiful, and nutritious food to support an increased global population, all while reducing environmental impacts and enhancing sustainability. Studies have estimated that the world's population will reach 10 billion people by 2050, and that we have already surpassed 6 out of 9 planetary boundaries in terms of safe operating zones for humanity [1]. Agricultural systems are facing mounting pressures due to the accelerating impacts of climate change, including more frequent and severe extreme weather events. Concurrently, consumer preferences are evolving, with growing demand for more ethical, sustainable, and animal-friendly livestock production practices. These converging trends underscore the urgent need to transform agrifood systems to enhance productivity while strengthening ecological resilience and advancing social responsibility. Without systemic innovation and the adoption of more sustainable practices, current agricultural models may become increasingly unsustainable, jeopardizing both global food security and environmental integrity.

These intersecting challenges indicate the urgent need to determine what types of innovations in sustainable food systems are most likely to address emerging global demands. Further, the adoption and assessment of such innovations are shaped by national contexts as cultural values, regulatory frameworks, and agrifood policy priorities vary across countries. Accordingly, this study centers on U.S.-based stakeholders to help inform how U.S. agencies and institutions might evaluate and govern emerging agrifood technologies in alignment with national interests. For example, in the U.S., federal agencies including the Food and Drug Administration (FDA), the United States Department of Agriculture's (USDA's) National Institute of Food and Agriculture (NIFA), and others proposed varying definitions and frameworks for sustainability in agrifood systems, often incorporating dimensions of environmental health, economic viability, and social equity to guide both public research funding and regulatory decisions while reflecting the complexity and context-dependence of sustainability as a concept [2]. In response, many innovations have been explored in recent years, including advances in alternative protein development, precision irrigation systems, climate-resilient crop breeding, and improved nutrient delivery mechanisms such as phosphorus efficiency technologies [3–9]. Amid this landscape, nanotechnology and genetic engineering have emerged as two particularly prominent and controversial approaches, each offering potentially transformative solutions for sustainable food production. Nanotechnology (nano) enables precision interventions at the molecular scale, while genetic engineering (GE) facilitates targeted modifications to plant and animal genomes. Both fields have been promoted by researchers and developers as means to improve agrochemical delivery, reduce waste, enhance food preservation, increase yields, foster climate resilience, and provide nutritional enhancements [10–14].

Despite their intended benefits, agrifood products using nano and GE have generated discussions and debates regarding their environmental, ethical, and societal implications as their value ultimately depends on how they are developed, deployed, and perceived by those who influence, regulate, and interact with them. Thus, it is critical to assess whether such products deliver on sustainability claims and how they are perceived by key stakeholders who influence the development, governance, and use of agrifood innovations. While numerous methodologies exist to evaluate technological innovation (e.g., life-cycle assessments, risk analyses, market forecasting), this study employs a systematic stakeholder engagement approach that relies on assessment of experts embedded in relevant sectors on developed nano- and GE-enabled agrifood products. These individuals shape scientific discourse, regulatory pathways, and public narratives, making their input vital for assessing the sustainability potential of GE and nano in agrifood systems. A robust evaluation framework to garner insights from stakeholders that accounts for multiple dimensions of sustainability can identify new opportunities for responsible innovation, improve legitimacy and trust in technological development, and contribute to broader scholarly and policy conversations around the governance of emerging agrifood technologies [15–19].

To navigate the complex tradeoffs and contested perceptions surrounding uses of GE and nano in agrifoods, our research is guided by the following overarching research question: *How do expert stakeholders perceive the sustainability of nano and GE when applied to agrifood systems?* To answer this question, we systematically evaluate and compare how expert stakeholders in the U.S. perceive the sustainability of real-world applications of GE and nano in agrifood products. We begin with a review of existing literature on how sustainability is defined in the context of agrifoods, synthesizing insights from diverse sources to identify ten distinct dimensions of sustainability: environmentally sustainable, economically sustainable, socially sustainable, responsible, useful, superior to alternatives, ethical, fosters a fair and just society, contributes to a collective good, and equitable. Next, we present a brief overview of the development and application of GE and nano in the agrifood sector. This review informs the selection of six real-world case studies: three involving GE and three involving nano. Using a multi-activity online engagement platform, we elicited evaluations of the sustainability of the six case studies from 42 U.S.-based stakeholders representing academia, industry, government, and civil society, which we then discuss within broader debates about responsible innovation, the governance of emerging technologies, and the future of sustainable food systems. By systematically examining stakeholder perspectives across a standardized framework of ten dimensions of sustainability, this study achieves two primary outcomes, (1) it advances understanding of how real-world applications of GE and nano are positioned within ongoing efforts to define and achieve sustainability in agrifoods; and (2) it provides novel theoretical and methodological contributions for sustainability assessment by offering a multidimensional framework that captures the complex and interrelated factors across diverse sustainability dimensions.

2. Literature Review

Understanding stakeholder perceptions of sustainability in emerging agrifood technologies requires grounding in two key areas of scholarship. First, we review how sustainability has been defined and operationalized in the context of agrifoods, drawing from this literature to develop a framework of ten distinct but interconnected dimensions of sustainability. This framework serves as the analytical foundation to guide expert stakeholders in evaluating novel technologies. Second, we examine the development and application of GE and nano within the agrifood sector. This includes a review of how these technologies have been used to date, as well as the extent to which their sustainability implications have been assessed.

2.1. Defining Sustainability in Agrifood Systems

Publications on sustainability have increased significantly over the past two decades, often structured around three foundational pillars: environmental, economic, and social sustainability [20]. These pillars are embedded in the United Nations' (UN) Sustainable Development Goals (SDGs) and the broader concept of "sustainable development", which spans domains from urban infrastructure to agrifood systems. In the U.S. agricultural context, these ideas are reflected in the USDA's 2011 consensus statement, which outlines sustainability as a balance of four goals: "satisfying human needs; enhancing environmental quality, the resource base, and ecosystem services; sustaining the economic viability of agriculture; and enhancing the quality of life for farmers, ranchers, forest managers, workers, and society as a whole" [2].

To define sustainability in our survey, we adopted definitions for each of the USDA's pillars of sustainable agriculture that align with established literature and policy frameworks. We defined environmental sustainability as the ability to preserve and protect the natural environment over time through appropriate practices, meeting present needs without compromising the availability of resources in the future. This reflects the Brundtland Report and longstanding U.S. environmental policy, such as the National Environmental Policy Act of 1969, which emphasizes achieving "productive harmony" between humans and nature [21,22]. We defined economic sustainability as the ability to preserve and promote long-term economic well-being, consistent with SDG 8, which calls for inclusive and sustainable economic growth and decent work for all [23]. Finally, we defined social sustainability as the ability to preserve and protect the well-being of people and communities, encompassing principles like equity, human rights, ethical labor, and community development. This aligns with the UN's framing of social sustainability as managing business impacts on people and reflects broader commitments to ethical, legal, and social implications (ELSIs) in responsible innovation.

While the UN's three pillars of sustainability—environmental, economic, and social—provide a foundational framework for understanding sustainable development, they are not always sufficient for evaluating the complexity of agrifood systems, particularly in the context of emerging technologies. The introduction of novel technologies into agrifoods is rarely judged on efficacy alone, as stakeholders bring to the table diverse concerns, values, and priorities that shape their views of what counts as "sustainable". As such, it is not sufficient to evaluate these technologies solely through environmental or economic lenses; rather, it is essential to incorporate a more granular and multidimensional view of sustainability that includes ethical, social, and justice-related considerations. A growing body of literature has argued for expanding the traditional sustainability framework to incorporate additional dimensions that better reflect the unique challenges and values across different sectors [24–26]. In the realm of agricultural biotechnology, scholars have increasingly emphasized the need for a more comprehensive and context-specific approach to sustainability assessment [27–30]. The latter studies surveyed U.S. stakeholders with subject matter expertise to better understand their attitudes towards additional parameters, including sustainability parameters that are important for assessing novel agrifood products. The studies revealed a clear demand for expanding conventional sustainability frameworks to include factors such as health impacts, ethical considerations, and long-term societal implications, underscoring the need for a more holistic and adaptive understanding of sustainability in this space.

To identify additional dimensions of sustainability that should be evaluated alongside environmental, economic, and social considerations, we reviewed existing literature on the tenets of sustainable agrifood systems. Sustainable agriculture is often linked to responsible and ethical behavior, especially when compared to conventional methods [31]. To be competitive with conventional agrifood products, it is also important that new products practical, useful, and offer some advantage over existing alternatives, such as agricultural products that produce higher yields while also minimizing chemical use in food production [32]. Finally, sustainable agriculture models like fair trade often promote values such as social justice, contributing to a collective good, fairness in labor practices, and equitable benefits, especially for marginalized producers and communities [33,34]. Hence, we added 7 additional dimensions to investigate the sustainability of agrifoods that include responsible, useful, superior to alternatives, ethical, fosters a fair and just society, contributes to a collective good, and equitable. These added dimensions are also supported by results from the previous research as well as considerations in several risk-benefit frameworks [30,35,36]. Together, we find our identified 10 dimensions of sustainability to fully encompass the USDA 2011 consensus statement on what constitutes sustainable agricultural systems.

2.2. Nano and GE as Approaches to Sustainable Agrifoods

The advancement of new and novel technologies, including the use of GE and nano, is largely seen as essential for driving scientific breakthroughs that promote sustainable futures, including those in agrifoods [37]. GE and nano are broad-reaching tools that can significantly enhance the precision with which we can modify, monitor, and optimize a wide variety of foods—from improving crop traits at the genetic level to delivering nutrients and detecting contaminants at the molecular scale. While many emerging technologies, such as alternative proteins, vertical farming, or synthetic biology, are reshaping sustainability in the agrifood landscape, GE and nano are examined in this study because they are already in active use across diverse agrifood applications, have well-established regulatory and scientific foundations, and continue to generate public and policy debate. These technologies are also broadly enabling, serving as foundational platforms for a wide array of downstream innovations in both plant and animal agriculture.

Genetic engineering (GE) refers to various techniques used to modify an organism's genetic makeup by adding, removing, or altering specific genes using modern biotechnology methods such as recombinant DNA, and in this paper, includes gene editing which uses these molecular methods to create the product (even if the foreign genes may be backcrossed out of the final product). By intentionally modifying or editing an organism's genetic material, GE can create crops with greater resistance to pests, diseases, and environmental stressors (such as impacts from climate change). GE foods and agricultural products are anticipated to enhance food safety, nutrition, and access to healthy foods, while promoting more sustainable and environmentally friendly agricultural practices [38,39]. Some GE crops are specifically engineered to withstand extreme weather conditions such as flooding and drought [11,40,41] and to facilitate more sustainable crop disease management [37]. In the context of livestock, gene editing may help accelerate the development of traits associated with improved welfare, such as polledness, while preserving other desirable genetic characteristics that might be lost through traditional selective breeding [42].

Nanotechnology (nano) refers to the field of science and engineering that creates, develops, and manipulates materials on the nanoscale, which is roughly 1–100 nm. At this very small scale, materials can have unique and different physical and chemical properties compared to the same materials on bulk scales. For over two decades, nano applications in agrifoods have been explored to improve sustainability through more efficient agrochemical delivery (e.g., nanopesticides, nanofertilizers), extend the shelf life of fresh-cut produce using nano-emulsion coatings to inhibit microbial growth, and enhance vaccine delivery for livestock via nano-vaccines [10,14,43–48].

Given the importance of understanding stakeholder perceptions and views of novel technologies in agrifood contexts, numerous studies have investigated the perceptions of GE and nano in agrifoods [19,49–53]. Among others, several studies have found that stakeholders raised concerns over the absence of multi-stakeholder collaborations in GE/agrifood development, the lack of transparent information about GE products, inadequate oversight, and potential gaps in regulatory systems [54]. For nano applications in agrifoods, previous studies have found stakeholder concerns related to uncertainties and data gaps related to understanding human health and environmental impacts of using engineered nanomaterials in agrifood systems [19,55,56], the lack of transparent information on their use in the food supply, issues of trust [57], and whether the product fulfilled a societal need [58]. Further, studies have found that the public is more willing to accept nanomaterials in packaging materials than inside the food and for purposes of preventing spoilage or enhancing nutrition than for taste or cheaper production [59–61]. Consumers also place strong value on labeling nanofood products, but do not prioritize nano labeling as high as GE food labeling [62]. When comparing nano applications to GE applications for food, consumers are willing to pay more to avoid GE foods than to avoid nanotech foods, but for both technologies, prefer applications directed towards nutrition and safety benefits over environmental or taste benefits [51,62–64]. For both technologies, trust in government to manage the technologies influences consumer willingness to accept nano and GE foods [64].

While many have begun to assess stakeholder perceptions of GE and nano in agrifood products, there is an absence of deeper and more inclusive assessments of the degree to which GE and nano contribute to sustainable agrifood systems. Hence, our study provides an important step forward by operationalizing a multidimensional evaluation of sustainability grounded in expert stakeholder input. Rather than relying solely on conventional metrics or top-down assessments, we designed a method that foregrounds how sustainability is interpreted and applied by those with lived experience and professional expertise across the agrifood landscape. By presenting stakeholders with detailed case studies of actual or near-market GE and nano applications, and by asking them to evaluate these across ten specific sustainability dimensions—including environmental, economic, social, ethical, and justice-based criteria—this research captures a more textured and realistic view of how emerging technologies are judged in context. The findings not only offer practical guidance for innovators, regulators, and funders, but also contribute conceptually to the growing literature on responsible innovation and sustainability science. In doing so, our study, described in detail below, helps to identify where opportunities and tensions lie, where greater clarity or communication may be needed, and where stakeholder priorities may diverge across technological approaches. It ultimately serves as a foundation for more responsive, trustworthy, and societally aligned pathways for advancing sustainable agrifood systems.

3. Methods

To investigate stakeholder perceptions of GE and nano-based agrifood products through the lens of sustainability, we employed a multi-phase tailored design survey presented via a custom-built online engagement platform [65,66]. This platform was designed to facilitate structured, multi-step interaction with expert stakeholders across a range of sectors, including academia, government, industry, and non-governmental organizations. Participants were presented with detailed case studies of six real-world products, three involving GE and three involving nano, and asked to evaluate each across ten distinct dimensions of sustainability. The convergent mixed methods study combined quantitative survey data with open-ended qualitative responses and online discussion forums, allowing for both standardized comparison and deeper insight into the reasoning behind stakeholder judgments [67]. This approach enabled us to capture both individual assessments and broader patterns across the sample, offering a comprehensive view of how sustainability is perceived in practice across diverse technological applications.

3.1. Respondent Sampling Procedure

Potential stakeholders were selected by identifying experts associated with topical peer reviewed literature (e.g., journal article publications, organizational reports), conferences and workshops, USDA's Current Research Information System (CRIS) database, as well as the research team's networks within GE, nano, food science, agriculture, veterinary medicine, and governance areas. The aim was to include participants from diverse affiliations and sectors in the U.S., including U.S. academic institutions, industry, non-governmental organizations (NGOs), think tanks, advocacy groups (including consumer and environmental advocacy groups), and government agencies. In total, the research team identified 570 individuals with the goal of having 60–75 stakeholder participants accept the invitation to participate in this study. Before reaching out to participants, the research team submitted the survey protocol to the research institutions under which the research took place (NC State: IRB ID 26701; Iowa State: IRB ID 24-038), which deemed the research IRB exempt.

Identified stakeholders were contacted via email with a brief description of the project and the reason for contacting them. Interested stakeholders were given a consent form that outlined an overview of the project, what they will be asked to do in the participation process, risks, benefits, aspects of confidentiality, compensation, and contact information. After consenting, participants were given a temporary username and password with instructions on how to access the website and complete the activities. 61 people consented to the study, of which 51 completed at least part of the study and 42 completed the study in full. Participants received an honorarium of $100 to complete all activities in the study.

3.2. Sample Composition

As the purpose of this study was to sample a diverse range of expert stakeholders from sectors in or adjacent to relevant fields, the priority of sampling was less on ensuring a representative sample and more on ensuring that the sample included representation from a significant proportion of fields. Of the 42 participants that completed the study 12 came from industry, 17 from academia, 2 from government, 9 from NGOs, 1 from "consulting" and one a member of a "think tank" (see Figure 1a). Participants ranged from 2 to 31 years of work within their sector with the average being 14.6 and a positive skew of a 10.5-year median (see Figure 1b). Participants were asked to report on their areas of expertise and disciplinary knowledge. The majority of participants selected more than one option with half identifying expertise in crop or agricultural sciences, half identifying expertise in natural sciences, one quarter identifying expertise in social sciences, and 2 participants reporting expertise in law. A compilation of the disciplines mentioned, mapped by their relation to the domains of science, technology, and society and size corresponding to the number of participants who listed the sector, is found in Figure 1c. Much of the population in this field is highly connected to each other so for the sake of protecting participant identity, we only collected demographic information on their areas of expertise and discipline, and their position within those areas.

While our sample of 42 U.S.-based stakeholders is comparable in size to similar studies in the literature, the modest sample limits the generalizability of the findings. We sought to ensure balance across stakeholder sectors, but representation from the government sector was notably lower, and the participant pool overall skewed toward individuals actively engaged in or supportive of biotechnology, potentially underrepresenting more critical or skeptical voices. Additionally, it should be noted that this study focused on U.S. agricultural systems and policy contexts, and therefore recruited stakeholder participants in the U.S. While this allowed for in-depth engagement with individuals familiar with domestic regulatory frameworks, markets, and cultural expectations, we recognize that perceptions of sustainability, particularly in relation to emerging technologies, may vary considerably across countries due to differences in governance structures, food systems, and public attitudes.

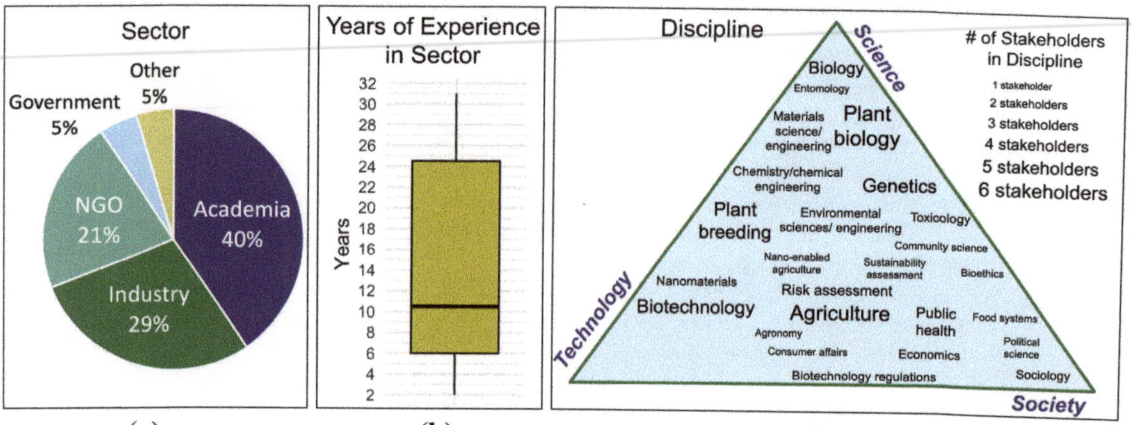

(a) (b) (c)

Figure 1. Demographic data for the 42 study participants, including (**a**) the sector they belong to, (**b**) years of experience in that sector, and (**c**) a conceptual representation of the depth and breadth of disciplines represented by the stakeholders.

3.3. Case Studies

To facilitate a structured and comparative assessment of stakeholder perceptions, we developed six detailed case studies, three focused on GE products and three on nano applications, each selected to represent different sectors within agrifoods. The goal was to present real-world or near-market examples that reflect the diversity of innovations being pursued under the umbrella of sustainable agrifood technologies. Case selection was guided by three primary criteria: (1) relevance to ongoing scientific and commercial development, (2) diversity in application across plant, animal, and agricultural innovations, and (3) variation in the type of sustainability benefit targeted (e.g., environmental protection, food security, animal welfare, nutrition). After initial selection, each case was subjected to extensive review and background research using peer-reviewed literature, regulatory reports, and public communications to ensure technical accuracy and contemporary relevance. How the case studies were presented to participants is described in the following section. This is followed by a summary of each case study, with full case study descriptions found in Supplementary Information.

3.3.1. Case Study Portrayal

Each case study document was carefully designed to be approximately two pages in length, following a standardized format that included: (1) the context and purpose of the product, (2) a plain-language explanation of how GE or nano was used, and (3) a summary of anticipated benefits and potential risks. This structure allowed for cross-case comparability while ensuring accessibility to a wide range of stakeholders with diverse disciplinary backgrounds. Relevant references were cited throughout to support factual claims and to encourage transparency and replicability. Importantly, we worked to minimize framing bias in the presentation of the cases. Information was presented in a neutral, evidence-based tone, and we carefully balanced descriptions of potential risks and benefits. The language avoided emotive or speculative phrasing and refrained from normative judgments. Each draft underwent multiple rounds of revision by the research team, including content experts and social scientists, to ensure the final versions were concise, balanced, and appropriately nuanced. The final cases included GE applications such as faster-growing Atlantic salmon, heat-tolerant beef cattle, and less pungent mustard greens;

and nano-enabled products such as phosphorus-efficient fertilizer, nanoencapsulated avian flu vaccines, and fruit coatings designed to increase shelf life and reduce waste. By presenting these case studies in a uniform, digestible, and unbiased manner, we aimed to elicit informed stakeholder evaluations across a range of sustainability dimensions in the subsequent phases of the study.

After stakeholders expressed interest in participating in the study, they were provided with anonymized, temporary login credentials to access a secure, custom-built online engagement platform. The platform guided participants through a series of five structured activities designed to gather both quantitative and qualitative insights. In Activity 1, participants completed a Qualtrics-embedded survey that collected demographic information, including their professional sector and areas of expertise or disciplinary background. This survey also included a previously validated set of scale items designed to measure participants' trust in science and technology [68]. In Activity 2, participants reviewed product overview documents for each of the six case studies, three involving GE and three involving nano—and then completed short surveys assessing their perceptions of each product's potential risks and benefits. These dimensions included human health, animal health, environmental impact, and ethical, legal, and social implications (ELSIs), as well as their level of certainty in those judgments. Participants repeated this process across all six case studies. In Activity 3, after completing all individual case assessments, participants completed a final survey capturing their holistic perceptions of each product's sustainability and potential for overall benefit or harm. This survey also included open-ended prompts about actions developers or regulators could take to address any concerns. Finally, Activity 4 consisted of an asynchronous discussion board where participants shared and responded to others' perspectives on two central prompts related to technology governance and the role of GE and nano in sustainable agrifoods.

This article focuses on stakeholder assessments of sustainability regarding Activity 3 of the platform. In this activity, participants were asked to evaluate each of the six case studies across ten dimensions of sustainability: environmentally sustainable, economically sustainable, socially sustainable, responsible, useful, superior to alternatives, ethical, fosters a fair and just society, contributes to a collective good, and equitable. These dimensions were presented using semantic differential scales, with three of the most conceptually significant terms (environmental, economic, and social sustainability) accompanied by hover-text definitions to ensure shared understanding. This design enabled a multidimensional and comparative assessment of each product's perceived sustainability profile, generating rich data that could be analyzed quantitatively and contextualized through earlier responses and discussion board contributions.

3.3.2. GE Case Studies

Case Study 1: Faster-Growing Atlantic Salmon

Faster-growing Atlantic salmon ("AquAdvantage Salmon") is an FDA-approved transgenic organism engineered by AquaBounty. It incorporates the Chinook salmon's growth hormone gene to grow Atlantic salmon faster compared to its non-GE counterpart [69,70]. The potential advantages of this GE salmon include an increase in seafood production to meet increased demand, a year-round growth, and potential to increase aquaculture sustainability through more efficient feed utilization as well as a prospected pressure reduction on wild salmon stocks. The FDA concluded that AquAdvantage Salmon and the derived foods are as safe to eat as food derived from non-GE Atlantic salmon since they have largely equivalent nutritional profiles [69]. Despite its approval, some have voiced concerns about the adequacy of the FDA assessment, particularly with regard to the risk of uncontrolled release [71]. These concerns persist despite several proactive measures taken to prevent escape: the GE salmon is only authorized to be grown in land-based facilities (one in Indiana, one in Canada) with highly sophisticated containment systems, and only sterile females are grown [69]. Additionally, a 2019 study reported a general more negative consumer perception for the GE salmon compared to GE vegetables [72]. Another study came to a similar conclusion, pointing to the fact that in general consumers appear to be more concerned with the use of biotechnologies in animals compared to plants, and concerns for animals include animal welfare and gene editing animals for productivity traits [73]. As of 2025, AquaBounty has ceased the fish farming operations due to lack of funds to maintain the operation [74].

Case Study 2: Heat-Tolerant Slick-Hair Beef Cattle

Heat-tolerant slick-hair beef (SLICK) cattle originate from two founder beef calves that were gene edited with Clustered Regularly Interspaced Short Palindromic Repeats (CRISPR) to achieve a haircoat that allows them to stay comfortable in warmer temperatures [75]. The founder cattle were modified by making a heritable edit to the prolactin receptor gene (i.e., PRLR gene), shortening the prolactin receptor protein in cattle to mimic a trait present in many breeds of cattle commonly found in tropical and subtropical areas [30,75]. Potential benefits of these GE cattle include faster introduction of the trait compared to traditional selective breeding, the ability to expand cattle production into previously unsuitable regions, and improved adaptability to rising temperatures—potentially enhancing animal welfare. PRLR-SLICK cattle are the first low-risk determination and decision to exercise enforcement discretion for an Intentional Genomic Alternation (IGA) in animals for food use. FDA risk evaluations concluded that there are no known safety risks from the PRLR-SLICK cattle food. Notably, the regulatory process for these GE cattle was less extensive than the GE salmon, as the SLICK cattle are not transgenic, as the edit did not involve DNA from other species [30]. The FDA also concluded that there is low environmental risk as many conventionally raised cattle in the US already have the trait, animal escape is unlikely (and easy to remedy if it occurs), and there are very few feral cattle populations in the US [75]. The agency therefore used its "enforcement discretion" in March 2022 to review the PRLR-SLICK cattle under a less extensive approval process [30]. Despite the overall low risks presented by the GE cattle to consumers and the environment, important considerations appear to be missing from FDA's risk assessment, including a lack of clear data concerning the actual welfare of the SLICK cattle. Additionally, the environmental risk section did not address potential land use and agrochemical impacts, and what a production expansion might imply for the environment should that occur (although there are no clear data on whether the SLICK cattle would lead to an increase in production and consumption). Lastly, similarly to the AquAdvantage salmon, studies point to the fact that the public perception of GE animals is generally more negative compared to GE plants due to concerns including animal welfare and misguided substitutes for conventional husbandry practices [42,73].

Case Study 3: Less Pungent Mustard Greens

Researchers at Pairwise (now overseen by Bayer) developed genetically engineered mustard greens (*Brassica juncea*) to alter the flavor to be less pungent and bitter than the conventional variety [76,77]. CRISPR was utilized to target and alter base pairs in the genes that are responsible for some of the pungent and spicy flavor that happens when eating the greens raw [76]. Multiple genes were targeted in the modification across seven chromosomes, including the deletion of two whole genes, blocking the conversion of glucosinolates to pungent oils. While normal cooking methods can break down the oils responsible for pungency [78], the anticipated benefit of the GE mustard greens is to increase the consumption of raw nutritious leafy greens. An anticipated risk of the GE mustard greens is related to change in glucosinolates, which are responsible as a plant defense mechanism against insects [79]. This decrease in glucosinolates may increase vulnerability to insects and in turn lead to an increase in pesticide use. Despite this risk, the GE mustard greens were found to be exempt from regulation, both in accordance with the Coordinated Framework for the Regulation of Biotechnology (CFRB) by USDA-Animal Plant Health and Inspection Service (APHIS) under the Plant Protection Act (PPA) [30] and under the now previous SECURE Rule with a Regulatory Status Review (RSR) [80]. Public opinion has shown that most people are more accepting of cisgenic modification rather than transgenic when it comes to GE techniques [81]. However, conventionally bred products are still preferred [82,83]. Concerns may center around the lack of labeling and regulation because of the use of gene editing and the National Bioengineered Food Disclosure Standards require foreign DNA in the final product for labeling [30]. The GE mustard greens were the first CRISPR agrifood product to be launched in North America.

3.3.3. Nano Case Studies

Case Study 4: More Efficient Phosphorus Fertilizer

Nano-based fertilizers have been developed and commercialized for agricultural applications to improve the efficiency of plant nutrient delivery, including phosphorus (P) [46]. Enhanced efficiency can increase crop yields while reducing nutrient runoff into the surrounding environment. To create a nanoscale encapsulated P fertilizer, engineered nanoparticles composed of polyamidoamine (PAMAM) dendrimers are mixed with conventional fertilizers [84,85]. The nutrient molecules in the fertilizer solution are attracted to the PAMAM dendrimer coating, forming an outer layer. The resulting mixture is then applied to crops, with electrostatic forces preventing nutrient leaching or premature binding to the soil, enabling efficient uptake by plants [86]. By improving the effectiveness of traditional fertilizers, nanoencapsulated P fertilizers could enhance plant growth and crop yields as well as help reduce economic losses for farmers [87]. Additionally, more efficient P delivery systems may benefit the environment and society by decreasing P runoff and agricultural pollution, thereby reducing the risk of eutrophication and harmful algal blooms, which can negatively affect human health, animal health, and ecosystems [46]. Currently, many uncertainties remain regarding the potential risks of nanoencapsulated P fertilizers to non-target environmental organisms and human health. While no toxicity or ecotoxicity studies have specifically examined PAMAM dendrimers loaded with P, research on PAMAM dendrimer nanoparticles and other nanomaterials suggests possible adverse effects on aquatic invertebrates, green algae, microorganisms, and zebrafish embryos [88–90]. Significant data gaps persist in understanding the environmental, health, and societal implications of nano-based fertilizers, and as of now, there are no nano-specific fertilizer regulations in the U.S. Although few studies have explored public perceptions and acceptance of nanoencapsulated fertilizers, broader research on nano applications in agriculture suggests moderately positive public support. Despite limited familiarity with nano products, many consumers perceive agricultural nano (e.g., nanopesticides) as offering greater efficiency and lower environmental impact compared to conventional agrochemicals. A majority also support continued development and use of nano in agriculture, as well as labeling nanoscale ingredients and additives in food products [62].

Case Study 5: Improved Avian Influenza Vaccines

Researchers are actively developing nanoencapsulated vaccines to combat poultry diseases, including avian influenza. This highly contagious disease poses significant economic challenges in the U.S. due to poultry mortality and trade restrictions imposed on infected flocks [91]. While several vaccines have been developed to address avian influenza, many have shown limited effectiveness. Polyanhydride nanoparticles (PANs) are synthetic, biodegradable copolymers with mucoadhesive properties, allowing them to adhere to cell membranes and enhance vaccine delivery. When used in vaccines, these nanoparticles enable a sustained release of antigens, potentially improving immune responses [47]. The use of PANs as a vaccine delivery system has been approved by the FDA for human applications, and the vaccine is considered safe for poultry. This is largely due to PAN's biodegradable nature, breaking down into nontoxic, metabolizable byproducts [92]. Ongoing research aims to assess the full range of benefits and potential impacts associated with nanoencapsulated vaccine delivery in poultry. While no toxicological risks of PANs have been reported in existing studies [47,93], their long-term effects on health and the environment remain largely unexplored. Available research suggests that stakeholders generally find their use in animal husbandry more acceptable than nanomaterials incorporated directly into food products or non-essential applications. While some consumers appreciate the increased oversight of animal health and welfare, others may express concerns about the presence of vaccine-related materials in food products.

Case Study 6: Fruit Coatings Designed to Increase Shelf Life and Reduce Waste

Researchers have developed and patented a nanocellulose-based coating designed to extend the shelf life of fresh produce, maintain freshness, and reduce food waste [48]. Nanocellulose is a natural, edible, biodegradable biopolymer derived from plant materials. To create a nanocellulose-based coating, cellulose nanofibers are extracted from plant biomass and then combined with other ingredients before being dispersed in water to form a coating solution [94]. The solution can be applied to fresh produce [48]. Extending the shelf life of fresh produce can significantly reduce food waste. Various forms of cellulose have been used in food for decades, and the FDA has classified bacterial cellulose and microcrystalline cellulose as Generally Recognized as Safe (GRAS). Unlike conventional food coatings, nanocellulose-based coatings provide an enhanced protective barrier against moisture loss, gas exchange, UV light, and microbial contamination, which can improve water retention and inhibit microbial growth [95]. However, the potential health and environmental risks of nanocellulose remain largely uncertain. While some studies have found no toxicity even at high concentrations, others have reported potential adverse effects, including inflammation, oxidative stress, and disruptions to gut microbiota [96–98]. Due to limited data, cellulose nanofibers have not yet received a GRAS designation. Few studies have examined consumer attitudes toward nanocellulose coatings on fresh produce. Existing research suggests that public acceptance is lower for applications where nanocellulose is directly ingested (e.g., coatings on fresh fruits) compared to non-ingested uses, such as removable food packaging.

3.3.4. Quantitative Comparison of the Sustainability of Case Studies

The data analysis presented in this paper is based on a set of questions from Activity 3 that asked stakeholder participants how they would rate each case study related to 10 dimensions of sustainability: environmentally sustainable, economically sustainable, socially sustainable, responsible, useful, superior to alternatives, ethical, fosters a fair and just society, contributes to a collective good, and equitable. For each item, stakeholders chose from five positions on a scale ranging from negative sentiment (e.g., unsustainable) to positive sentiment (e.g., sustainable). We categorized the responses into three groups: neutral (center of the scale), negative sentiment (the two positions closer to the negative sentiment word), and positive sentiment (the two positions closer to the positive sentiment word). To clearly capture the positive, neutral, or negative orientation of stakeholder views, we recorded participant responses from a 5-point scale to a 3-point scale to better align with overall sentiment toward sustainability. Specifically, responses of 4 or 5 were classified as sustainable (recoded as 3), 3 as neutral (recoded as 2), and 1 or 2 as unsustainable (recoded as 1). An example survey question can be found in Figure 2, with colored boxes exemplifying our grouping schema.

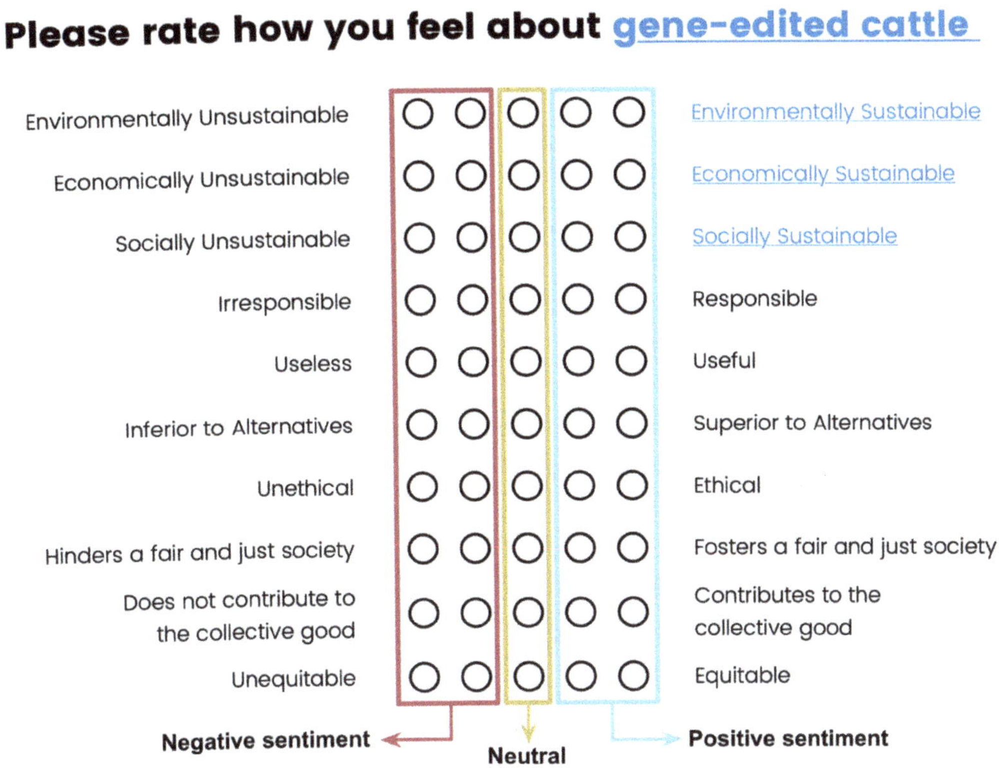

Figure 2. Example survey question with categorization. Blue text indicates terms that, when hovered over, displayed the definition used by our research team.

Using the new 3-point sentiment scale, we calculated the mean sustainability score for each case study across each of the ten dimensions, as shown in Table 1. These scores were used to generate 10 bar charts—one for each of the 10 dimensions of sustainability—that compare the proportion of positive, neutral, and negative sentiments about each of the six cases (Figure 3). These visualizations helped highlight specific strengths and weaknesses in each case study's perceived sustainability profile, offering a foundation for the more nuanced interpretation that followed. The raw data, including the frequency of each sentiment (sustainable, neutral, unsustainable) across all ten dimensions and six case studies, is available in the Supplemental Materials.

To explore whether the ten sustainability dimensions could be meaningfully combined into an overall sustainability factor, we examined how each dimension related to one another across the six case studies. Using the corrplot package in R, we generated a correlation matrix and plot (see Figure 4 in the Results and Table S11 in Supplementary Information), which visualized the strength and direction of linear relationships between each pair of dimensions based on their average scores on the original 1- to 5-point scale. The analysis revealed that all ten dimensions were positively correlated, with no correlation falling below 0.7, indicating that the dimensions tended to trend together and likely reflect a common underlying construct of sustainability.

Table 1. Mean sustainability scores for each dimension of sustainability across each of the 6 case studies (G = genetic engineering, N = nanotechnology). Scale = 1 to 3, with 1 being unsustainable and 3 being sustainable.

	Salmon (G)	Cattle (G)	Mustard Greens (G)	Fertilizer (N)	Flu Vaccine (N)	Fruit Coating (N)
Environmentally Sustainable	2.38	2.48	2.43	2.69	2.48	2.74
Economically Sustainable	2.64	2.67	2.62	2.62	2.62	2.74
Socially Sustainable	2.33	2.48	2.60	2.52	2.57	2.60
Responsible	2.50	2.60	2.50	2.64	2.79	2.67
Useful	2.71	2.76	2.52	2.83	2.91	2.86
Superior to Alternatives	2.38	2.45	2.29	2.57	2.83	2.64
Ethical	2.52	2.57	2.51	2.62	2.79	2.69
Fosters a fair and just society	2.38	2.42	2.32	2.37	2.45	2.57
Contributes to a collective good	2.52	2.49	2.49	2.69	2.81	2.69
Equitable	2.41	2.44	2.44	2.42	2.57	2.64

While the ten dimensions trend together and collectively represent an overall picture of sustainability, examining them individually offers valuable insights into specific areas where each product was assessed differently in each dimension. To better understand and visualize these differences, we developed the bar charts seen in Figure 3 by calculating the percentage of sampled stakeholders that rated their sentiment of a dimension as either sustainable, neutral, or unsustainable. Repeating this calculation for each of the ten dimensions across all six products, results in a measure of all three sentiments for all ten dimensions for each product that can be used to rank products by individual dimensions (See Figure 3) or overall assessed sustainability (See Figure 5). To support and contextualize findings about individual products, we triangulated our quantitative results with participants' open-ended responses collected in text-based sections of the surveys.

4. Results

This study aimed to answer the central research question: *How do expert stakeholders perceive the sustainability of nano and GE when applied to agrifood systems?* Drawing from quantitative ratings (and major claims supported by qualitative responses), our results offer a multidimensional view of how expert stakeholders evaluate these technologies across a range of real-world case studies. The findings illuminate key differences in perceived sustainability between nano and GE applications, as well as the specific sustainability dimensions—environmental, economic, social, ethical, and others—that shape stakeholder judgments. We begin by presenting the quantitative data, comparing stakeholder ratings of the six case studies across ten sustainability dimensions. Where appropriate, we incorporate qualitative insights from open-ended responses and online discussion board interactions to further contextualize these assessments. Together, these findings shed light on the factors that contribute to stakeholder evaluations and offer a deeper understanding of how sustainability is interpreted in the context of emerging agrifood technologies. An overview of the positive, neutral, and negative sentiments toward each of the 10 dimensions of sustainability is presented in Figure 3. Raw data is available in the Supplementary Information in Tables S1–S10.

First and foremost, most respondents viewed all six case studies as generally sustainable across the 10 dimensions. For both nano and GE applications, over half of stakeholders selected positive ratings for most sustainability dimensions, suggesting a broad openness to emerging technologies in agrifoods. This overall favorability coexisted with more nuanced differences; nanotech products were more consistently rated as environmentally and economically sustainable, socially beneficial, and ethically sound, while GE products, though still often rated positively, drew more mixed and uncertain responses, particularly around fairness, ethics, and social responsibility.

Consistent trends emerged across the dimensions with respect to nano- vs. GE-based products. Nano-based products were consistently rated more favorably than those developed through GE. When comparing stakeholder assessments across the ten dimensions of sustainability, nano products dominated the top rankings in several key areas. In five of the ten categories—responsible, useful, superior to alternatives, ethical, and contributes to a collective good—the three highest-rated products were all nano-enabled. The nano-based fruit coating received the highest scores in environmental, economic, and social sustainability, as well as in fostering a fair and just society. Meanwhile, the nanoencapsulated avian influenza vaccine ranked highest in the remaining six categories, including responsibility, usefulness, ethicality, and equity. In contrast, the three GE case studies were more uneven in their evaluations, with stakeholders expressing greater caution, particularly around issues of equity, environmental impact, and other ethical concerns.

The fairness and justice dimension revealed the most uncertainty overall, with no product receiving a clear endorsement. Still, the fruit coating stood out as relatively more favorable, while the GE products attracted the most skepticism. Similarly, the equity dimension produced mixed responses, with nanotech products again rated higher than their GE counterparts. The GE products, especially mustard greens and salmon, often prompted concerns or neutral stances regarding their ethical standing, societal value, and environmental or economic implications. Taken together, these results suggest that nano-based innovations in agrifoods are perceived as more aligned with sustainability goals than genetically engineered options.

Further analysis revealed that the ten sustainability dimensions included in this study were strongly positively correlated, with no pairwise correlation falling below 0.7 (see Figure 4 and Supplementary Information Table S11). This finding suggests that stakeholder perceptions of sustainability across these dimensions tend to move together, supporting the idea that they form a coherent and internally consistent construct. These results are particularly meaningful in light of calls from the literature to broaden traditional three-pillar models of sustainability—environmental, economic, and social—when evaluating agrifood technologies. Prior research has emphasized the need to incorporate additional considerations such as ethical responsibility, practical utility, fairness, and collective well-being [24–26]. The strong correlations observed here indicate that these expanded dimensions resonate with stakeholders as interconnected elements of sustainability, rather than isolated or conflicting concerns. This convergence reinforces the relevance of using a more holistic framework to assess the sustainability of novel agrifood products.

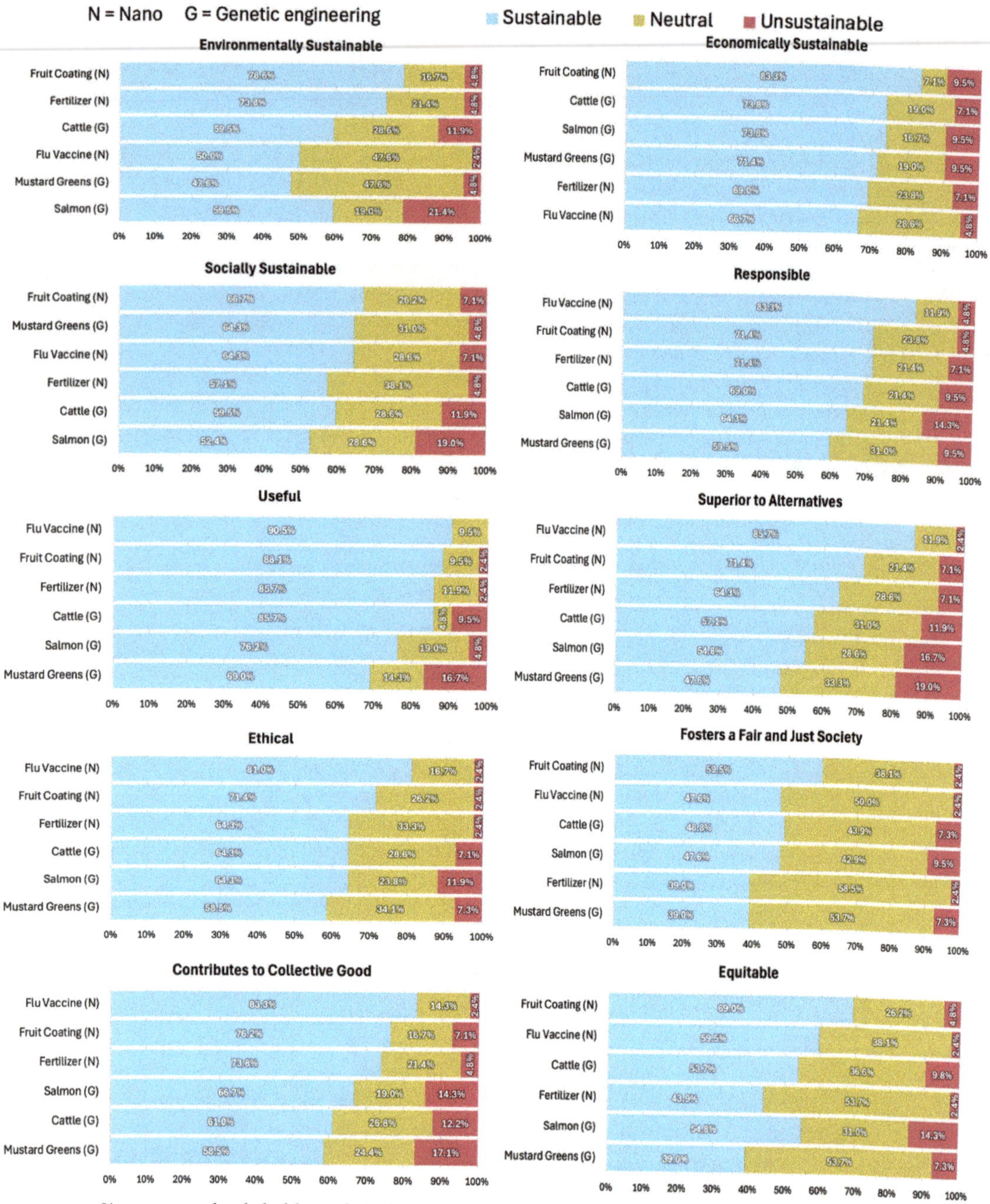

% = percent of stakeholders who selected each sentiment (sustainable, neutral, unsustainable)

Figure 3. Stakeholder sustainability assessments across ten dimensions. Each bar chart represents one dimension of sustainability, showing the percentage of responses rated as "sustainable" (positive), "neutral", or "unsustainable" (negative) for each of the six case studies. Cases are ordered independently for each dimension, from highest to lowest percentage of sustainable responses. In most dimensions, the nano cases appear closer to the top and GE cases appear at the bottom.

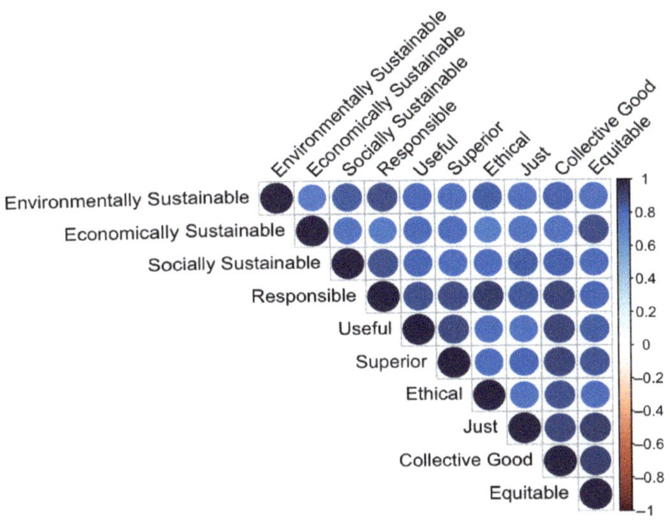

Figure 4. Correlation plot for the 10 dimensions of sustainability of all products. All 10 trend together, with no two dimensions having a correlation of less than 0.7. Raw correlation data can be found in Supplementary Information.

Because the ten sustainability dimensions trend together so closely, we calculated an overall sustainability profile for each case study by aggregating responses across all dimensions. For each case, we examined the total percentage of responses that were categorized as sustainable, neutral, or unsustainable across all ten categories. This aggregation offers a high-level summary of how each case study was perceived in terms of sustainability, providing a useful starting point for a comparison of the relative sustainability of the case studies, as seen in Figure 5.

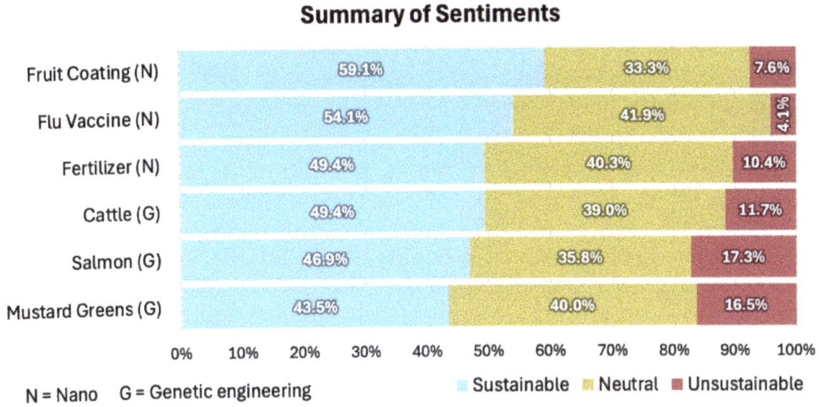

Figure 5. Summary of overall sustainability sentiment for all 6 case studies.

5. Discussion

The future of agrifoods depends on the development of systems that are both productive and sustainable, being capable of meeting global food demands while safeguarding environmental resources, promoting social equity, and building resilience to climate disruption. As the world population grows and ecological pressures intensify, innovations like

GE and nano are often positioned as promising tools to help achieve these goals. However, the true value of such technologies cannot be determined by technical performance alone. Their potential for contributing to sustainability must also be understood through the perceptions of stakeholders who influence development, governance, and public legitimacy. This study contributes to that understanding by offering the first empirical analysis of how expert stakeholders evaluate GE and nano-enabled agrifood products using a consistent, multidimensional sustainability framework.

It is first important to note that the majority of stakeholders viewed most of the products presented in this study as sustainable. For all six case studies, over 50% of respondents selected positive ratings for most sustainability dimensions, indicating a favorable overall orientation toward both nano and GE as tools for advancing sustainability in agrifoods. This widespread positive sentiment is encouraging, as it suggests that expert stakeholders see meaningful potential in these technologies to contribute to environmental, economic, and social goals. However, there were a few notable exceptions. Specifically, less than half of respondents rated the gene-edited mustard greens as environmentally sustainable or superior to conventional alternatives, and fewer than 50% rated either the nano-based phosphorus fertilizer or the mustard greens as equitable. Importantly, these lower scores were generally driven by an increase in neutral rather than negative sentiment. Across all ten dimensions and six products, negative ratings were uncommon—rarely exceeding 15% of responses. When they did occur, they were most likely to appear in evaluations of the genetically engineered salmon and the mustard greens. The salmon may elicit more polarized views due to its higher public visibility, regulatory controversy, and associations with transgenics and animal biotechnology. In contrast, the muted support for mustard greens may reflect stakeholder skepticism about whether improved taste alone constitutes a meaningful sustainability contribution, particularly when compared to products offering more clearly articulated environmental or health benefits. These findings highlight that while broad support exists, stakeholders apply discernment based on perceived impact, benefit type, and alignment with sustainability values—reinforcing the need for nuanced, context-specific engagement around emerging technologies.

Among the ten sustainability dimensions evaluated, "usefulness" stood out as the one with the greatest consensus and confidence among stakeholders. All six products were rated as useful by a clear majority of participants, with positive ratings above ~70% and neutral responses below 20% in every case. Four of the six products, three nano-enabled and one GE—were rated as useful by over 85% of stakeholders. This strong agreement suggests that expert stakeholders generally view these technologies as having a clear function or purpose within the agrifood system. The consistency of this response, regardless of whether a product was rated highly on other dimensions, indicates that "usefulness" may serve as a baseline or threshold consideration for innovation: if a product is not viewed as useful, it is unlikely to be considered sustainable, regardless of its other attributes.

Notably, several other sustainability dimensions tracked closely with "usefulness", offering additional insight into how stakeholders distinguish between products that are simply functional versus those worth pursuing. Dimensions such as "superior to alternatives", "responsible", "ethical", and "contributes to a collective good" followed nearly identical response patterns, with nano products ranking highest and GE products lagging behind. The nano-based flu vaccine emerged as the top-rated product across all five of these dimensions, with 80–90% of stakeholders selecting positive responses. Conversely, the genetically engineered mustard greens consistently received the lowest scores in this cluster, with positive ratings falling between 45% and 70%. These findings suggest that stakeholders may use "usefulness" as an initial practical measure but base their broader sustainability judgments on whether a product offers tangible improvements over existing options, upholds ethical standards, and contributes positively to society. Products perceived to meet these combined expectations—particularly the nanovaccine—are seen as not only viable but desirable innovations worthy of investment and continued development.

More broadly, our results echo a pattern in stakeholder preferences between nano- and GE-enabled products, with nano-based products generally perceived as more sustainable than their GE counterparts. When the sustainability sentiments for the six products were compared, the nano-enabled fruit coating and avian influenza vaccine received the most favorable ratings, followed by the phosphorus nano-fertilizer. In contrast, the genetically engineered mustard greens, beef cattle, and salmon were rated lower across most dimensions, with the mustard greens case receiving the lowest overall sustainability scores. These findings suggest that stakeholders distinguish between the two technology categories, not merely in technical terms but in relation to broader ethical, environmental, and societal considerations. Prior studies have also shown that consumers are more willing to purchase nano-enabled food products for multiple benefits than GE food products [62–64].

There are several possible explanations for the more favorable perception of nano. One contributing factor may be that nano applications, particularly those used in materials science and medicine—have been under development for several decades, leading to greater familiarity, more extensive safety evaluations, and more clearly articulated public benefits. However, GE foods have been in the market since the mid-1990s, albeit knowledge of their existence among consumers was low in earlier decades. The difference between nano and GE is mirrored in previous studies of consumer attitudes: research shows that while consumers express concern over both technologies, they are generally more averse to GE food products than to those containing nanomaterials. For instance, consumers are more likely to support labeling requirements for GE foods than for nano-enabled foods [62] and are often willing to pay more to avoid GE foods compared to nano-based alternatives [63,64]. Scholars have suggested that this divergence may be shaped by value-based predispositions—such as aversion to tampering with genes, preference for natural foods, or deeper associations between genetic modification and "unnaturalness"—in contrast to the perceived familiarity of chemical or material-based interventions in food [99]. They have also attributed past public controversies to GE foods and polarization surrounding them as factors in lower opinions of GE than nano [100]. These findings align with our results and suggest that perceived sustainability is shaped not only by technical or functional attributes, but by the cultural meanings, histories, and ethical associations that stakeholders bring to each technology.

The findings from this study offer actionable insights for innovation governance and policy development in agrifoods. Stakeholder assessments revealed clear distinctions not only between GE and nano-based products, but also among individual products within each category—highlighting that perceptions of sustainability are shaped by more than technical performance. Dimensions such as responsibility, equity, and contribution to the public good were central to stakeholder evaluations, underscoring the need for governance strategies that are nuanced and context-specific rather than guided solely by technology type. A one-size-fits-all regulatory approach may overlook important ethical and societal factors, whereas product-level evaluation informed by stakeholder input can support more responsible and widely accepted innovation pathways.

Nano-based products—particularly the nano-enabled avian influenza vaccine—emerged as strong candidates for future development. The nanovaccine received overwhelmingly positive ratings across multiple sustainability dimensions, including usefulness, ethicality, and contribution to a collective good, suggesting both high functionality and social value. However, while perceived as promising, some stakeholders also flagged barriers to commercialization, such as limited private-sector incentives and infrastructure readiness as well as expensive and uncertain regulatory hurdles. One stakeholder suggested that regulators should, "Allow for innovative, case-by-case flexibility in data requirements and assessments, [e.g.,] if a product is well understood and has numerous precedents with low risk—then find ways to expedite approval and minimize cost. Conversely do the opposite with products that pose potentially high uncertainty and risk. Rigid box-checking hurts all parties". These insights point to the need for proactive policy tools—such as targeted subsidies, public–private partnerships, or R&D investments focused on deployment—to help overcome market hurdles and bring high-value technologies to scale.

Crucially, our study also demonstrates the importance of incorporating stakeholder-informed sustainability assessments earlier and more systematically into the innovation pipeline. Traditional governance frameworks often emphasize risk assessment and regulatory compliance at later stages, typically after a product is nearing commercialization. However, such downstream evaluations may be too late to meaningfully influence product design or address public concerns before they solidify. By engaging stakeholders during earlier stages, especially between technology readiness levels (TRLs) 3 and 6, when product designs remain adaptable, developers can identify and respond to ethical, social, and practical concerns while there is still opportunity to refine the innovation trajectory. Researchers have argued that integrating ethical, legal, and social implication (ELSIs) assessments alongside traditional TRLs can enhance the legitimacy and viability of novel biotechnologies by surfacing concerns before they become barriers. For example, Trump et al. suggest evaluating ELSI concerns early in basic research stages in addition to assessing whether prototypes at more advanced research stages align with safety, security, and ethical expectations [101].

As with any expert elicitation study, several limitations should be acknowledged. First, while our sample of 42 U.S.-based stakeholders is comparable in size to similar studies in the literature, the modest sample limits the generalizability of the findings. We sought to ensure balance across stakeholder sectors, but representation from the government sector was notably lower, and the participant pool overall skewed toward individuals actively engaged in or supportive of biotechnology, potentially underrepresenting more critical or skeptical voices. While the generalization of findings may be more limited, transferability of qualitative results diminish the effect of this limitation by contextualizing individual stakeholder sentiments within repeatedly found concepts throughout the data. Second, the study focused exclusively on U.S. stakeholders to better inform U.S. policy; sustainability perceptions may differ significantly in other national or cultural contexts, and future cross-country comparisons could produce different, valuable results. Future studies should also consider the everyday immediate and adjacent consumers of products in stakeholder perceptions. Third, although the six case studies were carefully selected to reflect diversity across domains (e.g., environmental, human health, and animal applications), different case choices—particularly those perceived as highly controversial or unsustainable—could have led to different patterns of response. Further, our sample of stakeholders was more representative of plant biotechnology and agricultural systems expertise, with limited representation from individuals specializing in animal biotechnology or gene editing, despite three of the six case studies focusing on animal applications. However, we believe that the results of our data constitute a meaningful addition to the literature regardless of chosen products as each case study elicited both new and repeated results that contribute towards saturation of information related to more balanced products. Additionally, we focused our analysis on only one component of the engagement platform (Activity 3), meaning other potentially relevant data sources, such as participant characteristics and perspectives on science and technology gathered in other components, were not included here but may offer important explanatory context for future analyses. Despite these limitations, the study offers valuable insights into how a diverse set of stakeholders evaluates emerging food and agricultural technologies and provides a replicable framework for integrating sustainability considerations into technology governance.

The ten sustainability dimensions evaluated in this study also offer a novel theoretical and methodological framework for advancing sustainability assessment in emerging agri-food technologies and other fields. Theoretically, the multidimensional approach moves beyond conventional three-pillar models of sustainability by systematically incorporating additional ethical, social, and justice-based considerations that stakeholders actively apply when evaluating new technologies. Methodologically, the framework operationalizes these dimensions in a structured, stakeholder-centered evaluation platform that enables a more granular and holistic understanding of how diverse technologies are perceived across multiple domains of sustainability. While dimensions such as usefulness or technical efficacy elicited broad agreement, others revealed important divergence in stakeholder priorities and expectations, demonstrating the value of capturing multidimensional perspectives rather than relying on any single metric of sustainability. For example, while the gene-edited mustard greens were viewed as useful, they received lower ratings on environmental sustainability and equity, highlighting areas where further engagement or development may be needed to strengthen their sustainability profile. In contrast, the nanovaccine was broadly supported across both technical and normative dimensions, suggesting a clearer pathway for responsible advancement. Recognizing these differentiated patterns early in the innovation process allows policymakers, funders, and developers to better align product development with societal values, allocate resources more effectively, and design governance approaches that are responsive to stakeholder concerns. By embedding this multidimensional stakeholder evaluation framework into earlier phases of the innovation pipeline, the study demonstrates how sustainability can serve not only as an aspirational goal, but also as a practical, actionable tool for guiding technology design, governance, and investment decisions in ways that enhance legitimacy, trust, and societal benefit.

Finally, our findings carry broader implications for the future of sustainable agrifood innovation. If left unexamined, technological development risks becoming a narrow exercise in engineering efficiency rather than a collective project grounded in shared values. But the results of this study offer a more optimistic vision. They suggest that it is possible to design, assess, and govern technologies in ways that are not only technically effective but also socially legitimate and ethically informed. Whether that future becomes utopian or dystopian depends on our willingness to prioritize inclusive and anticipatory approaches to innovation. The tools to build a sustainable food system already exist—what remains is to ensure they are guided by the perspectives and priorities of the people they are intended to serve. As a next step, we feel that researchers should examine differences in stakeholder perceptions across sectors and disciplines, and how these perceptions evolve through structured dialog and deliberation. By exploring how conversations shape evaluations, we hope to deepen understanding of consensus, conflict, and learning within stakeholder communities. We also invite other researchers to adopt, adapt, and build upon this framework in future studies. The field urgently needs more transdisciplinary, stakeholder-engaged research to ensure that sustainability is not merely a label applied after the fact—but a guiding principle embedded at every stage of technological innovation.

6. Conclusions

This study provides the first systematic, multidimensional analysis of how U.S.-based expert stakeholders evaluate the sustainability of GE and nano-based products in agri-foods. By using real-world case studies and assessing them across ten key sustainability dimensions—environmentally sustainable, economically sustainable, socially sustainable, responsible, useful, superior to alternatives, ethical, fosters a fair and just society, contributes to a collective good, and equitable—we captured nuanced insights into how these emerging technologies are understood and judged by those with influence over their development, regulation, and adoption. Overall, nano-based products were viewed more favorably than their GE counterparts, with the nano-enabled flu vaccine and fruit coating receiving the highest ratings across multiple categories. At the same time, the GE mustard greens emerged as the least favored, raising questions about the types of benefits that are most valued when sustainability is at stake.

Notably, stakeholders evaluated sustainability holistically, tending to rate products consistently across dimensions. However, subtle variations in their responses reveal important distinctions that can inform more targeted and responsive innovation strategies. These findings underscore the value of multidimensional frameworks for sustainability assessment. Further, by capturing expert judgments on real-world products, our stakeholder engagement approach can help researchers, developers, and policymakers better anticipate which innovations are seen as both impactful and legitimate. Future work can extend this framework across additional products, sectors, and national contexts to build more responsive and inclusive strategies for sustainable agrifood innovation.

Supplementary Materials: The supplementary information file referenced in the text can be downloaded at: https://www.mdpi.com/article/10.3390/su17156795/s1.

Author Contributions: Conceptualization, M.D.H., C.L.C., J.K., K.D.G.; Methodology, M.D.H., C.L.C., K.D.G.; Software, M.D.H., C.L.; Validation, C.L.C., J.K., M.D., K.D.G.; Formal Analysis, M.D.H., C.L.C., C.L., K.D.G.; Investigation, M.D.H., C.L.C., J.K., M.D., I.C., M.C., C.L., N.L., K.D.G.; Resources, M.D.H., C.L.C., J.K., M.D., M.C., N.L., K.D.G.; Data Curation, M.D.H., C.L.; Writing—Original Draft Preparation, M.D.H., C.L.C., J.K., M.D., I.C., M.C., C.L., N.L., K.D.G.; Writing—Review and Editing, M.D.H., C.L.C., J.K., M.D., I.C., M.C., C.L., N.L., K.D.G.; Visualization, M.D.H.; Supervision, C.L.C., M.D., K.D.G.; Project Administration, K.D.G.; Funding Acquisition, K.D.G., J.K. All authors have read and agreed to the published version of the manuscript.

Funding: This research was funded by the Social Implications of Food and Agriculture Technologies Program Area, A1652, project award no. 2022-67023-36730, from the U.S. Department of Agriculture's National Institute of Food and Agriculture. Any opinions, findings, conclusions, or recommendations expressed in this publication are those of the authors and should not be construed to represent any official USDA or U.S. Government determination or policy.

Institutional Review Board Statement: The research protocol was reviewed by the Institutional Review Boards of North Carolina State University (IRB ID # 26701, date 2023-12-15) and Iowa State University (IRB ID 24-038, date 2024-03-01), both of which determined the research to be exempt from further ethical review.

Informed Consent Statement: Informed consent was obtained from all subjects involved in the study.

Data Availability Statement: The datasets presented in this article are not readily available because they are part of an ongoing study. Requests to access the datasets should be directed to the corresponding author.

Acknowledgments: The authors would like to thank Sharon Stauffer for her invaluable support in coordinating the engagement activities, as well as all of the participants who completed our study and provided valuable feedback.

Conflicts of Interest: The authors declare no conflicts of interest.

References

1. Richardson, K.; Steffen, W.; Lucht, W.; Bendtsen, J.; Cornell, S.E.; Donges, J.F.; Drüke, M.; Fetzer, I.; Bala, G.; von Bloh, W.; et al. Earth beyond Six of Nine Planetary Boundaries. *Sci. Adv.* **2023**, *9*, eadh2458. [CrossRef]
2. Definitions: Sustainability and Food Systems. Available online: https://www.usda.gov/about-usda/general-information/staff-offices/office-chief-economist/sustainability/definitions-sustainability-and-food-systems (accessed on 12 June 2025).
3. Saikanth, D.R.K.; Supriya; Singh, B.V.; Rai, A.K.; Bana, S.R.; Sachan, D.S.; Singh, B. Advancing Sustainable Agriculture: A Comprehensive Review for Optimizing Food Production and Environmental Conservation. *Int. J. Plant Soil Sci.* **2023**, *35*, 417–425. [CrossRef]
4. Singh, R.L.; Mondal, S. *Biotechnology for Sustainable Agriculture: Emerging Approaches and Strategies*; Woodhead Publishing: Sawston, UK, 2017; ISBN 978-0-12-812238-9.
5. Anderson, J.A.; Gipmans, M.; Hurst, S.; Layton, R.; Nehra, N.; Pickett, J.; Shah, D.M.; Souza, T.L.P.O.; Tripathi, L. Emerging Agricultural Biotechnologies for Sustainable Agriculture and Food Security. *J. Agric. Food Chem.* **2016**, *64*, 383–393. [CrossRef]
6. Hefferon, K.L.; De Steur, H.; Perez-Cueto, F.J.A.; Herring, R. Alternative Protein Innovations and Challenges for Industry and Consumer: An Initial Overview. *Front. Sustain. Food Syst.* **2023**, *7*, 1038286. [CrossRef]
7. Getahun, S.; Kefale, H.; Gelaye, Y. Application of Precision Agriculture Technologies for Sustainable Crop Production and Environmental Sustainability: A Systematic Review. *Sci. World J.* **2024**, *2024*, 2126734. [CrossRef]
8. Razzaq, A.; Kaur, P.; Akhter, N.; Wani, S.H.; Saleem, F. Next-Generation Breeding Strategies for Climate-Ready Crops. *Front. Plant Sci.* **2021**, *12*, 620420. [CrossRef]
9. Guelfi, D.; Nunes, A.P.P.; Sarkis, L.F.; Oliveira, D.P. Innovative Phosphate Fertilizer Technologies to Improve Phosphorus Use Efficiency in Agriculture. *Sustainability* **2022**, *14*, 14266. [CrossRef]
10. Nile, S.H.; Baskar, V.; Selvaraj, D.; Nile, A.; Xiao, J.; Kai, G. Nanotechnologies in Food Science: Applications, Recent Trends, and Future Perspectives. *Nanomicro. Lett.* **2020**, *12*, 45. [CrossRef] [PubMed]
11. Karavolias, N.G.; Horner, W.; Abugu, M.N.; Evanega, S.N. Application of Gene Editing for Climate Change in Agriculture. *Front. Sustain. Food Syst.* **2021**, *5*, 685801. [CrossRef]
12. Eş, I.; Gavahian, M.; Marti-Quijal, F.J.; Lorenzo, J.M.; Mousavi Khaneghah, A.; Tsatsanis, C.; Kampranis, S.C.; Barba, F.J. The Application of the CRISPR-Cas9 Genome Editing Machinery in Food and Agricultural Science: Current Status, Future Perspectives, and Associated Challenges. *Biotechnol. Adv.* **2019**, *37*, 410–421. [CrossRef] [PubMed]
13. Nguyen, T.H.; Ben Taieb, S.; Moritaka, M.; Ran, L.; Fukuda, S. Public Acceptance of Foods Derived from Genome Editing Technology: A Review of the Technical, Social and Regulatory Aspects. *J. Int. Food Agribus. Mark.* **2023**, *35*, 397–427. [CrossRef]
14. Hofmann, T.; Lowry, G.V.; Ghoshal, S.; Tufenkji, N.; Brambilla, D.; Dutcher, J.R.; Gilbertson, L.M.; Giraldo, J.P.; Kinsella, J.M.; Landry, M.P.; et al. Technology Readiness and Overcoming Barriers to Sustainably Implement Nanotechnology-Enabled Plant Agriculture. *Nat. Food* **2020**, *1*, 416–425. [CrossRef]
15. Sabio, R.P.; Lehoux, P. Responsible Research and Innovation in Food Systems: A Critical Review of the Literature and Future Research Avenues. *Agric. Hum. Values* **2024**. [CrossRef]
16. Testa, S.; Nielsen, K.R.; Vallentin, S.; Cicculló, F. Sustainability-Oriented Innovation in the Agri-Food System: Current Issues and the Road Ahead. *Technol. Forecast. Soc. Change* **2022**, *179*, 121653. [CrossRef]
17. Grobe, A.; Renn, O.; Jaeger, A. *Risk Governance of Nanotechnology Applications in Food and Cosmetics*; International Risk Governance Council: Geneva, Switzerland, 2008.
18. Selfa, T.; Lindberg, S.; Bain, C. Governing Gene Editing in Agriculture and Food in the United States: Tensions, Contestations, and Realignments. *Elem. Sci. Anthr.* **2021**, *9*, 00153. [CrossRef]
19. Grieger, K.D.; Merck, A.W.; Cuchiara, M.; Binder, A.R.; Kokotovich, A.; Cummings, C.L.; Kuzma, J. Responsible Innovation of Nano-Agrifoods: Insights and Views from U.S. Stakeholders. *NanoImpact* **2021**, *24*, 100365. [CrossRef]
20. Purvis, B.; Mao, Y.; Robinson, D. Three Pillars of Sustainability: In Search of Conceptual Origins. *Sustain. Sci.* **2019**, *14*, 681–695. [CrossRef]
21. Brundtland, G.H. *Our Common Future: Report of the World Commission on Environment and Development*; United Nations General Assembly: Geneva, Switzerland, 1987.
22. National Environmental Policy Act; 1969. Available online: https://www.energy.gov/sites/prod/files/nepapub/nepa_documents/RedDont/Req-NEPA.pdf (accessed on 27 June 2025).
23. United Nations Sustainable Development Goal 8. Available online: https://sdgs.un.org/goals/goal8 (accessed on 12 June 2025).
24. Seghezzo, L. The Five Dimensions of Sustainability. *Environ. Politics* **2009**, *18*, 539–556. [CrossRef]
25. Hahn, T.; Preuss, L.; Pinkse, J.; Figge, F. Cognitive Frames in Corporate Sustainability: Managerial Sensemaking with Paradoxical and Business Case Frames. *Acad. Manag. Rev.* **2014**, *39*, 463–487. [CrossRef]
26. Boyer, R.H.W.; Peterson, N.D.; Arora, P.; Caldwell, K. Five Approaches to Social Sustainability and an Integrated Way Forward. *Sustainability* **2016**, *8*, 878. [CrossRef]

27. Macken-Walsh, A.; Henchion, M.M.; Regan, A. 'Come Aboard' the Systems-Based Approach: The Role of Social Science in Agri-Food Research and Innovation. *Ir. J. Agric. Food Res.* **2022**, *61*, 168–183. [CrossRef]
28. Dalziel, P.; Saunders, C.; Tait, P.; Saunders, J.; Miller, S.; Guenther, M.; Rutherford, P.; Driver, T. Rewarding Responsible Innovation When Consumers Are Distant from Producers: Evidence from New Zealand. *Int. Food Agribus. Manag. Rev.* **2018**, *21*, 487–504. [CrossRef]
29. Wenzl, C.; Buddle, E.A.; Ankeny, R.A. Problematic Use of Sustainability Claims in Recent Scientific Literature on Crop Gene Technologies: Toward Improving Practices and Communication. *Plant J.* **2025**, *122*, e70137. [CrossRef] [PubMed]
30. Kuzma, J.; Grieger, K.; Cimadori, I.; Cummings, C.L.; Loschin, N.; Wei, W. Parameters, Practices, and Preferences for Regulatory Review of Emerging Biotechnology Products in Food and Agriculture. *Front. Bioeng. Biotechnol.* **2023**, *11*, 1256388. [CrossRef]
31. Paarlberg, R. The Ethics of Modern Agriculture. *Society* **2009**, *46*, 4–8. [CrossRef]
32. Giri, D.; Pokhrel, S. Organic Farming for Sustainable Agriculture: A Review. *Russ. J. Agric. Socio-Econ. Sci.* **2022**, *130*, 23–32.
33. Brown, S.; Getz, C. Towards Domestic Fair Trade? Farm Labor, Food Localism, and the 'Family Scale' Farm. *GeoJournal* **2008**, *73*, 11–22. [CrossRef]
34. Robinson, P.K. Responsible Retailing: Regulating Fair and Ethical Trade. *J. Int. Dev.* **2009**, *21*, 1015–1026. [CrossRef]
35. IRGC. *Introduction to the IRGC Risk Governance Framework*; EPFL International Risk Governance Center: Lausanne, Switzerland, 2017.
36. Assunção, R.; Alvito, P.; Brazão, R.; Carmona, P.; Fernandes, P.; Jakobsen, L.S.; Lopes, C.; Martins, C.; Membré, J.-M.; Monteiro, S.; et al. Building Capacity in Risk-Benefit Assessment of Foods: Lessons Learned from the RB4EU Project. *Trends Food Sci. Technol.* **2019**, *91*, 541–548. [CrossRef]
37. National Academies of Sciences, Engineering, and Medicine. *Science Breakthroughs to Advance Food and Agricultural Research by 2030*; National Academies Press: Washington, DC, USA, 2019; ISBN 978-0-309-47392-7.
38. Young, A.E.; Mansour, T.A.; McNabb, B.R.; Owen, J.R.; Trott, J.F.; Brown, C.T.; Van Eenennaam, A.L. Genomic and Phenotypic Analyses of Six Offspring of a Genome-Edited Hornless Bull. *Nat. Biotechnol.* **2020**, *38*, 225–232. [CrossRef]
39. Zhang, X.; Yao, G.; Vishwakarma, S.; Dalin, C.; Komarek, A.M.; Kanter, D.R.; Davis, K.F.; Pfeifer, K.; Zhao, J.; Zou, T.; et al. Quantitative Assessment of Agricultural Sustainability Reveals Divergent Priorities among Nations. *One Earth* **2021**, *4*, 1262–1277. [CrossRef]
40. Bailey-Serres, J.; Lee, S.C.; Brinton, E. Waterproofing Crops: Effective Flooding Survival Strategies. *Plant Physiol.* **2012**, *160*, 1698–1709. [CrossRef]
41. Esmaeili, N.; Shen, G.; Zhang, H. Genetic Manipulation for Abiotic Stress Resistance Traits in Crops. *Front. Plant Sci.* **2022**, *13*, 1011985. [CrossRef] [PubMed]
42. Yunes, M.C.; Osório-Santos, Z.; von Keyserlingk, M.A.G.; Hötzel, M.J. Gene Editing for Improved Animal Welfare and Production Traits in Cattle: Will This Technology Be Embraced or Rejected by the Public? *Sustainability* **2021**, *13*, 4966. [CrossRef]
43. Zaheer, S.; Shehzad, J.; Chaudhari, S.K.; Mustafa, G. The Application of Nanotechnology to Accelerate Plant Adaptation to Abiotic Stresses. In *Advanced Nanotechnology in Plants*; CRC Press: Boca Raton, FL, USA, 2023; ISBN 978-1-00-336857-1.
44. Sampathkumar, K.; Tan, K.X.; Loo, S.C.J. Developing Nano-Delivery Systems for Agriculture and Food Applications with Nature-Derived Polymers. *iScience* **2020**, *23*, 101055. [CrossRef] [PubMed]
45. Zambrano-Zaragoza, M.L.; González-Reza, R.; Mendoza-Muñoz, N.; Miranda-Linares, V.; Bernal-Couoh, T.F.; Mendoza-Elvira, S.; Quintanar-Guerrero, D. Nanosystems in Edible Coatings: A Novel Strategy for Food Preservation. *Int. J. Mol. Sci.* **2018**, *19*, 705. [CrossRef]
46. Basavegowda, N.; Baek, K.-H. Current and Future Perspectives on the Use of Nanofertilizers for Sustainable Agriculture: The Case of Phosphorus Nanofertilizer. *3 Biotech* **2021**, *11*, 357. [CrossRef]
47. Huntimer, L.; Ramer-Tait, A.E.; Petersen, L.K.; Ross, K.A.; Walz, K.A.; Wang, C.; Hostetter, J.; Narasimhan, B.; Wannemuehler, M.J. Evaluation of Biocompatibility and Administration Site Reactogenicity of Polyanhydride-Particle-Based Platform for Vaccine Delivery. *Adv. Healthc. Mater.* **2012**, *2*, 369–378. [CrossRef]
48. Jung, J.; Deng, Z.; Zhao, Y. A Review of Cellulose Nanomaterials Incorporated Fruit Coatings with Improved Barrier Property and Stability: Principles and Applications. *J. Food Process Eng.* **2020**, *43*, e13344. [CrossRef]
49. Kuzma, J.; Cummings, C.L. Cultural Beliefs and Stakeholder Affiliation Influence Attitudes Towards Responsible Research and Innovation Among United States Stakeholders Involved in Biotechnology and Gene Editing. *Front. Polit. Sci.* **2021**, *3*, 677003. [CrossRef]
50. Gómez-Llorente, H.; Hervás, P.; Pérez-Esteve, É.; Barat, J.M.; Fernández-Segovia, I. Nanotechnology in the Agri-Food Sector: Consumer Perceptions. *NanoImpact* **2022**, *26*, 100399. [CrossRef] [PubMed]
51. Liu, P.; Zheng, X.; Shangguan, S.; Zhao, L.; Fang, X.; Huang, Y.; Hermanowicz, S.W. Public Perceptions and Willingness-to-Pay for Nanopesticides. *Nanomaterials* **2022**, *12*, 1292. [CrossRef] [PubMed]

52. Woźniak-Gientka, E.; Agata, T.; Milica, P.; Anna, B.; Dennis, E.; Nick, V.; Godelieve, G.; Selim, C.; Naghmeh, A.; Tomasz, T. Public Perception of Plant Gene Technologies Worldwide in the Light of Food Security. *GM Crops Food* **2022**, *13*, 218–241. [CrossRef] [PubMed]
53. Kato-Nitta, N.; Tachikawa, M.; Inagaki, Y.; Maeda, T. Public Perceptions of Risks and Benefits of Gene-Edited Food Crops: An International Comparative Study between the US, Japan, and Germany. *Sci. Technol. Hum. Values* **2023**, *48*, 1360–1392. [CrossRef]
54. Kuzma, J.; Grieger, K. Community-Led Governance for Gene-Edited Crops. *Science* **2020**, *370*, 916–918. [CrossRef]
55. Baranowska-Wójcik, E.; Szwajgier, D.; Oleszczuk, P.; Winiarska-Mieczan, A. Effects of Titanium Dioxide Nanoparticles Exposure on Human Health—A Review. *Biol. Trace Elem. Res.* **2020**, *193*, 118–129. [CrossRef]
56. McClements, D.J.; Rao, J. Food-Grade Nanoemulsions: Formulation, Fabrication, Properties, Performance, Biological Fate, and Potential Toxicity. *Crit. Rev. Food Sci. Nutr.* **2011**, *51*, 285–330. [CrossRef]
57. EIT Food. *The Trust Report*; EIT Food: Madrid, Spain, 2020.
58. Grieger, K.; Merck, A.; Kuzma, J. Formulating Best Practices for Responsible Innovation of Nano-Agrifoods through Stakeholder Insights and Reflection. *J. Responsible Technol.* **2022**, *10*, 100030. [CrossRef]
59. Brown, J.; Fatehi, L.; Kuzma, J. Altruism and Skepticism in Public Attitudes toward Food Nanotechnologies. *J. Nanopart. Res.* **2015**, *17*, 122. [CrossRef]
60. Siegrist, M.; Cousin, M.-E.; Kastenholz, H.; Wiek, A. Public Acceptance of Nanotechnology Foods and Food Packaging: The Influence of Affect and Trust. *Appetite* **2007**, *49*, 459–466. [CrossRef]
61. Giles, E.L.; Kuznesof, S.; Clark, B.; Hubbard, C.; Frewer, L.J. Consumer Acceptance of and Willingness to Pay for Food Nanotechnology: A Systematic Review. *J. Nanopart. Res.* **2015**, *17*, 467. [CrossRef]
62. Kuzma, J. Society and Policy Maker's Responsibilities. In *Consumer Perception of Product Risks and Benefits*; Springer International Publishing/Springer Nature: Cham, Switzerland, 2017; pp. 547–566. ISBN 978-3-319-50528-2.
63. Yue, C.; Zhao, S.; Kuzma, J. Heterogeneous Consumer Preferences for Nanotechnology and Genetic-Modification Technology in Food Products. *J. Agric. Econ.* **2015**, *66*, 308–328. [CrossRef]
64. Yue, C.; Zhao, S.; Cummings, C.; Kuzma, J. Investigating Factors Influencing Consumer Willingness to Buy GM Food and Nano-Food. *J. Nanopart. Res.* **2015**, *17*, 283. [CrossRef]
65. Dillman, D.A.; Smyth, J.D.; Christian, L.M. *Internet, Phone, Mail, and Mixed Mode Surveys: The Tailored Design Method*, 4th ed.; John Wiley & Sons, Inc.: Hoboken, NJ, USA, 2014; pp. xvii, 509. ISBN 978-1-118-45614-9.
66. Ruzante, J.M.; Shumaker, E.T.; Holt, S.; Mayer, S.; Kokotovich, A.; Cuchiara, M.; Binder, A.R.; Kuzma, J.; Grieger, K. *Eliciting Stakeholder Perceptions Using a Novel Online Engagement Platform: A Case Study on Nano-Agrifoods*; RTI Press Occasional Papers; RTI Press: Research Triangle Park, NC, USA, 2022.
67. Creswell, J.E.; Plano Clark, V.L. *Designing and Conducting Mixed Methods Research*, 3rd ed.; SAGE: Thousand Oaks, CA, USA, 2018.
68. Cummings, C.L.; Kuzma, J. Societal Risk Evaluation Scheme (SRES): Scenario-Based Multi-Criteria Evaluation of Synthetic Biology Applications. *PLoS ONE* **2017**, *12*, e0168564. [CrossRef] [PubMed]
69. FDA. AquAdvantage Salmon Fact Sheet. Available online: https://www.fda.gov/animal-veterinary/aquadvantage-salmon/aquadvantage-salmon-fact-sheet (accessed on 12 June 2025).
70. FDA. Q&A on FDA's Approval of AquAdvantage Salmon. Available online: https://www.fda.gov/animal-veterinary/aquadvantage-salmon/qa-fdas-approval-aquadvantage-salmon (accessed on 12 June 2025).
71. Lindsey, B. The Potential Environmental Impacts of Genetically Modified (GMO) Salmon. Available online: https://sor.senate.ca.gov/sites/sor.senate.ca.gov/files/The%20Potential%20Environmental%20Impacts%20of%20Genetically%20Modified%20%28GMO%29%20Salmon.pdf (accessed on 27 June 2025).
72. Lefebvre, S.; Cook, L.A.; Griffiths, M.A. Consumer Perceptions of Genetically Modified Foods: A Mixed-Method Approach. *J. Consum. Mark.* **2019**, *36*, 113–123. [CrossRef]
73. Van Eenennaam, A.L.; Young, A.E. Public Perception of Animal Biotechnology. In *Animal Biotechnology 2: Emerging Breeding Technologies*; Niemann, H., Wrenzycki, C., Eds.; Springer International Publishing: Cham, Switzerland, 2018; pp. 275–303, ISBN 978-3-319-92348-2.
74. AquaBounty Technologies, Inc. AquaBounty Announces Plans to Cease Fish Farming Operations. Available online: https://investors.aquabounty.com/news-releases/news-release-details/aquabounty-announces-plans-cease-fish-farming-operations (accessed on 12 June 2025).
75. FDA. *PRLR-SLICK Cattle Risk Assessment Summary*; FDA: Silver Spring, MD, USA, 2022.
76. Karlson, D.; Mojica, J.P.; Poorten, T.J.; Lawit, S.J.; Jali, S.; Chauhan, R.D.; Pham, G.M.; Marri, P.; Guffy, S.L.; Fear, J.M.; et al. Targeted Mutagenesis of the Multicopy *Myrosinase* Gene Family in Allotetraploid Brassica Juncea Reduces Pungency in Fresh Leaves across Environments. *Plants* **2022**, *11*, 2494. [CrossRef]
77. Grinstein, J.D. Salad Days: Pairwise Gene Edits Food to Topple Nutrition Barriers. *GEN Biotechnol.* **2023**, *2*, 5–9. [CrossRef]
78. Rungapamestry, V.; Duncan, A.J.; Fuller, Z.; Ratcliffe, B. Effect of Cooking Brassica Vegetables on the Subsequent Hydrolysis and Metabolic Fate of Glucosinolates. *Proc. Nutr. Soc.* **2007**, *66*, 69–81. [CrossRef]

79. Hopkins, R.J.; van Dam, N.M.; van Loon, J.J.A. Role of Glucosinolates in Insect-Plant Relationships and Multitrophic Interactions. *Annu. Rev. Entomol.* **2009**, *54*, 57–83. [CrossRef]
80. Juarez, B. Confirmation of the Regulatory Status of Genome Edited Brassica Juncea Lines with Improved Flavor 2020. Available online: https://www.aphis.usda.gov/sites/default/files/20-168-07_air_response_signed.pdf (accessed on 27 June 2025).
81. Mielby, H.; Sandøe, P.; Lassen, J. Multiple Aspects of Unnaturalness: Are Cisgenic Crops Perceived as Being More Natural and More Acceptable than Transgenic Crops? *Agric. Hum. Values* **2013**, *30*, 471–480. [CrossRef]
82. Lusk, J.L.; Jamal, M.; Kurlander, L.; Roucan, M.; Taulman, L. A Meta-Analysis of Genetically Modified Food Valuation Studies. *J. Agric. Resour. Econ.* **2005**, *30*, 28–44.
83. Frewer, L.J.; van der Lans, I.A.; Fischer, A.R.H.; Reinders, M.J.; Menozzi, D.; Zhang, X.; van den Berg, I.; Zimmermann, K.L. Public Perceptions of Agri-Food Applications of Genetic Modification—A Systematic Review and Meta-Analysis. *Trends Food Sci. Technol.* **2013**, *30*, 142–152. [CrossRef]
84. Bunderson, L.D.; Hunter, B.L.; Isaacson, K.J. PAMAM Dendrimers for Fertilizer Delivery 2020. Available online: https://patents.google.com/patent/US11390571B2/en (accessed on 27 June 2025).
85. Bunderson, L.D.; Hunter, B.L.; Carpenter, J.O. Blended Nanoparticle Fertilizer Delivery 2020. Available online: https://patents.google.com/patent/WO2020210672A1/en (accessed on 27 June 2025).
86. Nano-Yield Nanotechnology. Available online: https://www.nano-yield.com/nanotechnology (accessed on 12 June 2025).
87. NanoPhos: Nanoliquid Technology for Delivering Phosphorus and Enhancing Nutrient Uptake. Available online: https://www.nano-yield.com/nano/nanophos (accessed on 12 June 2025).
88. Gonzalo, S.; Rodea-Palomares, I.; Leganés, F.; García-Calvo, E.; Rosal, R.; Fernández-Piñas, F. First Evidences of PAMAM Dendrimer Internalization in Microorganisms of Environmental Relevance: A Linkage with Toxicity and Oxidative Stress. *Nanotoxicology* **2015**, *9*, 706–718. [CrossRef]
89. Oliveira, E.; Casado, M.; Faria, M.; Soares, A.M.; Navas, J.M.; Barata, C.; Piña, B. Transcriptomic Response of Zebrafish Embryos to Polyaminoamine (PAMAM) Dendrimers. *Nanotoxicology* **2014**, *8*, 92–99. [CrossRef] [PubMed]
90. Petit, A.-N.; Eullaffroy, P.; Debenest, T.; Gagné, F. Toxicity of PAMAM Dendrimers to *Chlamydomonas reinhardtii*. *Aquat. Toxicol.* **2010**, *100*, 187–193. [CrossRef] [PubMed]
91. CDC. Current Situation: Bird Flu in Poultry. Available online: https://www.cdc.gov/bird-flu/situation-summary/current-bird-flu-situation-in-poultry.html (accessed on 12 June 2025).
92. Kingstad-Bakke, B.A.; Chandrasekar, S.S.; Phanse, Y.; Ross, K.A.; Hatta, M.; Suresh, M.; Kawaoka, Y.; Osorio, J.E.; Narasimhan, B.; Talaat, A.M. Effective Mosaic-Based Nanovaccines against Avian Influenza in Poultry. *Vaccine* **2019**, *37*, 5051–5058. [CrossRef]
93. Ojer, P.; de Cerain, A.L.; Areses, P.; Peñuelas, I.; Irache, J.M. Toxicity Studies of Poly(Anhydride) Nanoparticles as Carriers for Oral Drug Delivery. *Pharm. Res.* **2012**, *29*, 2615–2627. [CrossRef]
94. Pradhan, D.; Jaiswal, A.K.; Jaiswal, S. Emerging Technologies for the Production of Nanocellulose from Lignocellulosic Biomass. *Carbohydr. Polym.* **2022**, *285*, 119258. [CrossRef]
95. Perumal, A.B.; Nambiar, R.B.; Moses, J.A.; Anandharamakrishnan, C. Nanocellulose: Recent Trends and Applications in the Food Industry. *Food Hydrocoll.* **2022**, *127*, 107484. [CrossRef]
96. Brand, W.; van Kesteren, P.C.; Swart, E.; Oomen, A.G. Overview of Potential Adverse Health Effects of Oral Exposure to Nanocellulose. *Nanotoxicology* **2022**, *16*, 217–246. [CrossRef]
97. Shatkin, J.A.; Kim, B. Cellulose Nanomaterials: Life Cycle Risk Assessment, and Environmental Health and Safety Roadmap. *Environ. Sci. Nano* **2015**, *2*, 477–499. [CrossRef]
98. Lie, E.; Ålander, E.; Lindström, T. *Possible Toxicological Effects of Nanocellulose—An Updated Literature Study, No 2*; Inventia: Stockholm, Sweden, 2017.
99. Hekmat, S.; Dawson, L.N. Students' Knowledge and Attitudes towards GMOs and Nanotechnology. *Nutr. Food Sci.* **2019**, *49*, 628–638. [CrossRef]
100. Ineichen, C.; Biller-Andorno, N.; Deplazes-Zemp, A. Image of Synthetic Biology and Nanotechnology: A Survey among University Students. *Front. Genet.* **2017**, *8*, 00122. [CrossRef]
101. Trump, B.; Cummings, C.; Klasa, K.; Galaitsi, S.; Linkov, I. Governing Biotechnology to Provide Safety and Security and Address Ethical, Legal, and Social Implications. *Front. Genet.* **2022**, *13*, 1052371. [CrossRef]

sustainability

MDPI

Article

Evaluation of Sanitary and Environmental Impact of Plant Protection Practices in Vineyards of Southwestern France: Organic and Conventional/Integrated Agriculture

Chaima Grimene [1,2,3,*], Oussama Mghirbi [4], Philippe Le Grusse [1,3] and Jean-Paul Bord [2,3]

1 CIHEAM-IAMM Mediterranean Agronomic Institute of Montpellier, 3191 Route de Mende, Cedex 5, 34093 Montpellier, France; legrusse@iamm.fr
2 Paul Valéry Montpellier 3 University (UPVM), Route de Mende, Cedex 5, 34199 Montpellier, France; jpbord@gmail.com
3 LAGAM, Laboratory for Geography and Territorial Planning in Montpellier, Paul Valéry Montpellier University (UPVM), Site Saint Charles, Rue du Professeur Henri Serre, 34090 Montpellier, France
4 UniLaSalle Polytechnic Institute—Beauvais Campus, InTerACT (UP 2018.C102), 19 Rue Pierre Waguet—BP 30313, Cedex, 60026 Beauvais, France; oussama.mghirbi@unilasalle.fr
* Correspondence: chaymagrimen@gmail.com; Tel.: +33-6-62-98-92-69

Abstract: The French wine industry is spread across the country and represents 789,000 ha (2023). Over 20% of the plant protection products (PPPs) sold in France are used in viticulture on less than 4% of the French UAA (Utilized Agricultural Area). The share of wine estates with organic farming certification has risen sharply, reaching 9% of French vineyards in 2016. The position occupied by the wine sector on both the national and international scale confirms the need to examine the impacts of different management practices in viticulture on human health and the environment. This study presents an approach to the assessment of plant protection practices in vineyards based on indicators of plant protection pressure and risk. It was carried out on wine-growing farms in the southwest of France, surveyed according to the two farming systems: conventional/integrated and organic. The main objective of this study was to compare the health and environmental impact of the PPPs used in these two farming systems. The impact assessment result of wine-growing plant protection practices shows that some pesticides and molecules used in organic farming, especially those based on copper and sulfur, are more harmful than products used in conventional/integrated farming, in particular to the environment. For this reason, all stakeholders involved in pesticide management should recognize the health and environmental impact of PPPs in order to reduce and to control their toxicity risks to public health and the natural environment.

Keywords: pesticides; risk indicators; vineyards

1. Introduction

The agriculture industrialization phase between 1950 and 1980 saw the introduction of new technologies in terms of mechanization and the type of inputs used [1]. It marked the beginning of the evolution of practices with a view to increasing the profitability and productivity of farms. The use of chemical fertilizers, plant protection products, genetic development, and monocultures became widespread. Agriculture became intensive and less and less diversified, which led to the emergence of health and environmental concerns a few decades later.

Like most agricultural activities, wine-growing phytosanitary practices have impacts on many environmental systems due to the excessive use of plant protection products

(PPPs) [2–4]. Over 20% of the PPPs sold in France are used in viticulture on less than 4% of the French UAA (Utilized Agricultural Area) [5]. The effects of wine-growing practices were also highlighted in the 2004 report of the *Institut Français de l'Environnement* (IFEN, i.e., the French Institute for the Environment), which exposed the presence of pesticides in the surface and groundwater. Indeed, although these practices have made it possible to improve the sanitary quality of crops and produce quality wine, they are the source of diffuse pollution with a cumulative nature, which also has potential health impacts linked to the excessive use of plant protection products. Wine-growing plant protection practices are responsible for the pollution of surface and underground water in two forms: point-source pollution and diffuse pollution [6]. Diffuse water pollution can be significant if treatments with fungicides or pesticides are repeated, especially during rainy episodes following an application or if the topography of the vineyard is conducive to runoff [7]. In addition, active ingredients are easily carried away by runoff to surface water or to groundwater located below the vineyard by infiltration into the fault structure [8,9]. Point-source pollution occurs following errors in the handling of plant protection products or spray applications, such as emptying tank bottoms and discharging rinsing water loaded with chemical residues [10,11].

In addition, imposing vine varieties and using monocultures have a negative impact on animal biodiversity, which is known to depend on plant biodiversity. Indeed, the intensification of wine-growing practices, and particularly the use of plant protection products, leads to an imbalance because it reduces and fragments semi-natural habitats, which degrades biodiversity [12–15].

For several years now, the intensive paradigm that dominates viticulture has been actively called into question [16]. Although wine-growing practices have made it possible to improve the sanitary quality of harvests and to produce quality wine, they have caused damage to the image of wine-growing areas and to the sector. Indeed, the image of wine is strongly linked to its quality, but also to the conditions in which it is produced. The various consequences on human health [17–19] and the environment [20,21] linked to the use of plant protection products are now known. Consequently, the objective is no longer only to produce in large quantities, but also to respect the quality of the products [22,23] and to limit environmental impacts [24,25].

The evolution of viticulture towards more environmentally friendly practices is driven by regulatory and societal pressure [26]. The demand does not only come from consumers but also from other stakeholders involved in wine-growing, such as producers/farms, wine merchants, or cooperatives. In addition to guaranteeing the viability and sustainability of viticulture by protecting ecosystems and consumer health, an improved approach to environmental issues could provide economic value.

Indeed, integrated farming is an approach that considers environmental protection, human health, and animal welfare. It was regulated by the French Ministries of Agriculture and Ecology between 2002 and 2013 through certification but is no longer regulated [27].

Organic farming (AB, i.e., "*Agriculture Biologique*") was created by the Ministry of Agriculture and recognized by the law of 4 July 1980 [28]. Its practices were formalized by the 1994 regulation and aim to respect natural balances by avoiding the use of synthetic chemicals [29].

The same regulations apply throughout the European Union. EU Regulation 2018/848 and its implementing texts specify all the provisions to be complied with (see Regulation (EU) 2018/848 of the European Parliament and of the Council of 30 May 2018 relating to organic production and labeling of organic products. Consolidated version of 1 January 2022. https://eur-lex.europa.eu/legal-content/FR/TXT/?uri=CELEX:02018R0848-2022 0101 (accessed on 5 September 2024)). Organic agricultural production activities are based on specific principles such as the preservation and development of the natural fertility of the soil, the minimization of the use of non-renewable resources and external inputs, the preservation of plant health through preventive measures, the use of seeds and animals with greater genetic diversity, a high degree of disease resistance, and high longevity, etc.

The Common Agricultural Policy (CAP) has set up a system of conversion aid for AB. The share of wine estates under the organic farming (AB) certification has risen sharply, reaching 10% of French vineyards in 2018 [30].

The share of this area is greater in viticulture than in agriculture in general, since 6.5% of the French agricultural UAA is certified AB [30]. This organic farming (AB) certification is rapidly spreading in the Occitanie region. Organic viticulture in Occitanie is expanding from one year to the next, both in terms of surface area and in terms of the number of producers under organic certification. Indeed, the Occitanie wine industry covered 28,833 ha in 2020 (36% of the French organic wine industry) and presented notable evolution dynamics [31]. Occitanie and Nouvelle-Aquitaine, in particular, showed the most significant increases in surface area in 2017, with +14% and +11%, respectively [32].

These changes in practices towards more environmentally friendly practices are either part of sustainable development approaches or voluntary private sector initiatives. The stakeholders of this branch of the wine industry are investing in and looking for solutions to improve practices. Indeed, the environmental consequences of agricultural activities have become a major concern for society and for the CAP and environmental policy institutions, which are increasingly encouraging farmers to adopt environmentally friendly practices [33].

By using 20% of the total plant protection products used in French agriculture, even though it only represents 4% of the French agricultural area, viticulture is directly involved in all health and environmental issues [34]. Indeed, it is partly responsible for the pollution of surface and groundwater, but also for the risks to consumer health.

According to the OECD (2001), two families of indicators for pesticides can be distinguished in order to study and manage plant protection products:

- Pesticide pressure/use indicators describe trends in the use of pesticides over time. These types of indicators are the simplest, since they require less information;
- Risk indicators are associated with pesticides that relate to potential polluting pressure. They are characterized by a more complex construction, since they integrate the characteristics of active substances and their toxicities.

Current policies for reducing the use of protection products, such as the Ecophyto 2018 plan, use pressure indicators, including the three main indicators to monitor the evolution of the use of plant protection products in France: the Amount of Active Substances (QSA), the Number of Dose Units (NODU), and the treatment frequency indicator (TFI).

Pressure indicators do not consider the specific characteristics of each plant protection product, such as its behavior in the environment, toxicity to non-target organisms, ecotoxicity to the environment, or the effects on the applicator's health [35]. In the context of a risk study on the use of plant protection products, it is essential to adopt indicators that provide additional information on health and environmental impacts.

In addition to pressure indicators, impact/risk indicators have been developed to allow for the assessment of the environmental and health risks of pesticides. There is a multitude of methods and indicators developed in the literature to study pesticide risks.

Our research work also seeks to assess the risk of diffuse pollution related to plant protection practices in vineyards of southwest France. The main purpose is to assess the toxicity risk linked to plant protection products applied during vineyard plot treatment based on a combination of existing tools [36–39] and to establish a comparison between the risk associated with organic wine-growing and conventional/integrated farms. Through this result, it will be possible to identify the pesticides with the highest risk in order to improve farmers' choices in terms of phytosanitary treatment. This study is an assessment of plant protection practices in viticulture based on indicators of pressure (TFI: treatment frequency indicator) and risk (IRSA: indicator of risk to applicator health; IRTE: indicator of toxicity risk to the environment). Several indicators were developed to assess the impact of pesticides on health and the environment [40–43]. Due to the lack of global indicators to assess the toxicity risk of plant protection practices on health and the environment, the CIHEAM-IAMM team has developed risk indicators (IRSA and IRTE) of pesticide use on health and the environment, allowing for the consideration of the ecotoxicological and toxicological impact of molecules and their physico-chemical properties [38,39]. The results of the indicators are derived from the EToPhy software (2020, APP deposit n°: IDDN.FR.001.090003.000. S.P.2020.000.31500. https://www.dephyto.com/ (accessed on 12 September 2024)) developed by the CIHEAM-IAMM research team [38,44–46]. This approach therefore requires the mobilization of a database of plant protection practices in the vineyards of southwestern France.

Despite the significant role of viticulture in the French agricultural sector, the health and environmental impacts of plant protection practices remain a critical and insufficiently explored issue. Organic farming is often perceived as more sustainable and environmentally friendly. However, its reliance on copper- and sulfur-based products raises questions about its actual environmental impact compared to conventional or integrated farming systems. This study addresses the following research problem: To what extent do organic and conventional/integrated farming systems differ in their health and environmental impacts, and how do specific phytosanitary practices contribute to these outcomes? We hypothesize that organic farming, while reducing certain health risks, may present significant environmental risks due to the use of copper and sulfur. Conversely, we expect conventional/integrated farming to pose higher health risks due to synthetic products but potentially lower environmental risks in specific contexts. By examining these hypotheses, this research aims to provide a nuanced understanding of the trade-offs and inform stakeholders about strategies for sustainable viticulture.

This study is essential because it addresses a significant gap in the existing literature: the lack of a comparative and localized assessment of phytosanitary practices in organic and conventional/integrated farming systems. Previous research has often been limited in geographical scope or focused on broader trends without examining specific farming practices in detail. By focusing on the Gironde department, a key wine-producing region, this study provides a unique perspective that combines methodological rigor with practical relevance. It aims to support stakeholders in adopting more sustainable practices, thus contributing to both scientific understanding and practical solutions in sustainable viticulture.

This article is structured into three main sections (except for the Introduction), each addressing key aspects of the health and environmental impact assessment of plant protection practices in vineyards. The first section introduces the approach used in this assessment, including a presentation of the study area and an analysis of phytosanitary practices in organic and conventional/integrated farming systems. It details the methodological framework, the sample of wine-growing farms surveyed, and the indicators developed to evaluate the impact of these practices. The second section focuses on the analysis of wine-growing phytosanitary practices, comparing them across farming systems and within the Gironde department. This includes descriptive statistical analyses and specific impact assessments of cropping treatments for conventional/integrated and organic farming plots. Finally, the article concludes by summarizing the findings and their implications for sustainable wine-growing practices.

2. Approach to the Health and Environmental Impact Assessment of Plant Protection Practices in Vineyards

2.1. Presentation of the Study Area

In order to study plant protection practices in viticulture and to assess their associated risk level to human health and the environment, various field surveys were carried out to collect information on cropping treatment schedules at the wine-growing plot level. This work was carried out in two departments in southwestern France: Gironde and Hérault (Figure 1).

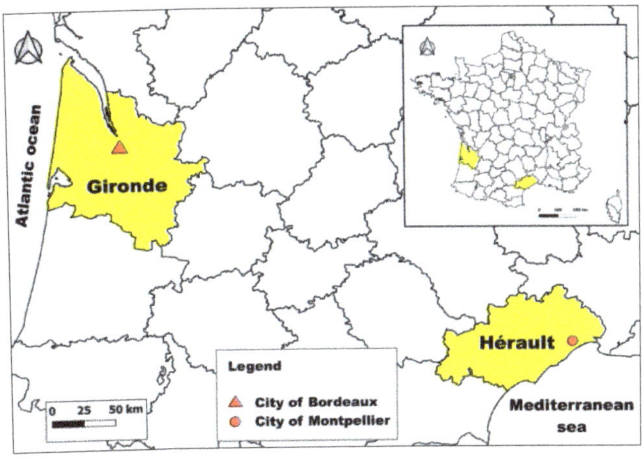

Figure 1. Location of the study area (source: BDD Geofla (https://www.data.gouv.fr/fr/datasets/geofla-r/ (accessed on 18 September 2024)) 2016).

In the Hérault department, the farms surveyed are located across 9 communes in the *Etang de l'Or* watershed, a few kilometers east of Montpellier, and in the commune of Combaillaux (north of Montpellier). In the Gironde department, the farms are located in the experimental catchment area of Marcillac in the Blayais area (north of Bordeaux), a wine-growing region located on the right bank of the Gironde estuary.

The southern and southwestern French departments, especially the Gironde department, offer significant territorial diversity across the Great South-West of France. However, agriculture occupies a large part of each department: nearly half of the department area (Figure 2).

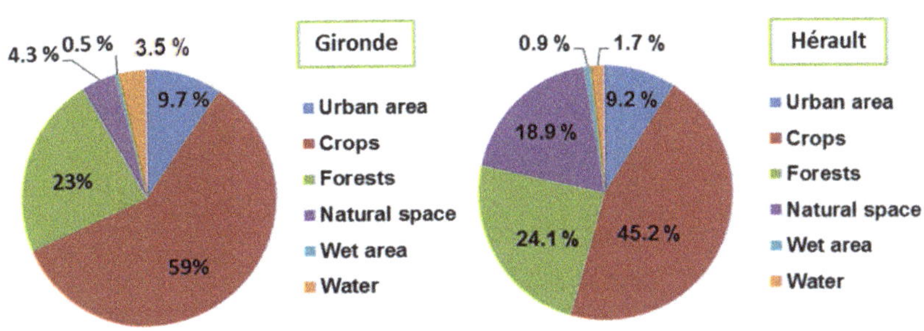

Figure 2. Land use in the Gironde and Hérault departments (source: OSO 2017 data of Theia cnes (https://theia.cnes.fr/atdistrib/rocket/#/search?collection=OSO (accessed on 28 September 2024))).

The Gironde department is dominated by permanent crops, which accounts for its high pesticide consumption [47]. It is ranked as the leading pesticide consumer in France, with over 3400 tons [48]. It is characterized by a strong wine-growing footprint (it is the largest French wine-growing department), and its wine industry extends over 120,120 ha, i.e., almost half of the departmental agricultural area (272,062 ha), of which 13,909 ha are certified organic [49]. Gironde is the country's largest organic wine-growing department, followed by Hérault.

The Hérault department is the second largest wine-growing department in France, with 84,945 ha of vineyards (45% of his UAA) [50]. The share of wine estates under AB organic certification has increased significantly in this department. Organic wine-growing areas cover 12,255 ha [49]. The Occitanie region is the emblem of organic wine in France, as it covers 38% of the French organic wine-growing area [30].

2.2. Approach to Analysis of Wine-Growing Phytosanitary Practices in Organic and Conventional/Integrated Farming Systems

2.2.1. Methodological Approach

Figure 3 below presents the methodological approach to the analysis of plant protection practices in the vineyards of southwestern France. It shows the initial database and its use in the process of calculating the pressure and risk indicators TFI, IRSA, and IRTE [37,38], as well as the analysis of the results obtained according to the organic and conventional/integrated farming systems.

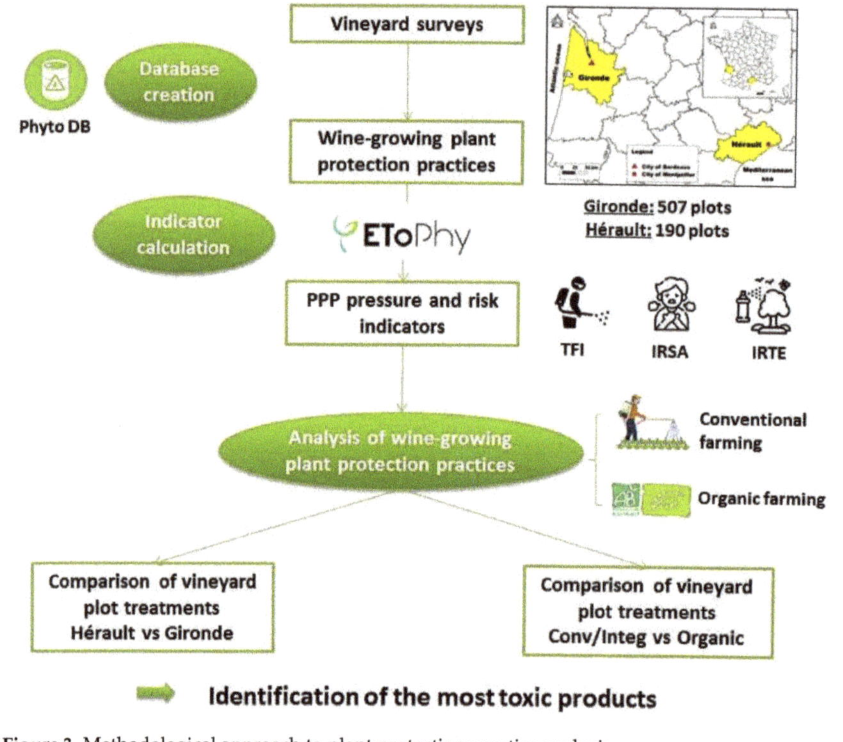

Figure 3. Methodological approach to plant protection practice analysis.

2.2.2. Presentation of the Sample of Wine-Growing Farms Surveyed

Since 2009, the research team at the Mediterranean Agronomic Institute of Montpellier (CIHEAM-IAMM) has been building a database of plant protection practices collected from farmers and agricultural cooperatives in two departments in the south-west of France: Hérault and Gironde. A total of 49 representative wine-growing farms were surveyed during only one year, 2015/2016. In the Gironde department, 507 cropping treatment schedules were collected (which corresponds to the number of plots). These surveys in the Gironde department were carried out in collaboration with INRAE of Bordeaux, ETBX research unit [51,52]. A total of 190 cropping treatment schedules were collected from winegrowers in the Hérault department. The sampling of vineyard plots in this department was carried out within the framework of the Tram research project (Plan Ecophyto 2018) [38,39,44,53]. (The Tram (2010–2014) research project was approved in September 2010 and was funded by ONEMA. Its objectives were to develop a methodology for testing the agro-environmental and technical-economic impact of an integrated reduction in the use of pesticides, taking into account the different levers of action from field level to catchment area level with weightings to take account of environmental specificities. https://ecophytopic.fr/recherche-innovation/concevoir-son-systeme/projet-tram (accessed on 12 September 2024)) These farms in the departments of Gironde and Hérault were divided into conventional/integrated and organic vineyards (Table 1).

Table 1. Distribution of the wine-growing farms surveyed by department.

Department	Crop	Number of Farms	Number of Plots	Area (ha)
Gironde	Conv/integrated vineyard	30	467	726.60
	Organic vineyard	9	40	195.83
Hérault	Conv/integrated vineyard	9	180	348.74
	Organic vineyard	1	10	19.82
	Total	49	697	1291

This sampling will be used to assess the health and environmental impact of plant protection agricultural practices using pressure (TFI) and risk (IRSA and IRTE) indicator outcomes from the EToPhy tool on the surveyed wine-growing farms in the Hérault and Gironde departments and according to conventional/integrated and organic farming systems.

2.2.3. Indicators for Assessing the Plant Protection Impact of Wine-Growing Phytosanitary Practices

The assessment of plant protection practices is based on the complementarity between the TFI, IRSA, IRTE, and risk sub-indicators (acute IRSA; chronic IRSA; terrestrial IRTE; bird IRTE; aquatic IRTE), which makes it possible to determine the degree of toxicity of the practices to human health and to the three environmental systems: soil, air, and water [38,39].

- The treatment frequency indicator (TFI): Plant protection pressure varies from one region to another and depends on soil and climatic conditions, agricultural practices, sanitary pressure, and the crops concerned. Because of their large surface area or their particular sensitivity to one or more pest(s), some crops, particularly fruit trees and vines, accumulate a high proportion of the pesticides used. The treatment frequency indicator (TFI) corresponds to the number of registered doses applied to a plot during a crop year. The registered dose is defined as the effective application dose of a product according to the pair (crop/pest).

$$TFI = \frac{\text{applied dose}}{refrence\ dose} \times \frac{treated\ surface}{\text{plot area}} \tag{1}$$

This indicator is calculated at different levels depending on the need for analysis (product, plot, crop, farm, and region) [54]. Consequently, the TFI reflects the intensity of

PPPs use and therefore the plant protection pressure exerted at the different levels, and it also describes the dependence of farmers on these products.

- Agri-environmental indicators (IRSA and IRTE): In this study, the choice of parameters was based on the risk indicators IRSA (indicator of risk to applicator health) and IRTE (indicator of toxicity risk to the environment), both calculated using the EToPhy software. These indicators are generic and modular, and they can be calculated at different levels, from plot to farm [38,39,55]. They are subsequently used to analyze the health and environmental risk of plant protection practices by crop. The calculation of IRSA and IRTE indicators is performed for each active ingredient (AI) according to the following equations:

$$IRSA = \text{IRT AI} \times \text{FPf} \times \text{FCP} \quad (2)$$

$$IRTE = (1.75 \times (T + O) + A + M + P + 1)^2 \quad (3)$$

IRSA and IRTE are composite indicators that assess the acute and chronic toxicity of plant protection products by taking into account several critical variables such as the characteristics of the active ingredient (physicochemical and ecotoxicological properties), the commercial preparation (concentration of the active substance, applied dose, ...), the place of application (full field, greenhouse cultivation, ...) and the type of crop (market gardening, arboriculture, ...).

The indicator of risk to applicator health (IRSA) is a scoring indicator. It assesses the acute and chronic toxicities of plant protection products by considering the physicochemical and toxicological properties of active ingredients. Furthermore, this indicator is broken down into sub-indicators: acute toxicity (acute IRSA), which is related to skin and eye irritation, inhalation, etc., and chronic toxicity (chronic IRSA), which represents the risks related to cancer, reproduction, neurotoxicity, and endocrine disruption [38,39,41]. This indicator is based on the calculation of the Toxicity Risk Index (IRT), which takes into consideration the acute and chronic toxicity of active ingredients with their persistence factor (bioaccumulation in living tissues).

The indicator of toxicity risk to the environment (IRTE) assesses the eco-toxicological impacts on non-target living organisms (terrestrial invertebrates, birds, aquatic organisms), as well as the physico-chemical behavior of molecules in the receiving environment (mobility, persistence in the soil, bioaccumulation). Its calculation is based on physicochemical parameters, eco-toxicity, interception factors, drift, runoff, and drainage potential [38,39,41,56]. This indicator is broken down into three sub-indicators: terrestrial IRTE (IRTE T), bird IRTE (IRTE B), and aquatic IRTE (IRTE A). They allow decision-makers and researchers to implement strategies for protecting target organisms, mainly bees and pollinating insects, and reducing toxicity in aquatic environments.

3. Analysis of Wine-Growing Phytosanitary Practices According to Farming Systems

3.1. Overall Analysis of Plant Protection Practices in Vineyards and Comparison Between Departments and Farming Systems

The results of the global analysis of the plant protection practices applied on the surveyed farms in the Gironde and Hérault departments are illustrated in the figure below (Figure 4).

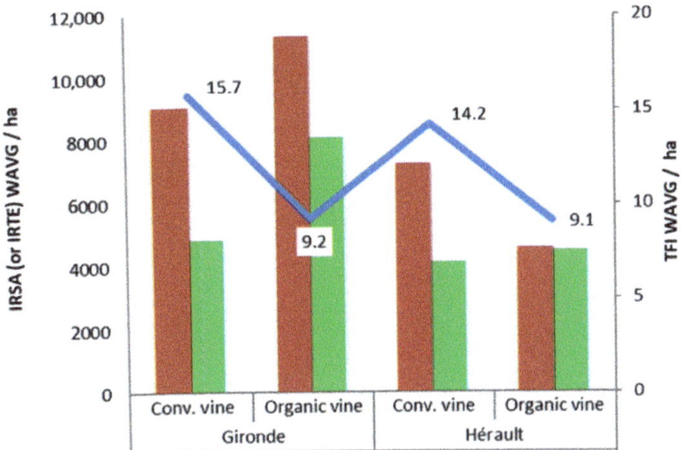

	Gironde Conv. vine	Gironde Organic vine	Hérault Conv. vine	Hérault Organic vine
IRSA/ha	9039	11,305	7243	4589
IRTE/ha	4852	8117	4146	4515
TFI/ha	15.7	9.2	14.2	9.1

Figure 4. Variability of indicators according to conventional/integrated and organic farming between departments (values expressed as weighted average per hectare).

The graph presents the results of the phytosanitary pressure and risk indicators (TFI, IRSA, and IRTE) in the two departments for conventional/integrated and organic farming systems.

This illustration shows a difference between the two departments in terms of risk. The risk to the applicator's health and to the environment is higher in Gironde than in Hérault. However, it is important to bear in mind that the data collected in the Gironde department only concern one year (2015–2016), which shows that the climate effect is not negligible. This effect acts indirectly on the choice of plant protection products, which changes from a dry year to a wet year, requiring more interventions and more effective products against certain diseases and pests.

According to farming systems, the average TFI/ha in organic farming is lower than in conventional/integrated farming (in Gironde, TFI conventional/integrated vine = 15.7; TFI organic vine = 9.2). This explains why farmers tend to decrease the treatment frequency when switching from conventional/integrated to organic farming. By comparing the average TFI/ha values of our sampling with the average TFI values on the different wine-growing areas in France based on surveys of wine-growing phytosanitary practices during the year 2016 calculated and published by Agreste (The Agriculture Ministerial Statistical Department in France) in 2019, we find that the value of TFI in the Bordelais wine-growing area is 17.2 (Figure 5) against a value of 15.7 calculated on our sampling. In the Languedoc wine-growing area, the average TFI/ha according to the Agreste report is 13.8 (Figure 5), against a value of 14.2 in our study (Figure 4). The values are close, which confirms the results of a comparison study [57] which aims to analyze the phytosanitary pressure variability between the different wine-growing areas but without taking into account organic wine-growing practices. Our study complements these results while also emphasizing the comparison between farming systems (conventional/integrated vs. organic).

According to results presented in Figure 4, the risk level is much higher for organic farming in our study sample from the Gironde department. This result shows, firstly, that there is no correlation between treatment frequency and risk. The risk to the applicator's health is not correlated with the TFI of plant protection pressure. Therefore, an increase in the TFI cannot lead to a direct increase in risk. However, this increase is mainly due to the products used and the acute and chronic toxicity degree of the active ingredients chosen by the farmers. Secondly, it can be concluded that even if the plant protection products in organic farming are not used very frequently, they present a significant risk to health and the environment compared to conventional methods in the Gironde department. This is therefore due to poor choices on the part of the farmers, choices based only on the efficiency of the products yet ones which do not take into account their eco-toxicological and toxicological characteristics.

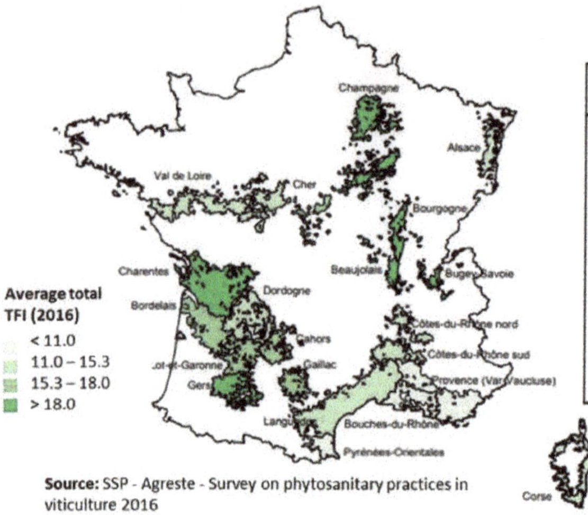

	2010	2013	2016
Alsace	10.4	11.1	14.9
Beaujolais	16.4	19.1	18.7
Bordelais	14.2	17.0	17.2
Bourgogne	15.2	20.0	19.4
Champagne	17.8	21.5	23.5
Charentes	15.3	18.1	18.1
Languedoc hors PO	11.8	13.5	13.8
Pyrénées-Orientales	8.9	9.3	10.1
Provence (Var-Vaucluse)	11.3	9.4	10.3
Val de Loire	8.9	13.7	15.1
Bouches-du-Rhône	8.8	9.3	9.4
Dordogne	12.4	15.0	15.1
Gers	17.2	19.7	18.9

Source: Agreste - Surveys on phytosanitary practices (2010, 2016) in viticulture

Figure 5. Average total TFI of wine-growing areas in France [57].

In order to test the dependence of the indicators on each other, a correlation analysis was conducted in order to test the shape of the correlation curve between two indicators. This analysis was performed using RStudio software (Version 1.2.5042), with the indicators of risk and phytosanitary pressure values as input data. The graphs below are the output result (Figure 6). This presentation illustrates a scatter plot of indicator values in order to analyze the correlation between phytosanitary pressure and risk indicators.

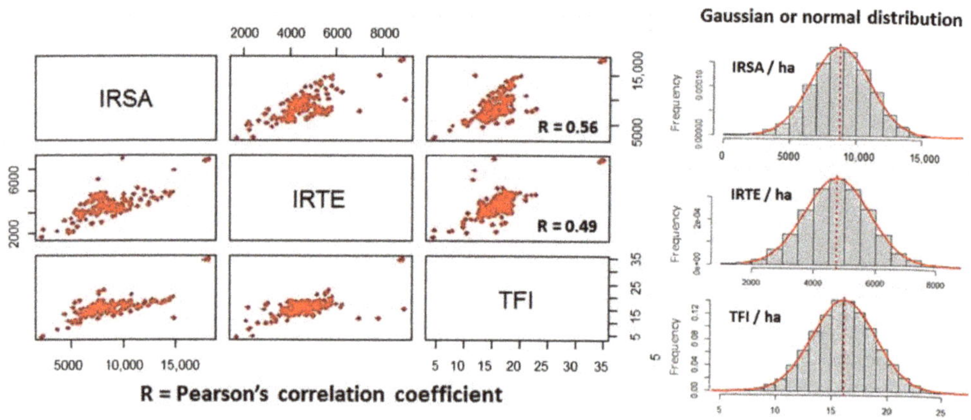

Figure 6. Correlation analysis and Gaussian distribution of phytosanitary pressure and risk indicators.

The scatter plots in the correlation graph between the two risk indicators (IRSA, IRTE) and the TFI barely take on the appearance of a straight line through the origin. Points are distributed randomly. Therefore, the phytosanitary pressure indicator (TFI) is moderately correlated (R value is between 0.4 and 0.6) with the risk indicators IRSA and IRTE. This result shows that the phytosanitary treatment frequency cannot indicate the toxicity expressed by the risk indicators. Even at low frequencies of phytosanitary treatment, the risk indicators appear with very high values. This high toxicity risk is related to the eco-toxicological characteristics of the products and active ingredients and it is not directly linked to the dose and frequency of the applied treatment. This is sometimes the case of products applied at a low dose but which induce a very high toxicity risk. This analysis clearly shows the usefulness of risk and phytosanitary pressure indicators and the complementarity between these indicators in order to provide an exhaustive analysis of the health and environmental impact of agricultural phytosanitary practices to the different stakeholders involved in pesticide management.

An assessment of the toxicity degree of plant protection practices was carried out using sub-indicators (acute and chronic IRSA; terrestrial, bird, and aquatic IRTE) to obtain a deepened analysis of their health and environmental impact (Figure 7). The graph below shows a comparison of the toxicity share between the two wine-growing farming systems (conventional/integrated, conv.; organic) and between the two departments (Gironde, Hérault). The sub-indicators of risk to human health are presented in figure A and the sub-indicators of risk to the environment are presented in figure B. All values were expressed as weighted average per hectare.

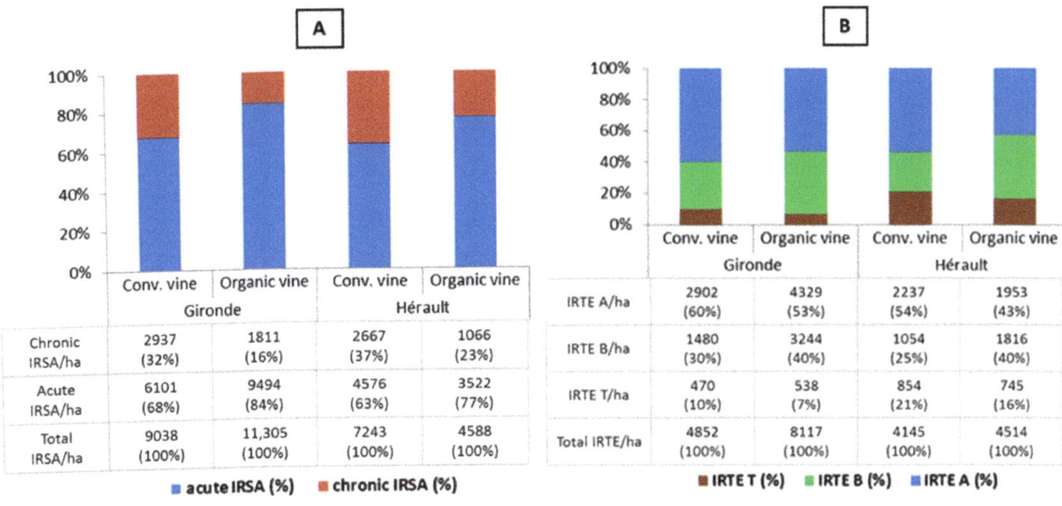

Figure 7. The toxicity share of plant protection practices between farming systems and between departments ((**A**): share of acute and chronic toxicity; (**B**): share of toxicity on each environmental system).

Regardless of the farming system and department, the share of acute toxicity risk related to plant protection practices is greater than 60% (Figure 7A). Most of the products used on the wine-growing farms surveyed have a health risk that is more acute (risk of irritation and risk due to inhalation, skin, or ingestion exposure) than chronic (carcinogenic, mutagenic, toxic to reproduction, neurotoxic, and endocrine-disruptive).

Figure 7B illustrates the impact on non-target organisms in the three environmental systems: water (aquatic IRTE), air (bird IRTE), and soil (terrestrial IRTE). The share of toxicity risk to the aquatic environment and birds represents over 80% of the overall risk, regardless of the farming system.

In order to better understand the risk values calculated according to the two farming systems, it is necessary to analyze the plant protection products used that are responsible for this toxicity, whether to human health or to the environment. The following tables illustrate the products and active ingredients that were used the most in our sample of vineyard plot treatments in Gironde and Hérault, including the active ingredients that present the highest risk to human health and the active ingredients that present the highest risk to the environment (Tables 2 and 3).

Table 2. Classification of Top 5 plant protection products and active ingredients used in the two departments according to TFI and the quantity applied in kg/ha.

Most Used Products (High TFI/ha)				Most Used Active Ingredients (High AI Quantity/ha)			
Gironde		Hérault		Gironde		Hérault	
Product	Active Ingredient	Product	Active Ingredient	Active Ingredient	AI Qtity (kg/ha)	Active Ingredient	AI Qtity (kg/ha)
Chaoline	Fosetyl-aluminum	Abilis	Triadimenol	Sulphur *	10.0	Sulphur *	10.0
Steward	Indoxacarb	Bouillie bordelaise RSR disperss	Copper sulfate	Potassium bicarbonates *	4.2	Potassium phosphonates *	2.9
Ysayo	Cyazofamid	Kavea DG	Mancozeb	Potassium phosphonates *	3.0	Oryzalin	2.9
Jokari	Acrinathrin	Turkoise	Fenazaquin	Copper sulfate *	3.0	Metiram	2.8
Consist	Trifloxystrobin	Clameur	Alpha-cypermethrin	Metiram *	2.8	Mancozeb	2.6

Fungicide , insecticide , acaricide . * Active ingredient used in organic wine-growing plots.

Table 3. Classification of the Top 5 plant protection products used in the two departments according to risk level.

AIs with Higher Risk to Human Health (High IRSA/ha)				AIs with Higher Risk to Environment (High IRTE/ha)			
Gironde		Hérault		Gironde		Hérault	
Active Ingredient	IRSA/ha	Active Ingredient	IRSA/ha	Active Ingredient	IRTE/ha	Active Ingredient	IRTE/ha
Diquat *	3880	Copper oxychloride	1768	Diquat *	900	Dimethoate	1469
Fluazinam *	1167	Chlorothalonil	1353	Chlorpyrifos-methyl *	756	Chlorpyrifos	1024
Maneb *	837	Fluazinam	1247	Cyfluthrin	650	Chlorpyrifos-methyl *	711
Alpha-cypermethrin *	820	Chlorpyrifos	879	Sulfur *	506	Copper oxychloride	676
Meptyldinocap *	774	Meptyldinocap *	853	Emamectine Benzoate *	473	Cyfluthrin	652

Fungicide , insecticide , acaricide . * Active ingredient used in organic wine-growing plots.

The plots surveyed in the Gironde department used 171 products (with 74 active ingredients). In the Hérault department, 155 products (with 91 active ingredients) were used in all the analyzed plant protection treatments.

The five products with the highest TFI in the Gironde department are fungicides and insecticides (Table 2). The five products with the highest TFI in the Hérault department are a mix of fungicides, insecticides, and an acaricide. They are totally different from those used in Gironde.

The most used active ingredients (AIs) in both departments are sulfur and potassium phosphonates, with different quantities per hectare (10 kg/ha in the Gironde department;

3.2 kg/ha in the Hérault department). These AIs are used in both conventional and organic wine-growing plots. Overall, the results in this table indicate that most of AIs identified in both departments with high quantities are used in organic farming.

The products that represent the highest risk to human health and the environment are classified in the table above (Table 3). Diquat represents the AI with the highest risk to human health (IRSA/ha = 3880) and the environment (IRTE/ha = 900) in the Gironde department.

The active ingredients used in the Gironde department that represent the highest risk of toxicity to human health and the environment are used in organic farming. Copper oxychloride represents the highest risk to human health in the Hérault department (IRSA/ha = 1768).

3.2. *Results of the Analysis of Plant Protection Practices in Wine-Growing Plots in the Gironde Department*

3.2.1. Descriptive Statistical Analysis of Plant Protection Practices in Wine-Growing Plots

The descriptive statistical analysis of phytosanitary pressure and risk indicators was carried out using RStudio software (Version 1.2.5042) to define a set of statistical parameters in order to assess the variability in plant protection practices at the wine-growing plot level in the Gironde department. The results are presented for the Gironde department as it has more vineyard plot samples.

This analysis was carried out for each farming system (conventional/integrated and organic farming) separately in order to compare the health and environmental impact of the plant protection practices. Table 4 presents a descriptive analysis of the phytosanitary pressure and risk indicators for plots in conventional/integrated farming.

Table 4. Descriptive statistics of phytosanitary pressure and risk indicators of conventional/integrated wine-growing plots in the Gironde department.

Indicators	Min.	Max.	Median	Mean	STDEV	CV
IRSA/ha	2274	18,097	8346	8786	2206	0.25
IRTE/ha	1693	8983	4737	4745	1038	0.22
Acute IRSA/ha	1463	11,730	5766	5853	1423	0.24
Chronic IRSA/ha	539	6500	2893	2933	974	0.33
IRTE T/ha	0	1494	559	512	278	0.54
IRTE B/ha	304	4383	1415	1439	565	0.39
IRTE A/ha	938	6661	2772	2789	645	0.23
TFI/ha	4.3	34.9	15.9	16.1	2.8	0.17

The results show a wide variability in indicator values between the minimum and the maximum values. The risk indicator for human health ranges from 2274 to 18,097. Likewise, the environmental risk indicator ranges from 1693 to 8983. The treatment frequency indicator ranges from 4.3 to 34.9 (Table 4); we know that the average TFI in the Bordeaux wine-growing area was 17.2 in 2016 [57]. These results represent the toxicity risk and phytosanitary pressure values calculated at the plot scale (weighted per hectare) using the EToPhy software (Version 1.2.5042). Although the treated plots were occupied by the same crop and the same farming system, the indicators are highly variable. This variability can be explained by the great differences in farmers' treatment strategies and their choices of plant protection products.

Table 5 presents a descriptive analysis of phytosanitary pressure and risk indicators for plots in organic farming.

The results show a wide variability in indicator values between the minimum and the maximum values. The risk indicator for human health ranges from 8065 to 14,669, with a mean value of 11,469. Likewise, the environmental risk indicator ranges from 6256 to

10,273, with a mean value of 8227. The treatment frequency indicator ranges from 7.3 to 11.3, with a mean value of 9.3.

Table 5. Descriptive statistics of phytosanitary pressure and risk indicators of organic wine-growing plots in the Gironde department.

Indicators	Min.	Max.	Median	Mean	STDEV	CV
IRSA/ha	8065	14,669	11,048	11,469	3047	0.26
IRTE/ha	6256	10,273	9289	8227	1426	0.17
Acute IRSA/ha	6677	12,920	9403	9839	2848	0.29
Chronic IRSA/ha	1118	3189	1709	1630	329	0.20
IRTE T/ha	0	890	415	294	218	0.74
IRTE B/ha	2265	4509	3678	3567	440	0.12
IRTE A/ha	3242	5684	5111	4343	980	0.22
TFI/ha	7.3	11.3	9.9	9.3	1.0	0.10

In order to better present the distribution of the risk indicators calculated for our sample, we present them using box plots, which represent the most suitable method to display our data (Figure 8). Boxes are drawn with ends at quartiles Q1 and Q3. The statistical median Q2 is represented as a horizontal line in the box; there are as many values above this value as there are below it in the sample (Figure 9).

Figure 8. Variability analysis of risk indicators for the surveyed wine-growing plots (conventional/integrated and organic farming).

The graph in Figure 8 shows the variability in the risk indicators and sub-indicators calculated for plots in conventional/integrated and organic farming in the Gironde department. This graph shows more or less symmetrical data, which indicates that the results of the indicators were normally distributed.

The range of the acute IRSA indicator shows good symmetry, with a wide distribution presented by the large difference between the min and max risk values. In contrast, the terrestrial IRTE indicator shows a narrow distribution of values, which indicates that the variability in terrestrial risk toxicity is very low from one cropping treatment schedule to another. The products applied in this farming system (conventional/integrated) present approximately the same level of toxicity.

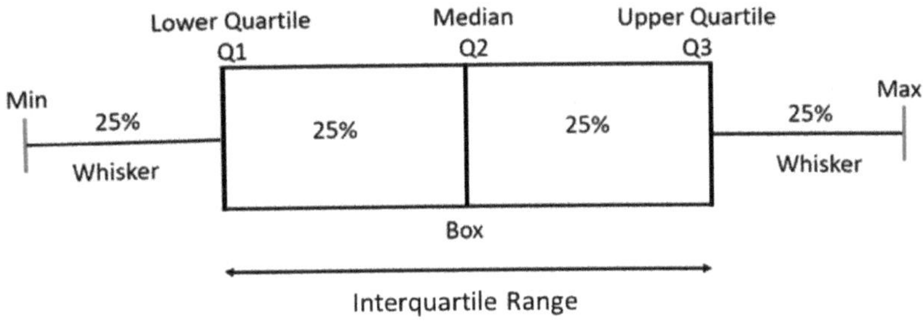

Figure 9. Explanation of the data.

The green boxes show the distribution of the risk indicators and sub-indicators calculated for the organic plots in Gironde (Figure 8). The risk toxicity indicators on the environment show low asymmetry with a narrow distribution. The minimum and maximum risk values are not too far apart, except for acute and human health risk s(acute IRSA and IRSA). So, acute risk represents the indicator with the most variability between minimum and maximum values. Acute risk depends largely on the formulation of the phytosanitary products applied, specifically on the toxicological properties of active ingredients, although the attenuation of human health risk can be achieved by choosing less toxic active ingredients during phytosanitary treatments.

A comparison of the distribution of the two farming systems' risk values shows that the variability within conventional/integrated farming is much higher, as the acute risk indicator varies within a range of 10,000, while that for organic farming does not exceed 6000.

3.2.2. Analysis of the Impact of Cropping Treatments on a Wine-Growing Plot in Conventional/Integrated Farming

This case study of wine-growing plot treatments will allow us to identify the plant protection products used during the cropping season, their treatment frequency, and the toxicity risk level associated with each product. This will be used to select the products that most contribute to the overall risk level for plots.

A conventional/integrated wine-growing plot in the Gironde department was chosen from the group of plots with a medium input of phytosanitary treatments, as it represents values close to the average risk and pressure values (as determined through a cluster analysis of plant protection practices based on pressure and risk indicators as classification criteria). This vineyard plot has an area of 85 ha. On this plot, the farmer chose to treat his vineyard with 22 products (Figure 10 and Table A1 in Appendix A).

The figure shows the risk indicators and the TFI calculated for each product. First of all, it can be noted that there is no correlation between treatment frequency (TFI) and risk (IRSA and IRTE). Fungicide 6 contributes more to risk than to pressure (low TFI), while fungicide 11 has a high TFI and a low level of risk to human health and the environment. The TFI/ha in this plot is low (12.6) if we compare it with the average TFI value for the Bordelais wine-growing area [57].

Six of the products used contribute to more than 50% of the plot's overall risk to human health. These products include five fungicides used against downy mildew and one herbicide. Herbicide 1 is made from ammonium glufosinate, which was withdrawn from the French market by the Anses (French Agency for Food, Environmental, and Occupational

Health and Safety) in 2017 because the risks to human health related to exposure to this product could not be ruled out for the farmers using it as well as for people in the vicinity of the treated areas.

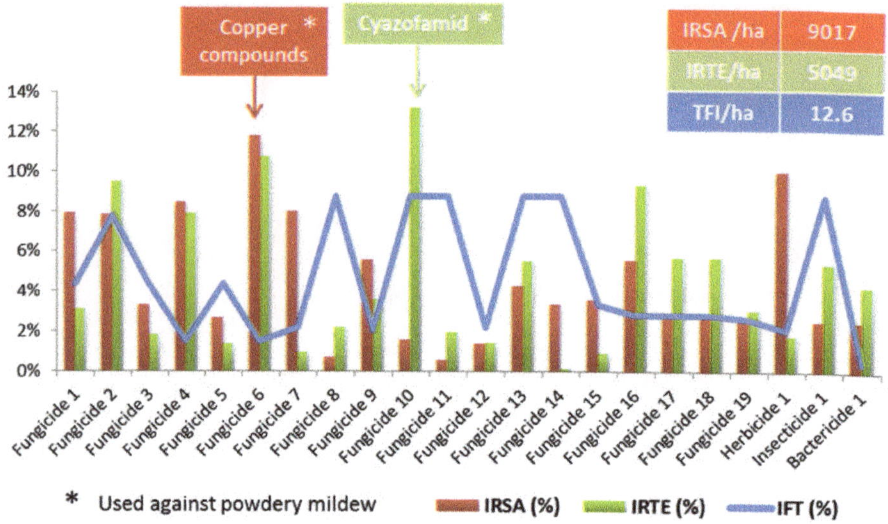

Figure 10. The contribution of the plant protection products applied on a wine-growing plot to risk and plant protection pressure (conventional/integrated farming).

In terms of risk to the environment, we found that four fungicides (fungicides 2, 4, 6, and 10) used against downy mildew contribute the most to environmental risk. Fungicide 10 (active ingredient: Cyazofamide) presents the highest level of environmental risk (14% of the overall plot risk level).

Figure 11 shows the share of acute and chronic toxicity for each product used in the treatment of the wine-growing plot studied in the previous graph (Figure 10). Acute IRSA is equal to 6291 (70% of total IRSA) and chronic IRSA is 2726 (30% of total IRSA). The risk of toxicity to human health on this wine-growing plot mainly involves acute toxicity, which exceeds 50% of the overall toxicity level of all fungicides, except fungicide 7 and 11. On the other hand, the share of toxicity is more chronic rather than acute only in the case of herbicide 1 and insecticide 1. The chronic IRSA of insecticide 1 represents 70% of the overall risk to applicator health; this product generates neurotoxicity, impacts reproduction and organ development, and has endocrine-related effects. In addition, fungicide 7 (active ingredient = Meptyldinocap) and fungicide 11 (active ingredient = Trifloxystrobin) present a high chronic risk.

Breaking down the IRSA into two sub-indicators (acute and chronic IRSA) makes it easier for the farmer to recognize the toxicological characteristics of each product used to avoid products with a high chronic toxicity and to improve pesticide management with a better choice of plant protection products.

Figure 12 presents the share of toxicity risk of each product used for the different environmental systems: air (birds), water (aquatic organisms), and soil (invertebrate terrestrial organisms). The value of aquatic IRTE is equal to 3178 (63% of total IRTE), aerial IRTE is 1674 (33% of total IRTE), and terrestrial IRTE is 197 (4% of total IRTE). We can observe that most of the products have a high aquatic toxicity level, which for some products can represent 100% of their IRTE value, as is the case for fungicides 7 and 8, made with the active ingredients Meptyldinocap and Quinoxyfen.

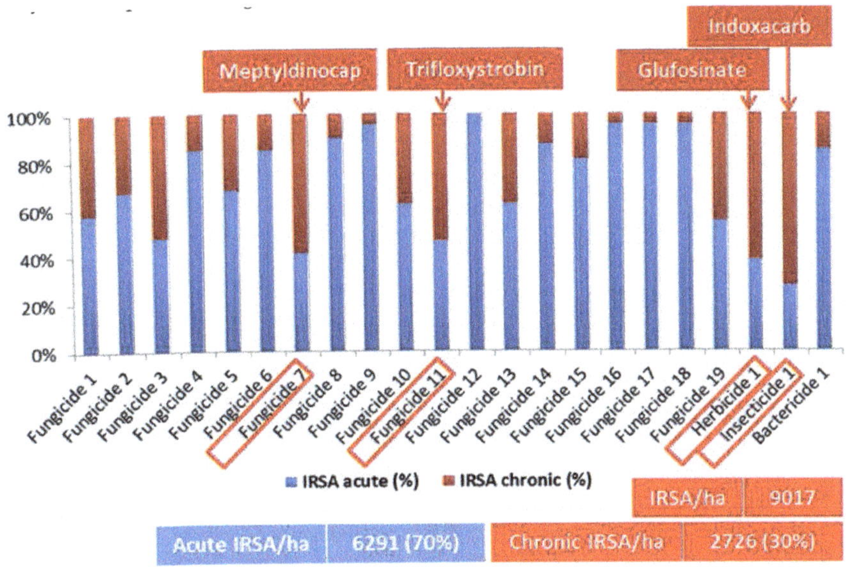

Figure 11. The contribution of the plant protection products used on the conventional/integrated wine-growing plot to acute and chronic toxicity.

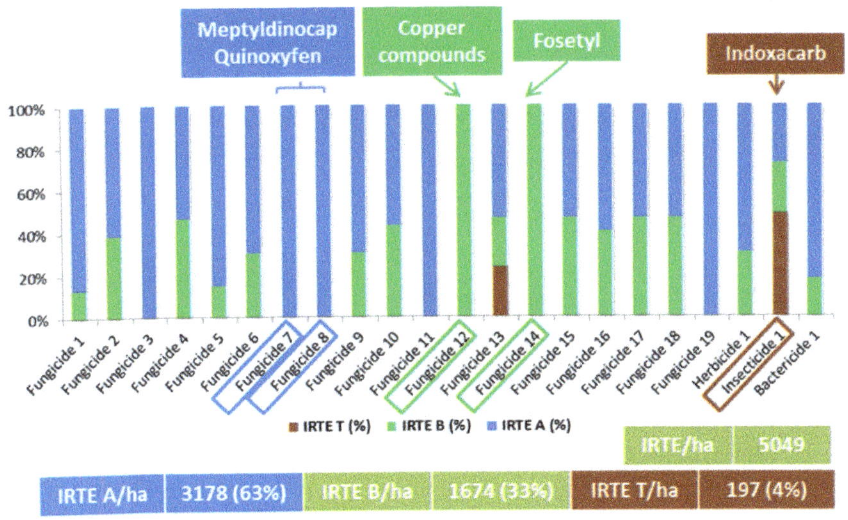

Figure 12. The contribution of the plant protection products used on the conventional/integrated wine-growing plot to toxicity risk for each environmental system.

Fungicide 12 (active ingredient = copper compounds) and fungicide 14 (active ingredient = Fosetyl) represent the highest risk for air. Fungicide 12 is copper-based which gives us an indication of the toxicity of copper-based products to the air environment, in particular birds.

Despite the low risk to the terrestrial environment of all the products used, air toxicity represents more than 50% of the total IRTE for insecticide 1 (active ingredient = Indoxacarb). It is a toxic product for bees that must be used outside the flowering stage.

3.2.3. Analysis of the Impact of Cropping Treatments on a Wine-Growing Plot in Organic Farming

In this part, we will present the results of the plant protection treatment of an organic wine-growing plot in the Gironde department. The plot was chosen from a group of plots which represents the average indicator values for organic farming (as determined through a cluster analysis of plant protection practices based on pressure and risk indicators as classification criteria). The area of this plot is 3.59 ha. On this plot, the farmer used five plant protection products, which are all sulfur- and copper-based fungicides (Figure 13 and Table A2 in Appendix A).

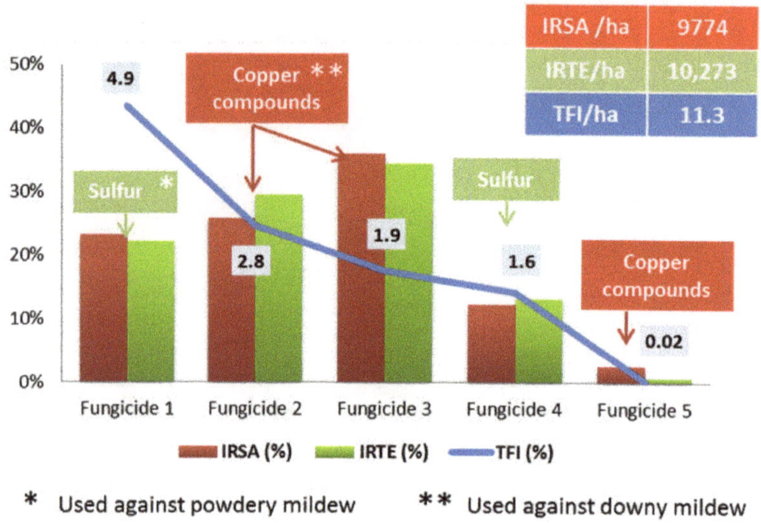

Figure 13. The contribution of the plant protection products applied on the wine-growing plot (organic farming) to risk and plant protection pressure.

These illustrations show the risk indicators and the TFI calculated for each product used in this wine-growing plot. The values of the phytosanitary pressure (TFI) and risk (IRSA and IRTE) indicators do not show a correlation relationship, as in the case of the previously studied conventional/integrated farming plot. Indeed, the decrease in toxicity risk is not associated with a reduction in treatment frequency (TFI), as shown by the comparison between fungicides 1 and 3.

All of the fungicides used in this plot are based on sulfur or copper and thus present a risk to human health and to the environment, particularly fungicides 1, 2, and 3 (Figure 13).

Fungicide 2 and fungicide 3 contribute the most to human health and environmental risk. Nevertheless, they are not used with the highest TFI. These two fungicides are based on the same active ingredient, "copper compounds" made of copper, and used against downy mildew.

Fungicide 1, made of 80% sulfur, is used with the highest TFI (4.9) to control powdery mildew.

Figure 14 presents the share of acute and chronic toxicity for each product used in the treatment of the organic wine-growing plot. IRSA per hectare is equal to 9774; acute IRSA represents the most significant share, with 83% at the plot level. A more detailed analysis shows that the share of toxicity varies from one product to another. Fungicide 2 and fungicide 5 (copper-based) represent more than 70% of the chronic risk to overall human health risk. This type of risk is related to long-term effects such as neurotoxicity, reproduction, and endocrine effects. These two fungicides are used against downy mildew. However, fungicides 1 and 4 (sulfur-based) present a high acute toxicity risk related to short-term effects such as skin and eye irritation and respiratory tract impact via inhalation.

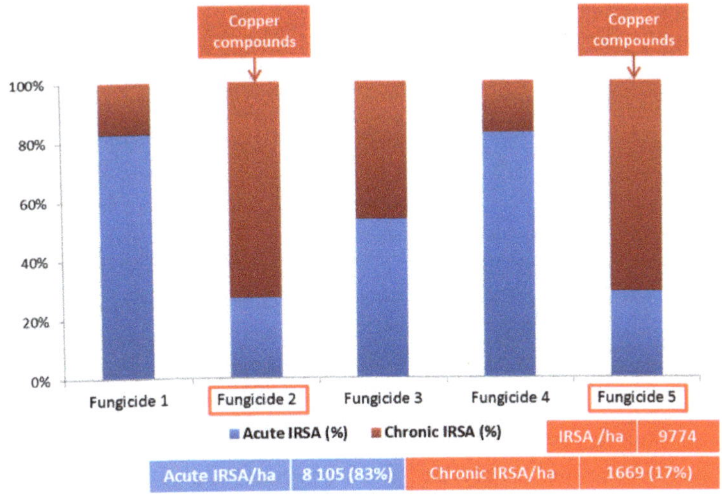

Figure 14. The contribution of the plant protection products used on the organic wine-growing plot to acute and chronic toxicity.

Figure 15 presents each product's share of toxicity risk to the different environmental systems: air (birds), water (aquatic organisms), and soil (invertebrate terrestrial organisms). The value of aquatic IRTE is equal to 5684 (55% of total IRTE), aerial IRTE is 5509 (44% of total IRTE), and terrestrial IRTE is 80 (1% of total IRTE). Copper-based products (fungicide 2, 3, and 5) are almost as toxic to air and aquatic environments as fungicide 2 and fungicide 5. Fungicide 3 contributes the most to aquatic toxicity risk, with more than 60% of the total IRTE. It is made of 35% copper and it is used against downy mildew.

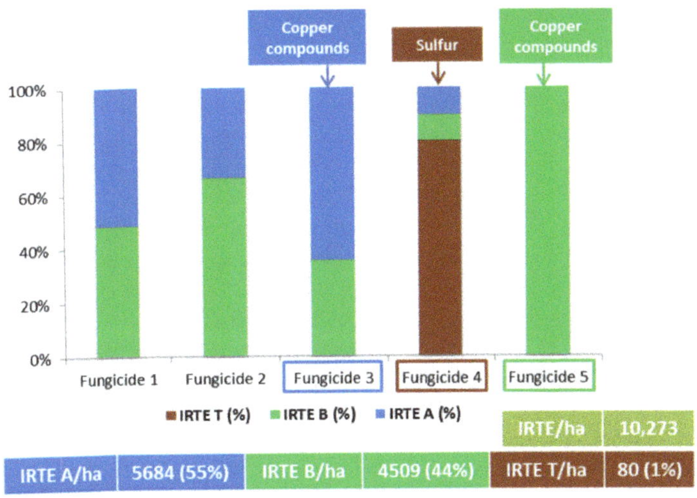

Figure 15. The contribution of the plant protection products used on the organic wine-growing plot to toxicity risk for each environmental system.

A deeper analysis of the environmental impact of plant protection products using sub-indicators can help land management authorities to develop biodiversity protection plans for the fauna and flora and to manage toxicity to natural environments, especially aquatic ones. Moreover, according to the soil type and plot location in relation to watercourses, risk levels can be weighted and more appropriate and targeted action plans can be drawn up in order to limit the impact of some plant protection products that are potentially harmful to human health and the environment.

Figures 12 and 15 illustrate the contribution of each product used to the toxicity risk for different environmental systems: air (birds), water (aquatic organisms), and soil (terrestrial invertebrates), in both conventional/integrated and organic farming systems.

Comparing Figures 12 and 15, 22 products were applied on the conventional vineyard plot, and only 5 products were used on the organic vineyard plot. However, the environmental toxicity risk of the pesticides used on the organic vineyard plot is higher compared to those used on the conventional/integrated vineyard plot (IRTE/ha for organic farming = 10,273; IRTE/ha for conventional/integrated farming = 5049). In contrast, the toxicity risk to applicator health is similar between phytosanitary practices in conventional/integrated and organic vineyards (Figures 11 and 14).

This difference mainly arises from the composition and formulation of the phytosanitary products used. In organic farming, most of the plant protection products are based on sulfur and copper (Appendix A, Tables A1 and A2). Consequently, these substances have a negative impact on aquatic, terrestrial, and aerial environments [58,59] and, in some cases, can be more toxic to living organisms than the substances used in conventional/integrated farming (Figure 15). This observation contradicts the common perception that plant protection products used in organic farming are free from risks to human health and the environment.

As a result, this analysis of the toxicity risk of plant protection products, based on various risk (sub-)indicators (IRSA, IRTE, acute IRSA, chronic IRSA, IRTE A, IRTE B, and IRTE T), helps farmers improve decision-making and choose active ingredients according to their physicochemical characteristics as well as their toxicological and eco-toxicological properties. In both conventional or organic farming, there are several approved products and sub-stances against a given pest or disease. However, farmers typically lack information on the toxicological and eco-toxicological properties needed to identify the substance or product that is least toxic to human health and the environment.

This strategy for improving plant protection practices aims to reduce the impact of pesticides on the environment and human health while designing sustainable farming systems with low inputs, combined with other alternative practices such as biological control and the use of natural pest control methods or environmentally friendly substances.

4. Conclusions

This work demonstrates the value of risk indicators such as IRSA, IRTE, and the treatment frequency indicator (TFI) as essential decision support tools for assessing and managing plant protection practices at the plot level. These indicators help farmers make informed choices to minimize the risks associated with plant protection products (PPPs) that pose significant threats to human health and the environment. By utilizing these tools, farmers can select better alternatives to high-risk products, thus contributing to more sustainable agricultural practices.

Through our analysis, we identified a novel approach for managing the selection of plant protection products based on their potential impact on human health and various environmental components, including air, soil, and water. This study highlights that the risk associated with plant protection practices is primarily determined by the formulation of the products used, with the active ingredient playing a key role. This finding underscores that the risk level is contingent upon the specific molecule applied, rather than the farming system employed.

The variability between farming systems (conventional/integrated and organic) in terms of toxicity risk to both human health and the environment arises from differences in the toxicity profiles of the products used. Conventional farming relies on synthetic products, while organic farming uses naturally derived substances, which, despite being deemed less harmful, still present significant toxicity risks. In organic farming, the high application rates of certain products, particularly sulfur- and copper-based compounds, contribute to environmental toxicity, especially when applied in large quantities. In contrast, the human health risks associated with organic products tend to be more acute and less chronic compared to conventional products.

This study also reveals how assessing the toxicity of different molecules used by farmers can improve the management of phytosanitary treatments at the plot level. The results of this analysis can help refine the monitoring efforts of chemical concentrations in rivers conducted by water agencies in France. By identifying the most commonly used molecules, it is possible to predict which of them are likely to be found in higher concentrations in water sources.

However, this study is not without limitations. The analysis was conducted at the plot level and focused on specific farming systems within a defined geographic area. A broader, territorial-scale approach, such as mapping phytosanitary pressures at the watershed level, would allow for more comprehensive monitoring of the impact of plant protection practices across larger areas. This approach could provide critical insights into the cumulative risks associated with high-intensity agricultural areas and inform future management practices at a regional scale.

Future studies should extend this work to explore the long-term effects of specific plant protection practices on both human health and the environment. A more extensive analysis at a territorial or watershed level, incorporating diverse agricultural landscapes, would allow for a better understanding of the spatial distribution of phytosanitary risks. Furthermore, investigating alternative pest management strategies, such as Integrated Pest Management (IPM), could provide valuable insights into reducing the dependency on high-risk products while maintaining agricultural productivity.

Author Contributions: All authors contributed to the study conception and design. Material preparation, data collection, and analysis were performed by C.G., O.M., P.L.G., and J.-P.B. The first draft of the manuscript was written by C.G. and all authors commented on previous versions of the manuscript. All authors have read and agreed to the published version of the manuscript.

Funding: This work is funded by the Occitanie region as part of a doctoral thesis entitled "Caractérisation spatiale de la vulnérabilité des ressources naturelles et gestion des risques de pollution diffuse liée aux pratiques phytosanitaires agricoles". Part of the funding is provided by the LAMES laboratory of CIHEAM-IAMM.

Institutional Review Board Statement: Not applicable.

Informed Consent Statement: Not applicable.

Data Availability Statement: The plant protection practice data have been collected from the Qualisol cooperative via many field surveys over 10 years.

Acknowledgments: This work was carried out as part of the research project SPAGER_Phy (Spatialization and Management of Plant Protection Risks) funded by the Occitanie region (call for projects entitled "Allocations doctorales 2019 de la région Occitanie") as part of a doctoral thesis entitled "Caractérisation spatiale de la vulnérabilité des ressources naturelles et gestion des risques de pollution diffuse liée aux pratiques phytosanitaires agricoles". We gratefully acknowledge the financial support provided by the LAMES laboratory of CIHEAM IAMM. This work was conducted in collaboration with the Qualisol cooperative, which participated by organizing investigations with farmers and providing databases of agricultural practices.

Conflicts of Interest: The authors declare that they have no competing interests.

Appendix A

Table A1. List of plant protection products used on a wine-growing plot (conventional/integrated farming).

Category	Name of Product	Active Ingredient [60]
Fungicide 1	AMALFI	Benalaxyl + Folpet
Fungicide 2	AMALINE FLOW	Copper compounds + Zoxamide
Fungicide 3	FIANAKY	Tebuconazole
Fungicide 4	FUNGURAN OH	Copper (II) hydroxide
Fungicide 5	GRIP TOP	Dimethomorph + Metiram
Fungicide 6	HELIOCUIVRE	Copper (II) hydroxide
Fungicide 7	KARATHANE 3D	Meptyldinocap
Fungicide 8	LEGEND	Quinoxyfen
Fungicide 9	MICROTHIOL SP LIQ	Sulfur
Fungicide 10	MILDICUT	Cyazofamid
Fungicide 11	NATCHEZ	Trifloxystrobin
Fungicide 12	NORDOX 75 WG	Copper (I) oxide
Fungicide 13	PROSPER	Spiroxamine
Fungicide 14	SERVAL	Fosetyl
Fungicide 15	SILLAGE	Fosetyl
Fungicide 16	SOUFREBE DG	Sulfur
Fungicide 17	SULFOJET DF	Sulfur
Fungicide 18	TRILOG	Sulfur
Fungicide 19	TSAR	Myclobutanil + quinoxyfen
Herbicide 1	BASTA F1	Glufosinate
Insecticide 1	STEWARD	Indoxacarb
Bactericide 1	COPERNICO HI BIO WG	Copper (II) hydroxide

Table A2. List of plant protection products used on a wine-growing plot (organic farming).

Category	Name of Product	Active Ingredient [60]
Fungicide 1	AMODE DF	Sulfur
Fungicide 2	BOUILLIE BORDELAISE RSR DISPERSS	Copper sulfate
Fungicide 3	KOCIDE 35 DF (ANCIEN)	Copper (II) hydroxide
Fungicide 4	PENNTHIOL	Sulfur
Fungicide 5	STYROCUIVRE DF	Copper oxychloride

References

1. Meynard, J.M.; Messéan, A.; Charlier, A.; Charrier, F.; Fares, M.; Le Bail, M.; Magrini, M.B.; Savini, I. Freins et leviers à la diversification des cultures. Etude au niveau des exploitations agricoles et des filières. In *Synthèse du Rapport d'étude*; INRA: Paris, France, 2013; p. 52.
2. Karimi, B.; Cahurel, J.Y.; Gontier, L.; Charlier, L.; Chovelon, M.; Mahé, H.; Ranjard, L. A meta-analysis of the ecotoxicological impact of viticultural practices on soil biodiversity. *Environ. Chem. Lett.* **2020**, *18*, 1947–1966. [CrossRef]
3. Renaud, C.; Benoît, M.; Thiollet-Sholtus, M.; Jourjon, F. Evaluation globale des impacts environnementaux des itinéraires techniques viticoles: L'Analyse du Cycle de Vie (ACV). *Rev. Suisse D'arboriculture Vitic. Arboric. Hortic.* **2011**, *43*, 184–189.
4. Litskas, V.; Mandoulaki, A.; Vogiatzakis, I.N.; Tzortzakis, N.; Stavrinides, M. Sustainable viticulture: First determination of the environmental footprint of grapes. *Sustainability* **2020**, *12*, 8812. [CrossRef]
5. Butault, J.P.; Delame, N.; Jacquet, F.; Zardet, G. L'utilisation des pesticides en France: État des lieux et perspectives de réduction. *NESE* **2011**, *35*, 7–26. Available online: https://www.researchgate.net/publication/309050944_L%E2%80%99utilisation_des_pesticides_en_France_Etat_des_lieux_et_perspectives_de_reduction (accessed on 14 September 2024).
6. Marsala, R.Z.; Capri, E.; Russo, E.; Bisagni, M.; Colla, R.; Lucini, L.; Gallo, A.; Suciu, N.A. First evaluation of pesticides occurrence in groundwater of Tidone Valley, an area with intensive viticulture. *Sci. Total Environ.* **2020**, *736*, 139730. [CrossRef]
7. Labite, H.; Holden, N.M.; Richards, K.G.; Kramers, G.; Premrov, A.; Coxon, C.E.; Cummins, E. Comparison of pesticide leaching potential to groundwater under EU FOCUS and site specific conditions. *Sci. Total Environ.* **2013**, *463*, 432–441. [CrossRef]

8. Pompermaier, A.; Carolina Cole Varela, A.; Fortuna, M.; Mendonça-Soare, S.; Koakoski, G.; Aguirre, R.; Oliveira, T.; Sordi, E.; Moterle, D.; Pohl, A.; et al. Water and suspended sediment runoff from vineyard watersheds affecting the behavior and physiology of zebrafish. *Sci. Total Environ.* **2021**, *757*, 143794. [CrossRef]
9. Yuan, L.; Sinshaw, T.; Forshay, K.J. Review of watershed-scale water quality and nonpoint source pollution models. *Geosciences* **2020**, *10*, 25. [CrossRef]
10. Boulanger-Fassier, S. La viticulture durable, une démarche en faveur de la pérennisation des territoires viticoles français? *Géocarrefour* **2008**, *83*, 181–190. [CrossRef]
11. Harrison, S.; McAree, C.; Mulville, W.; Sullivan, T. The problem of agricultural 'diffuse'pollution: Getting to the point. *Sci. Total Environ.* **2019**, *677*, 700–717. [CrossRef]
12. Agathokleous, E.; Feng, Z.; Iavicoli, I.; Calabrese, E.J. Nano-pesticides: A great challenge for biodiversity? The need for a broader perspective. *Nano Today* **2020**, *30*, 100808. [CrossRef]
13. Le Roux, X.; Barbault, R.; Baudry, J.; Burel, F.; Doussan, I.; Garnier, E.; Herzog, F.; Lavorel, S.; Lifran, R.; Roger-Estrade, J.; et al. Agriculture et biodiversité. Valoriser les synergies. Expertise scientifique collective INRA. *Versailles Quae* **2008**, 177.
14. Li, Y.; Miao, R.; Khanna, M. Neonicotinoids and decline in bird biodiversity in the United States. *Nat. Sustain.* **2020**, *3*, 1027–1035. [CrossRef]
15. Paiola, A.; Assandri, G.; Brambilla, M.; Zottini, M.; Pedrini, P.; Nascimbene, J. Exploring the potential of vineyards for biodiversity conservation and delivery of biodiversity-mediated ecosytem services: A global-scale systematic review. *Sci. Total Environ.* **2020**, *706*, 135839. [CrossRef]
16. Blouet, A.; Pervanchon, F.; Pervanchon, M. L'agriculture raisonnée. Limites et alternatives de modèle agricole dominant. *Futuribles* **2003**, *283*, 27–42.
17. Bukalasa, J.S.; Brunekreef, B.; Brouwer, M.; Vermeulen, R.; de Jongste, J.C.; van Rossem, L.; Vonk, J.M.; Wijga, A.; Huss, A.; Gehring, U. Proximity to agricultural fields as proxy for environmental exposure to pesticides among children: The PIAMA birth cohort. *Sci. Total Env.* **2017**, *595*, 515–520. [CrossRef]
18. Carles, C.; Bouvier, G.; Esquirol, Y.; Piel, C.; Migault, L.; Pouchieu, C.; Fabbro-Peray, P.; Lebailly, P.; Baldi, I. Residential proximity to agricultural land and risk of brain tumor in the general population. *Environ. Res.* **2017**, *159*, 321–330. [CrossRef]
19. Coste, A.; Goujon, S.; Faure, L.; Hémon, D.; Clavel, J. Agricultural crop density in the municipalities of France and incidence of childhood leukemia: An ecological study. *Env. Res.* **2020**, *187*, 109517. [CrossRef]
20. Gatignol, C.; Etienne, J.C. Pesticides et Santé. 2010. Available online: https://www.assemblee-nationale.fr/13/pdf/rap-off/i2463.pdf (accessed on 12 September 2024).
21. Tang, F.H.M.; Lenzen, M.; McBratney, A.; Maggi, F. Risk of pesticide pollution at the global scale. *Nat. Geosci.* **2021**, *14*, 206–210. [CrossRef]
22. Alem, H.; Rigou, P.; Schneider, R.; Ojeda, H.; Torregrosa, L. Impact of agronomic practices on grape aroma composition: A review. *J. Sci. Food Agric.* **2019**, *99*, 975–985. [CrossRef]
23. Medina-Meza, I.G.; VanderWeide, J.; Torres-Palacios, C.; Sabbatini, P. Quantitative Metabolomics Unveils the Impact of Agricultural Practices in the Grape Metabolome. *ACS Agric. Sci. Technol.* **2021**, *1*, 253–261. [CrossRef]
24. Parmentier, B.; Pisani, E. *Nourrir L'humanité: Les Grands Problèmes de L'agriculture Mondiale au XXIème Siècle*; La Découverte: Paris, France, 2004; 294p.
25. Tissot, C.; Neethling, E.; Rouan, M.; Barbeau, G.; Quénol, H.; Le Coq, C. Modeling Environmental Impacts on Viticultural Ecosystems. *Environ. Inf. Syst.* **2019**, *8*, 1403–1422.
26. Bélis-Bergouignan, M.; Saint-Ges, V. Quelle trajectoire environnementale pour la viticulture ? L'exemple du vignoble girondin. *Rev. D'économie Régionale Urbaine* **2009**, *3*, 491–516. [CrossRef]
27. Rochard, J. Evolution des pratiques viticoles françaises: De la viticulture raisonnée à l'agroécologie. In Proceedings of the 40th World Congress of Vine and Wine, Sofia, Bulgarie, 29 May–2 June 2017; pp. 1–8.
28. Bélis-Bergouignan, M.; Cazals, C. Démarches environnementales volontaires, conflit d'usage et proximité. *Développement Durable Et Territ.* **2006**, *7*, 1–16. [CrossRef]
29. Alim'agri. L'agriculture Biologique (AB). 2017. Available online: http://agriculture.gouv.fr/lagriculture-biologique-ab (accessed on 13 September 2024).
30. Agence Bio. Chiffres de la bio en France. 2018. Available online: http://www.agencebio.org/les-chiffres-cles/ (accessed on 28 August 2024).
31. Agence Bio. La Production et le Marché bio en France en. 2020. Available online: http://www.agencebio.org/les-chiffres-cles/ (accessed on 28 August 2024).
32. InterBio Occitanie. L'observatoire Régional de L'agriculture Biologique D'Occitanie: Les Fiches Filière—Viticulture. 2018. Available online: https://www.interbio-occitanie.com/content/uploads/2019/12/fiche_filiere_viti_2018_vd.pdf (accessed on 13 September 2024).

33. Meissner, G.; Athmann, M.E.; Fritz, J.; Kauer, R.; Stoll, M.; Schultz, H.R. Conversion to organic and biodynamic viticultural practices: Impact on soil, grapevine development and grape quality. *Oeno One* **2019**, *53*. [CrossRef]
34. Defer, A.; Gautier, M.; Georges, A.; Girard, D.; Hadrane, S.; Irrien, A.; Julliand, B.; Pigeon, H.; Pineau, N.; Commune par Commune, la Carte de France des Pesticides. Médiapart. 2019. Available online: https://www.mediapart.fr/journal/france/040719/commune-par-commune-la-carte-de-france-des-pesticides?onglet=full (accessed on 18 September 2024).
35. Zahm, F. Méthodes de Diagnostic des Exploitations Agricoles et Indicateurs: Panorama et cas Particuliers Appliqués à L'évaluation des Pratiques Phytosanitaires. *Sci. Eaux Territ.* **2003**, *33*, 13–34. Available online: https://hal-sde.archives-ouvertes.fr/hal-00465539/document (accessed on 17 September 2024).
36. Ayadi, H. Outils de Gestion de la Pollution Phytosanitaire Diffuse au Niveau d'un Territoire: Cas D'application Zone Humide Ramsar de la Merja Zerga au Maroc. Doctoral Dissertation, Université Paul Valéry-Montpellier III, Montpellier, France, 2013.
37. Ayadi, H.; Le Bars, M.; Le Grusse, P.; Mandart, E.; Fabre, J.; Bouaziz, A.; Bord, J.P. SimPhy: A simulation game to lessen the impact of phytosanitaries on health and the environment-the case of Merja Zerga in Morocco. *Environ. Sci. Pollut. Res.* **2014**, *21*, 4950–4963. [CrossRef]
38. Mghirbi, O.; Ellefi, K.; Le Grusse, P.; Mandart, E.; Fabre, J.; Ayadi, H.; Bord, J.P. Assessing plant protection practices using pressure indicator and toxicity risk indicators: Analysis of the relationship between these indicators for improved risk management, application in viticulture. *Environ. Sci. Pollut. Res.* **2015**, *22*, 8058–8074.
39. Mghirbi, O. Résilience des Exploitations Agricoles Face au Changement des Pratiques Phytosanitaires: Conception D'outils De gestion des Risques Liés Aux Pesticides—Cas du Bassin Versant de L'étang de L'or en France. Ph.D. Dissertation, Université Paul Valéry Montpellier 3, Montpellier, France, 2016.
40. Devillers, J.; Farret, R.; Girardin, P.; Rirvière, J.L.; Soulas, G. *Indicateurs Pour Evaluer les Risques liés à L'utilisation des Pesticides*; Tec et Doc: New York, NY, USA, 2005; p. 278.
41. Samuel, O.; Dion, S.; ST-Laurent, L.; April, M.H. *Indicateur de Risque des Pesticides du Québec*, 2nd ed.; IRPeQ: Québec, QC, Canada, 2012; 48p, (Santé et environnement). Ministère de l'Agriculture, des Pêcheries et de l'Alimentation/Ministère du Développement Durable, de l'Environnement et des Parcs/Institut National de Santé Publique du Québec; Available online: https://www.inspq.qc.ca/publications/1504 (accessed on 12 September 2024).
42. Kniss, A.R.; Coburn, C.W. Quantitative evaluation of the environmental impact quotient (EIQ) for comparing herbicides. *PLoS ONE* **2015**, *10*, e0131200. [CrossRef]
43. Bockstaller, C.; Pierlot, F.; Marks-Perreau, J.; Réal, B.; Constant, T.; Lioeddine, A.; Carluer, N.; Miralles, A.; Morin, A.; Villerd, J. Evaluation de la qualité prédictive d'indicateurs pesticides de transfert vers les eaux: Le projet EQUIPE. *Innov. Agron.* **2017**, *59*, 25–39.
44. Le Grusse, P.; Mandart, E.; Bouaziz, A.; Le Bars, M.; Bord, J.P.; Fabre, J. Gestion de la Toxicité en Zone Ramsar (TRam): Rapport Final. 2014. Available online: https://www.researchgate.net/publication/334049279_Gestion_de_la_Toxicite_en_zone_Ramsar_TRam_rapport_final (accessed on 16 June 2024).
45. Mghirbi, O.; Le Grusse, P.; Fabre, J.; Mandart, E.; Bord, J.P. OptiPhy, a technical economic optimisation model for improving the management of plant protection practices in agriculture: A decision-support tool for controlling the toxicity risks related to pesticides. *Environ. Sci. Pollut. Res.* **2017**, *24*, 6951–6972. [CrossRef]
46. Juan, G.; Barataud, F.; Billy, C.; Bouchet, L.; Carpentier, A.; Gouy, V.; Le Hénaff, G.; Voltz, M. Référentiel sur les Outils de la Recherche Pour Réduire les Pollutions de L'eau par les Pesticides. Rapport final 2018, Convention INRA-AFB. Available online: https://odeliane.com/wp-content/uploads/2019/11/Referentiel_outils_pesticide_version_finale_230218.pdf (accessed on 14 August 2024).
47. Moreau, S.; Janin, L.; Carrière, C.; Patin, F.; Janvier, F.; Vénus, S. Plan de Réduction des Produits Phytopharmaceutiques et Sortie du Glyphosate: Etat des Lieux des Ventes et des Achats en France. Paris (France): Datalab. 2019. Available online: https://www.statistiques.developpementdurable.gouv.fr/sites/default/files/2019-04/datalab-essentiel-172-plan%20de%20reductionavril2019.pdf (accessed on 14 August 2024).
48. SDES—Service des Données et Etudes Statistiques. 2020. Available online: https://www.statistiques.developpement-durable.gouv.fr/etat-des-lieux-des-ventes-et-des-achats-de-produits-phytopharmaceutiques-en-france-en-2020?rubrique=&dossier=211 (accessed on 15 June 2024).
49. Millesime Bio. Mondial du vin et des Autres Boissons Alcolisées Biologiques: Dossier de Presse. 2021. Available online: https://www.millesime-bio.com/files/download/documents/Dossier%20de%20presse%20Mill%C3%A9sime-Bio%202021-%20DIGITAL.pdf (accessed on 10 September 2024).
50. FranceAgriMer. La Filière vin. 2017. Available online: https://www.franceagrimer.fr/filieres-Vin-et-cidre (accessed on 5 September 2024).
51. Aouadi, N.; Macary, F.; Ugaglia, A.A. Évaluation multicritère des performances socio-économiques et environnementales de systèmes viticoles et de scénarios de transition agroécologique. *Cah. Agric.* **2020**, *29*, 19. [CrossRef]

52. Aouadi, N.; Macary, F.; Delière, L.; Roby, J.P. New Scenarios for a Shift towards Agroecology in Viticulture. *Agric. Sci.* **2021**, *12*, 1003–1033. [CrossRef]
53. Mghirbi, O.; Bord, J.P.; Le Grusse, P.; Mandart, E.; Fabre, J. Mapping for the management of diffuse pollution risks related to agricultural plant protection practices: Case of the Etang de l'Or catchment area in France. *Environ. Sci. Pollut. Res.* **2018**, *25*, 14117–14137. [CrossRef] [PubMed]
54. Pingault, N.; Pleyber, E.; Champeaux, C.; Guichard, L.; Omon, B. Produits Phytosanitaires et Protection Intégrée des Cultures: L'indicateur de Fréquence de Traitement. *Notes Et Études Socio-Économiques* **2009**, *32*, 61–94.
55. Grimene, C.; Mghirbi, O.; Louvet, S.; Le Grusse, P. Caractérisation spatiale de la Vulnérabilité des Ressources Naturelles et Gestion des Risques de Pollution Diffuse liés aux Pratiques Phytosanitaires Agricoles. In Proceedings of the 49. Congrès du Groupe Français des Pesticides: Pesticides et Transition Agroécologique: Enjeux et Avancées, Montpellier, France, May 2019; pp. 103–105. Available online: https://hal.science/hal-03046570 (accessed on 17 June 2024).
56. Grimene, C.; Mghirbi, O.; Louvet, S.; Bord, J.P.; Le Grusse, P. Spatial characterization of surface water vulnerability to diffuse pollution related to pesticide contamination: Case of the Gimone watershed in France. *Environ. Sci. Pollut. Res.* **2022**, *29*, 17–39. [CrossRef]
57. Agreste Agreste les Dossiers. Enquetes Pratiques Phytosanitaires en Viticulture en 2016: Nombre de Traitement et Indicateurs de Fréquence de Traitement. N° 2019-2—Février 2019. Available online: https://agreste.agriculture.gouv.fr/agreste-web/download/publication/publie/Dos1902/Dossier2019-2.pdf (accessed on 26 June 2024).
58. Lamichhane, J.R.; Osdaghi, E.; Behlau, F.; Köhl, J.; Jones, J.B.; Aubertot, J.N. Thirteen decades of antimicrobial copper compounds applied in agriculture. A review. *Agron. Sustain. Dev.* **2018**, *38*, 28. [CrossRef]
59. Pesce, S.; Mamy, L.; Sanchez, W.; Artigas, J.; Bérard, A.; Betoulle, S.; Chaumot, A.; Coutellec, M.-A.; Crouzet, O.; Leenhardt, S. The use of copper as plant protection product contributes to environmental contamination and resulting impacts on terrestrial and aquatic biodiversity and ecosystem functions. In *Environmental Science and Pollution Research*; Springer: Berlin/Heidelberg, Germany, 2024; pp. 1–17.
60. Lewis, K.A.; Tzilivakis, J.; Warner, D.; Green, A. An international database for pesticide risk assessments and management. *Human. Ecol. Risk Assess. An. Int. J.* **2016**, *22*, 1050–1064. [CrossRef]

sustainability

MDPI

Article

Quince (*Cydonia oblonga* Mill.) Waste By-Product Characterization as a Potential Functional Ingredient

Janira Romero [1], Alexandra Díez Méndez [2], María José Castro-Alija [1], Jorge Poveda [3] and Irene Albertos [1,*]

[1] Recognized Research Group: Assessment and Multidisciplinary Intervention in Health Care and Sustainable Lifestyles of the University of Valladolid, 47197 Valladolid, Spain; janira.romero22@estudiantes.uva.es (J.R.); mariajose.castro@uva.es (M.J.C.-A.)

[2] Universidad Católica de Ávila (UCAV), Calle Canteros s/n, 05002 Avila, Spain; alexandra.diez@ucavila.es

[3] Recognised Research Group AGROBIOTECH, UIC-370 (Junta de Castilla y León), Department of Plant Production and Forest Resources, Higher Technical School of Agricultural Engineering of Palencia, University Institute for Research in Sustainable Forest Management (iuFOR), University of Valladolid, Avda. Madrid 57, 34004 Palencia, Spain; jorge.poveda@uva.es

* Correspondence: irene.albertos@uva.es

Abstract: Currently, the production of waste in the food industry is increasing, which is a serious problem. However, most of these residues, especially those derived from fruits and vegetables, have great unknown properties that are not used. The main objective of this article is the analysis and characterization of the waste from quince after its processing to observe its properties and its potential use in different industries as a functional ingredient, thus favoring the circular economy and sustainability. Quince by-product nutritional parameters such as proteins, fibers, sugars, vitamins, and minerals were analyzed. Also, the antioxidant capacity was measured by various methods: 1,1-diphenyl-2-picrylhydrazyl radical scavenging activity (DPPH), antioxidant capacity in Trolox equivalent antioxidant capacity (TEAC/ABTS), and total polyphenol content (TPC). Finally, the antimicrobial capacity against different postharvest-pathogenic fungi was measured in direct sample and extract. The nutritional results showed a nutritional profile rich in soluble and insoluble fiber, potassium, calcium, and magnesium, and low in fat. The antioxidant results from the extract showed significant levels of phenols and higher antioxidant capacity from the extracted sample. No positive results were found in the antimicrobial capacity study. Quince by-products could be a potential ingredient in the industry due to their nutritional composition and antioxidant content.

Keywords: antioxidant capacity; food waste; nutritional composition; circular economy

1. Introduction

Currently, the population growth is increasing, having a forecast of 8.9 billion people by 2050, which will translate into an increase in demand for food between 56 and 98% for this year [1]. This increase in demand directly causes an increase in the production of waste and by-products that cause serious environmental and socio-economic problems [2]. The fruit and vegetable sector is one of the most affected in terms of waste generation due to the sensitivity and properties of these products. The most common reasons for discarding fruits and vegetables are injuries, bumps, excessive ripening, appearance, and freshness [3]. It is estimated that about 1300 million tons of food are wasted every year, with the fruit and vegetable sector accounting for 60% of these losses [4]. Food waste poses a significant environmental threat, representing a major source of greenhouse gas (GHG) emissions that exacerbate global warming and climate change, potentially contributing to the extinction of numerous species [5].

Highlighting this aspect, the food industry contributes to GHG emissions, generating 26% of global emissions, including methane (CH_4), dioxins, and ammonia (NH_3), which pollute air and water, impacting both environmental and human health [6]. Therefore, the implementation of effective waste management strategies within the food industry, particularly in the fruit and vegetable sector, is paramount to mitigating these severe consequences [6].

The growing demand for food is also driven by per capita consumption since there is greater access to energy-rich foods, which often require significant resources for their production and consumption [1], causing a change in dietary habits and increasing the interest in fresh and nutritious foods.

The United Nations, through the Sustainable Development Goals (SDGs), has set a target of reducing per capita food waste by 50% by 2030 [7]. There is a total of 17 SDGs and 7 of them, such as the end of poverty, health, and well-being, zero hunger, sustainable production and consumption, clean water and sanitation, underwater life, climate action, and life of terrestrial ecosystems, are directly related to the food supply chain and its sustainability [8].

For the reasons mentioned above, there has been a growing interest in recent years to find ways to utilize this waste to promote the circular economy and sustainability. Research has revealed that food waste by-products from fruits and vegetables, such as seeds, peels, leaves, and stems, are rich in antioxidants, fibers, bioactive compounds, and enzymes, making them valuable resources for the food, pharmaceutical, and cosmetic industries [3]. Likewise, these products possess low toxicity and high efficacy, generating an added value [9]. Studies have also shown antifungal and antimicrobial activity in plant by-products. For instance, several works have described the antioxidant and antimicrobial capacity of pomegranate peel and its results against various food pathogens, which can be applied in the food industry as a natural additive and as a natural antimicrobial [10–13]. Other authors analyzed its cytotoxicity and effect on human health such as Lai et al. (2013) [14] and Rodríguez-Gonzalez et al. (2017) [15] who described the anti-cancer and anti-diabetic properties of citrus and mango residues, respectively. Therefore, these residues can be used as functional foods that are described as foods that have a positive impact on a person's health, physical performance, or mood, in addition to their nutritional value [16].

Among fruits, the quince (*Cydonia oblonga* Mill.) is a great unknown despite its different properties, becoming one of the most prized pitted fruits despite its low interest as a fresh fruit [17]. Quince is part of the Rosaceae family and grows from a deciduous tree. The immature fruit has a green color that becomes more yellowish as the fruit matures. Quince has a shape like a pear or an apple and inside there are numerous brown seeds covered by a whitish mucilage. The fresh fruit is not very appreciated due to its hard and astringent pulp [18]. It is a climacteric fruit; therefore, it is harvested after it reaches physiological maturity, which usually occurs from October to November [18]. The most common way of fruit preservation and storage is refrigeration at temperatures between 0 and 5 °C helping to slow microbial growth and the generation of ethylene, a hormone responsible for fruit ripening, slowing it down [18]. The origin of the quince fruit dates to the Transcaucasian zone (Iran, Armenia, Azerbaijan, etc.), and it is estimated that it began to be cultivated in the year 4000 B.C. As it can tolerate climatic changes, it allowed its extension to China and Europe [18]. The introduction of quince into the Mediterranean is directly linked to the invasion of the Middle East by Alexander the Great, so its scientific name, *Cydonia*, derives from a city in Crete called Cydonea [18]. In ancient Greece, quince was used at wedding banquets or to make wine. This custom was continued until the Middle Ages and was considered a protector against the black death [18].

The worldwide production of quince between 2017 and 2019 was 674,894 tons, of which 77% was grown on the Asian continent while 6.8% of the production was on the European continent [17]. Turkey is the main country producing quince, followed by others such as China, Iran, and Morocco [19]. In 2022, 82,941 ha were dedicated to the cultivation of quince, producing 702,015 tons of fruit [20]. Bayav and Sahin (2023) estimated that Turkey would increase its production to 208,112 tons and export to 18,685 in 2023 [17].

In terms of its composition, quince is a highly nutritious fruit of great interest to various industries. It is one of the best sources of pectin and phytochemicals [18], rich in carbohydrates, fibers, proteins, vitamins, minerals, and organic acids [21]. Notably, quince contains a high level of phenolic compounds, which contribute to its antioxidant capacity [22].

Pectins, which are an important component in quince, are polysaccharides present in the cell wall and can act as a gel, binding cells together. After ingestion, it has various physiological benefits, such as reducing glucose and cholesterol levels and acting as a prebiotic [23]. Pectins are also used in various industries due to their gelling properties. Quince is a notable source of pectin, the content of which will vary depending on the degree of maturity [18].

Quince also boasts a high content of flavonoids, phenolic acids, and lignin among other active ingredients. The main phenolic compound is 5-O-caffeoylquinic acid, the primary substrate for the enzyme polyphenol oxidase, which makes quince highly susceptible to enzymatic browning [24]. Additionally, quince leaves have various medicinal uses, including antifungal properties, protection of the liver system, antioxidant properties, the treatment of skin lesions, and antitussive, sedative, and antipyretic properties [18,25].

Typically, quince is consumed cooked in the form of jams, marmalades, or jellies [25–27]. However, the industrial manufacturing process of these products generates by-products, around 50,000 kg per year in the quince paste manufacturing process (Figure 1) [20]. This by-product is formed by the central area of the fruit with seeds and mucilage, small parts of skin, and leaves that are not used and discarded. As mentioned above, quince is rich in numerous components and this by-product could have a rich nutritional profile and components that generates a functional ingredient for food that is not used in any way. Therefore, it is important to characterize the by-product to know its properties and possible applications and uses.

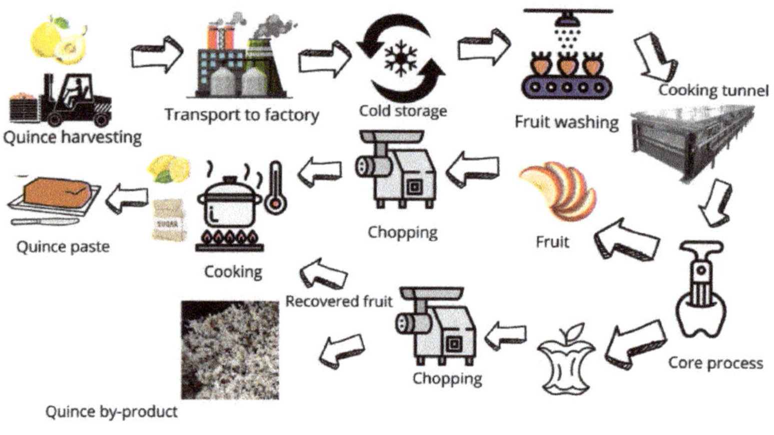

Figure 1. Quince manufacturing process.

Thus, the main objective of this study is to analyze and characterize the waste generated during quince product production, exploring its properties and potential applications as a functional ingredient in the food industry.

2. Materials and Methods

2.1. Raw Material

The quince residue (the central part of the fruit) obtained after the manufacturing process was kindly provided by the company Yemas de Santa Teresa S.L. (Ávila, Spain). To obtain the by-product, the fruit undergoes a specific processing consisting of several processes. First, the fruit is washed with water and subjected to a cooking process. Once cooked, it is cored to separate the central area of the pulp. The pulp obtained will be used for the manufacture of the final product. On the other hand, the quince core is passed through a chopper equipped with a sieve to recover as much pulp as possible, which will be used in the manufacturing process. The discarded surplus is the generated by-product that is wasted. Once obtained, it was stored in deep-freezing at −80 °C until use.

2.2. *Nutritional Parameters*

The nutritional parameters, such as moisture, fats, ash, carbohydrates, fibers, vitamins, and minerals, were analyzed in triplicate. The moisture content was determined gravimetrically [28]: a total of 5 g of by-product were prepared to constant weight in a hot air oven at 80 °C for about 24 h.

The fat was analyzed using petroleum ether (40–60 °C) for 4 h by means of a Soxtec System 2055 Tecator extractor (FOSS, Hillerød, Denmark) and subsequently measured gravimetrically [29]. The ash content was also determined by by-product incineration to constant weight in a muffle at 550 °C for 5 h [30]. The result was expressed in g per 100 g of by-product. Carbohydrates were calculated by difference. The method described by Dumas [31] was used to measure the protein content using a CN-2000 Analyzer (Leco Corp., St. Joseph, MO, USA). Based on the results, the protein content was calculated from nitrogen by the conversion factor 6.25.

Soluble and insoluble fibers were analyzed by the method described by the Association of Official Analytical Chemists (AOAC) [32] using the TDF-100 kit (Sigma-Aldrich, St. Louis, MO, USA). The samples were measured together with blank samples.

The total sugar content was analyzed by high-performance liquid chromatography (HPLC) according to the AOAC 1990, method 982.14 [33]. A 50% ethanol dilution was prepared for the extraction of sugars. Once obtained, the extract was passed through a Sep-Pak C18 cartridge and filtered through a 0.45 mm nylon disk. The quantification and separation were performed by means of a column with amino bonds and detection with a refractometer 1260 Infinity II (Agilent Technologies, Waldbronn, Germany). The results were expressed as g per 100 g of sample.

Vitamin C was measured using a standard vitamin C measurement kit Vitafast (R-Biopharm AG, Darmstadt, Germany). High-performance liquid chromatography was used to analyze the sugars [30], and the sample was homogenized with distilled water and filtered. The filtrate was injected into HPLC 1100 VWD (Agilent Technologies, Waldbronn, Germany). Standard sugars were purchased from Sigma-Aldrich. The minerals calcium, magnesium, and potassium were measured by means of ICP-MS Avio, 220 Max (PerkinElme, Waltham, MA, USA) in triplicate based on the method described by Santos et al., 2022 [34]. Proximal composition was expressed in g per 100 g of sample.

2.3. *Antioxidant Capacity*

2.3.1. Sample Processing

The by-product was subjected to a lyophilization process by a lyophilizer (LYOQUEST-55, Azbil Telstar Technologies S.L.U., Terrassa, Spain) for 24 h and stored at 5 °C until use.

2.3.2. Extraction

After lyophilizing the sample, it was ground by an analysis mill (Tube Mill control) of IKA Works S.L (Barcelona, Spain) until a fine and homogeneous powder was obtained. Once the powder was obtained, 5 g were dissolved in 100 mL of methanol and water (1:1) at pH 2. It was kept under stirring for 24 h at room temperature (20 °C) and covered to avoid solvent evaporation. After resting, the extract was centrifuged for 20 min at 6000 rpm. The supernatant was filtered with qualitative filter paper (Whatman, UK), made up to 100 mL with methanol and water (1:1) pH 2, and stored at −80 °C until use.

2.3.3. Antioxidant Capacity Determination

The total polyphenol content (TPC) was measured by the Folin–Ciocalteu method [35] with the use of a spectrometer at 760 nm. The results were expressed as mg gallic acid equivalent (GAE)/100 g weight using a gallic acid calibration curve (9.8 μm–70 Mm). The TP was measured only in an exact sample.

The antioxidant capacity in the Trolox equivalents (TEAC) method was used to evaluate the antioxidant capacity both during extraction and directly in the lyophilized sample following the method reported by Re et al., 1999 [36]. The reduction in the absorbance at 730 nm was recorded by a spectrophotometer (Thermo Fisher Scientific, Genesys 150, Madison, WI, USA). Trolox (7.5–240 μM) was used as a standard.

The effect of antioxidant activity on DPPH was analyzed based on the method described by Brand-Williams, Cuvelier, and Berset 1995 [37]. The results were expressed as a percentage of the inhibition of the DPPH radical. The TEAC and DPPH methods were both analyzed in an extract and direct pre-lyophilized sample.

All the antioxidant activity test samples were diluted 1:10 with miliQ water.

2.4. Antimicrobial Capacity

To measure antimicrobial capacity, the by-product extract was previously lyophilized. To evaluate the antimicrobial activity of quince, an assay was conducted against two phytopathogenic fungi, *Botrytis cinerea* (CECT 20973) and *Colletotrichum acutatum* (CECT 21009), obtained from the Spanish Type Culture Collection (CECT) in Valencia, Spain. Fungal disks (approximately 6 mm) were excised from fully grown 7-day-old cultures and placed at the center of a PDA medium. Regarding the quince extract solution, a concentration of [1:1] was prepared using distilled, autoclaved water, and sterilized using 0.22 μm syringe filters. In total, 5 μL of the extract was inoculated alongside four equidistant points surrounding the fungal disks, with each point containing the quince extract residue. The plates were incubated at room temperature for seven days and observed routinely for fungal growth inhibition. The experiment was repeated in triplicate, with a total of five plates per fungal species. The area of inhibition was measured by the ImageJ photographic analyses software 1.53k (NIH, National Institutes of Health, Bethesda, MD, USA).

3. Results and Discussion

3.1. Nutritional Composition

The results of the nutritional analysis of the quince by-product are shown in Figure 2.

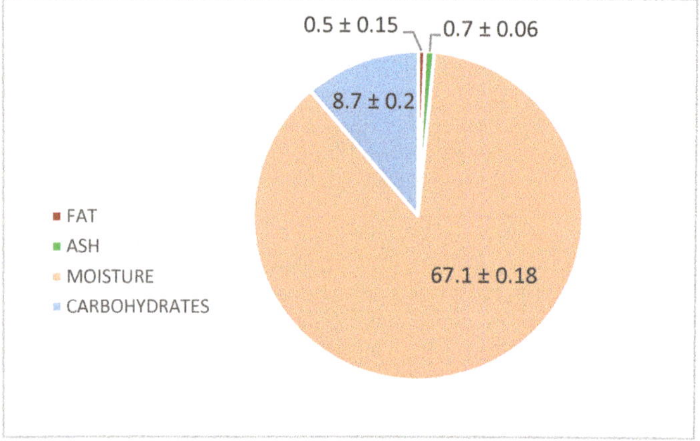

Figure 2. Nutritional profile of quince by-product. Fat, ash, moisture, and carbohydrates. Data expressed in g/100 g (Wet Base) as mean values ± standard deviation (n = 3).

The fat content showed significantly low values. According to the current legislation [38], it can be indicated as a "low-fat" product that does not contain more than 3 g of harrow per 100 g of solid matter, so according to the results obtained, this claim can be applied to our product. "Low-fat" could also be established with the results obtained in other studies where fat values range from 0.20 g to 0.6 g per 100 g [26,39,40].

Dimitriu et al. (2023) [41] characterized the composition of quince pulp without treatment and with ethanol treatment, where significantly higher values of untreated carbohydrates were obtained (27.53 ± 2.46 mg of glucose/xylose equivalent/100 g of sample) than in the present work (8.7 ± 0.2 g/100 WB sample). On the other hand, other authors reported higher levels (75.80 ± 0.28 g per 100 g sample) [42]. This difference in results may be due to various factors such as part of the fruit as a whole fruit or only the central area as in our test, the origin of cultivation, the ripeness index of the fruit, etc.

Additional results on the nutritional profile of the sample are shown in the following Table 1.

Table 1. Quince by-product nutritional composition.

Soluble Fiber (g/100 g WB)	Insoluble Fiber (g/100 g WB)	Total Sugars (g/100 g WB)	Vitamin C (g/100 g WB)	Calcium (g/100 g WB)	Magnesium (g/100 g WB)	Potassium (g/100 g WB)
2.2 ± 1.23	12.6 ± 0.32	7.4 ± 2.47	<0.02 ± 0.01	0.056 ± 0.01	0.024 ± 0.08	0.21 ± 0.52

Data expressed in g/100 g (Wet Base) as mean values ± standard deviation (n = 3).

The sugar content obtained is low compared to the results previously reported by Coimbra et al. (2023) [40], where they characterized the powder obtained from the peel of the quince and observed that it was a product rich in sugars, especially reducing sugars. Previous studies coinciding with Coimbra et al. (2023) [40] obtained greater reduction than non-reducing sugar, although not with a great difference [43]. The by-product sample consists of the central part of the fruit which includes seeds and mucilage but also skin and small portions of pulp from the central area which can be found in smaller proportions. By working with this by-product, numerous advantages are obtained, such as waste reduction and the optimization of resources rich in various nutritional components.

The most abundant mineral element found was potassium with 210 mg per 100 g, followed by calcium with 56 mg per 100 g followed by calcium. Potassium is one of the main intracellular elements in the body and is indispensable for the normal functioning of cells in processes such as ATP synthesis. Magnesium and calcium are of great importance in bone composition since 99% of the body's calcium is present in bones and teeth. Magnesium is involved in various enzymatic processes and in the maintenance of body levels of potassium and calcium. For its part, potassium is one of the main body elements and is indispensable in the normal functioning of cells in processes, such as ATP synthesis [39,44].

Calcium, magnesium, and potassium were analyzed in the present work and in previous works. The main component obtained was potassium, followed by magnesium and to a lesser extent calcium. In other studies that also analyzed these minerals together, potassium was also the majority element followed by magnesium [39,43,45]. However, in their paper, Rather et al. (2023) [21] described a higher content of calcium than magnesium in the nutritional parameters of quince. The values obtained vs. the previous literature are similar. However, the potassium levels recorded by Krzepiłko and Prażak (2023) [45] were much higher (887.17 mg and 781.02 mg) versus the 210 mg obtained. This difference may be since in our work the residue was composed not only of seeds, but also of mucilage and a very small part of pulp remains that could have been left while Krzepiłko and Prażak (2023) [45] focused their research only on seeds. Another possible cause of this difference could be due to the different species analyzed. *C. oblonga* is the species from which the analyzed residue comes, while the seeds of the revised work [45] belonged to two species, *Chaenomeles japonica* and *C. superba*. However, Byczkiewicz et al. (2021) [43] also analyzed the components in *C. japonica* and obtained a lower amount of potassium, resulting in values relatively more like our work. The relative variations in fiber content, and in mineral content can be altered by cultivation and even by the environmental conditions in which these are developed [46].

Vitamin C or ascorbic acid is an important antioxidant and its amount in fruit ranges from 2.5 to 11.6 mg per 100 g, its reference intake value according to the USDA being 15 mg/100 g [46]. Less than 20 mg/100 g were obtained in the study, being within the range reported in the literature. However, other studies reported higher values, such as 15.46 mg/100 g or 50 to 80 mg/100 g in cultivars from the Czech Republic [46]. It seems that the vitamin content is also affected by the location of the cultivar [46]. Likewise, vitamin C is thermolabile, and in the case of the study, the sample has previously undergone a cooking heat treatment necessary for the manufacturing process and obtaining the studied by-product, which could affect the final vitamin C content. Therefore, optimizing the treatment and process could help to better maintain vitamin C both in the final product and in the derivative product.

The results showed a high level of dietary fiber in the quince by-product, especially soluble fiber followed by carbohydrates and insoluble fiber. On the other hand, the fat content is very low. The quince by-product nutritional profile suggests that the sample may be suitable for diets that require a high intake of fiber with a low fat content. Fiber is defined as plant components that cannot be digested but can be fermented by the intestinal microbiome. The recommended daily intake of fiber is estimated to be 38 g/day in men and 25 g/day in women [47]. Foods rich in fiber can improve insulin sensitivity, metabolic profile, and weight control, as well as reduce blood pressure [47]. The non-soluble fiber consists of non-cellulosic polysaccharides such as pectin, which agrees with the results of the known high pectin content of quince. On the other hand, insoluble fiber constitutes the cell wall [48]. Soluble dietary fiber has a great water retention capacity and viscosity that dilutes nutrients in the intestine and causes a feeling of satiety allowing a lipid reduction mechanism [49,50]. Several studies describe the ability of soluble dietary fiber to absorb and sequester cholesterol, which reduces triglyceride levels. Pectin cannot be degraded by intestinal enzymes but is degraded by bacteria and various studies have described that pectin improves the intestinal population, with bacteria such as Lactobacilli and Bifidobacteria [48].

Other authors carried out an analysis of the nutritional composition of quince, where lower fiber results were obtained than those obtained in our study (1.9 g of fiber per 100 g of fruit) [51]. This difference in fiber can be due to two issues: the first is that the authors performed the analysis on fresh fruit while in the present analysis it was performed on already cooked fruit, which reduces the insoluble fiber. On the other hand, this work focuses on the residue obtained from industrial processing, that is, the heart of the fruit, while Khan and Ahmad (2021) [51] reported results on the complete fruit including pulp and skin.

As far as we know, there are no previous studies that indicate the nutritional content of quince by-product, that is, not only the seeds, although there are others that have described the skin and pulp content [39,52,53]. Quince peel fiber levels (20.2 g/100 g) were reported [39], being more like the present work, while the fiber content in the pulp was less than 1 to 6 g per 100 g [39,52,53].

3.2. Antioxidant Capacity

As mentioned above, except for the total polyphenols, the parameters selected to measure the antioxidant capacity were performed on the extract obtained from the by-product and directly (by-product previously lyophilized).

The TPC of the extract was 22 mg of GAE (Table 2). Polyphenols are compounds that have antioxidant capacity and have hydroxyl groups in the para or ortho position, which facilitates redox-type reactions; this allows them to be oxidized easily since they can transport protons [44].

Table 2. Quince residue total polyphenols and antioxidant capacity.

Sample	TCP (mg of GAE)	DPPH (% of Inhibition)	TEAC (µmol TE 100 g [1])
Direct sample	n.m [1]	47.88 ± 3.57 [2]	31.49 ± 6.12 [2]
Extract	22.62 ± 1.34 [2]	72.39 ± 2.77 [2]	377.10 ± 48.09 [2]

[1] not measured. [2] the results are expressed by media ± standard deviation.

Phenolic compounds are related to the health properties of the fruit and quince is considered a good source of these compounds [26]. Phenols are related to beneficial effects in diseases and are one of the most important antioxidants present in fruits and vegetables. Ibrahim Anber and Asadi-Gharneh (2024) [46] reported a phenolic content of 32.4–143.1 mg/100 g of GAE in the studied genotypes. Lower values were obtained in the previous study by Byczkiewicz et al. (2021) [43] on three varieties of quince which turned out to contain 17.10 mg GAE/g, 18.14 mg GAE/g, and 17.35 mg GAE/g, respectively. A study reported the presence of 16 phenolic compounds in the quince peel and it was determined that the extraction method quantitatively affected the phenolic content [43]. For its part, Silva et al. (2023) [3] analyzed the phenolic content of quince leaves (209.78 ± 14.28 µg/mg), but they saw that the phenol content in the seed extracts were the ones with the lowest phenolic content presented, 12.54 ± 1.09 µg/mg [3]. The results obtained in our trial agree with the previous literature, although, like the nutritional composition, there are variations. It is known that the total content of phenols can be affected by crop conditions since agents such as light, nutrients, and soil temperature [3,54]. Benahmed et al. (2021) [54] reported a phenolic content of 23.3 mg/100 g GAE in the fruit, like the content obtained in the test in the extract (22.62 mg/100 g GAE).

The highest antioxidant capacity is found in the extract with the TEAC method (377.10 ± 48.09 µmoles Trolox/g) followed by the % inhibition of DPPH. The opposite occurs in the direct residue where we find the highest value (Table 2) of the % inhibition of DPPH. However, in both cases, the values obtained are higher in the extract compared to the lyophilized residue. This may be because the extract was made with a methanol–water mixture at pH 2, which facilitates the extraction of components.

Two methods of measuring antioxidant capacity were used to obtain a more reliable assessment by combining both. The antioxidant activity is closely related to the phenolic content since they are responsible for this in large part [55]. Authors such as Silva et al. (2023) [3] corroborated this statement when they saw that quince leaves presented the highest phenolic content, followed by the peel and finally the seeds. The antioxidant activity was equated to being the highest in the part that had the highest phenolic content—the leaves followed by the peel and the seeds. Aguayo-Rojas et al. (2024) [56] analyzed the antioxidant capacity in the methanol extracts of quince (peel and pulp) with the combination of two methods (DPPH and ABTS/TEAC), as did the authors. In the same way, they registered a higher antioxidant capacity in the extracts by the ABTS/TEAC method (11,050 µmol Trolox equivalents/100 g). However, our test has higher values of antioxidant capacity. On the other hand, the opposite is true in the study by Byczkiewicz et al. (2021) [43] where a higher antioxidant capacity was reported in the DPPH method instead of in ABTS analysis in the three varieties of quince analyzed. Other authors also showed a relatively higher value with ABTS in different ethanol extracts and in two varieties of quince [49]. Krzepilko and Prazak. (2023) [45] reported a higher antioxidant capacity in the extracts with ethanol and water in *C. japonica* (3.38 mmol TE 100 g^{-1}), while in the *C. superba*, there were no significant differences in the antioxidant capacity between the different extracts.

3.3. Antimicrobial Capacity

The resistance of microorganisms to drugs is increasing worldwide, assuming a very important problem in human health. For this reason, the research and development of new components and antibiotics are becoming increasingly important [3].

The antimicrobial activity in this assay was measured by the disk diffusion method against two fungi (*B. cinerea* and *C. acutatum*) and was performed in pre-lyophilized extract dilution. There are numerous studies on the antimicrobial activity of quince on human pathogens. However, as far as the authors are aware, there is not much literature on its activity against fungal phytopathogens. For this reason, the two strains selected for the antimicrobial test were important fungal pathogens in post-harvest diseases. The results obtained are shown in Figure 3.

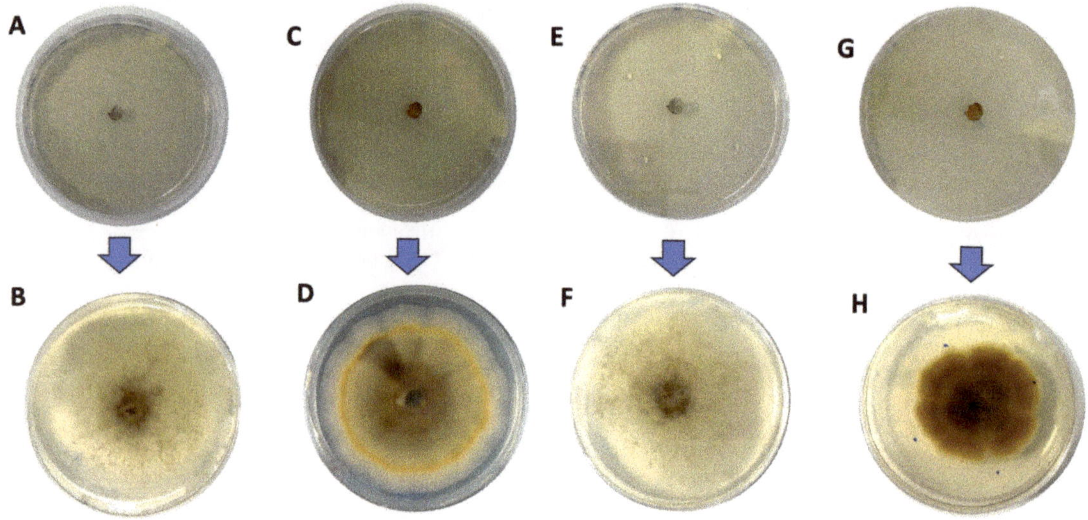

Figure 3. Quince by-product antimicrobial activity assay. (**A**) Control PDA plate with *B. cinerea* disk at time 0. (**B**) Control PDA plate with *B. cinerea* disk at time 7 of incubation at room temperature. (**C**) Control PDA plate with *C. acutatum* disk at time 0. (**D**) Control PDA plate with *C. acutatum* disk at time 7 of incubation at room temperature. (**E**) PDA plate inoculated pre-lyophilized extract dilution (1×) and *B. cinerea* disk at time 0 days. (**F**) PDA plate inoculated pre-lyophilized extract dilution (1×) and *B. cinerea* disk after 7 days of incubation at room temperature. (**G**) PDA plate inoculated pre-lyophilized extract dilution (1×) and *C. acutatum* disk at time 0 days. (**H**) PDA plate inoculated pre-lyophilized extract dilution (1×) and *C. acutatum* disk after 7 days of incubation at room temperature.

The results in the antifungal activity against pathogens were negative (Figure 3) since in none of the cases inhibition halos were observed at the points of application of the extract and the areas measured with the software presented non-existent values (400–650 μm^2). Higher concentration may be required to achieve antifungal activity. Similarly, the content of total phenolic compounds can interfere with the antimicrobial capacity of the extract [3]. Although phenols are generally positively correlated with antimicrobial ability, this was not the case with quince extract, which may be related to the TPC values obtained.

As mentioned above, not much literature has been found to study the antifungal properties of the fruit or by-product of quince against phytopathogens. However, Tarihi and Nejad. (2023) [57] studied the antibacterial capacity of silver nanoparticles obtained from the extract of the petals of the quince flower against *Erwinia amylovora*, responsible for the bacterial fire blight disease that devastates fruit tree crops such as pear trees, obtaining positive results dependent on the size and applied dose. Lykholat et al. (2022) [58] focused their study on knowing the endophytic population of quince (*Chaenomeles speciosa*) and its antifungal properties against various phytopathogens. The endophytic communities were isolated from both the skin and the pulp of the fruit. The species *Penicillium expansum*, *P. viridicatum*, and *P. hirsutum* were identified in the skin, while the species *P. chrysogenum*, *P.*

cyclopium, and *P. purpurogenum* were identified in the pulp [58]. The antifungal capacity was studied against the pathogens of the genus *Fusarium*, specifically *F. culmorummycelium* and *F. oxysporum*, where positive inhibition results were obtained without finding significant differences between the skin and pulp isolates [58].

Altuntas and Korukluoglu (2024) [59] studied the antifungal activity against *Saccharomyces cerevisiae*, *Candida albicans*, *Aspergillus flavus*, and *Penicillium roqueforti*. They obtained very positive results in the extracts made with the leaves of quinces. The extract with a concentration of 6.25 mg/mL resulted in 93.8% inhibition against *S. cerevisiae* while a concentration of 25 mg/mL inhibited *C. albicans* by 90%. However, if the concentration of the extract was increased, the inhibition was reduced. No 90% inhibition results were obtained with any concentration of any of the other quince parts used (peel, pulp, seeds, and juice). However, the highest concentration of seeds (100 mg/mL) managed to inhibit *A. flavus* and *P. roqueforti* [59]. It has been found that the number of phenolic compounds is not as important in antioxidant activity as the type of compound. The fungi could use the sugars available in the extract as nutrients, which could have affected the analysis carried out in this trial [59]. Likewise, the phenolic compounds, as indicated above, are closely related to the antimicrobial capacity [60] and there is a possibility that the extraction method of the phenolic compounds has affected the antimicrobial properties [3].

In the quince antimicrobial capacity from human pathogen studies, Anna et al. (2011) [61] studied the effect of plant extracts on *Helicobacter pylori* and described that quince extract had the greatest potential against this pathogen followed by blueberry extract. Other authors reported the activity of quince phenolic components against *Escherichia coli*, *Candida albicans*, *Pseudomonas aeruginosa*, and *Staphylococcus aureus*, obtaining positive results with different components, such as chromogenic acid [21]. Silva et al. (2023) [3] also studied the antimicrobial capacity of the extract of the different parts of quince against Gram+ and Gram− bacteria, where a stronger inhibition was observed against Gram+ bacteria since they are more susceptible to phenolic compounds. Gram− bacteria have a higher negative electrical charge that reduces the interactions between the membrane and phenolic compounds [62]. No action against any of the Gram-tested bacteria was observed with the quince extracts [3].

4. Conclusions

Quince by-product presents a very interesting nutritional profile which is high in fiber and low in fat, which makes it a potential ingredient for the formulation of products with specific properties. It also has a high antioxidant capacity, making it a natural antioxidant and a potential functional ingredient. This fact is very important because of the increasing demand for natural antioxidants versus chemical antioxidants by consumers. It does not have antifungal properties for the specific strains selected. The use and revalorization of the by-products also promote sustainability, circular economy, waste reduction, and resource optimization.

However, despite the results obtained in the assay, it is necessary to encourage research on the properties of this fruit since studies are relatively scarce. In future lines of research, it is important to study the behavior of the antioxidant capacity of this by-product in food. These results show that quince by-products could be used in the short future as a functional ingredient in the food industry such as a source of fiber for snacks or fiber bars for athletes. It also could be used as a natural preservative that delays oxidation thanks to its antioxidant properties.

Author Contributions: Conceptualization, J.R., I.A., M.J.C.-A. and A.D.M.; methodology, I.A., M.J.C.-A. and A.D.M.; validation, I.A., A.D.M. and J.P.; formal analysis, I.A.; investigation, J.R.; resources, I.A.; data curation, I.A.; writing—original draft preparation, J.R.; writing—review and editing, I.A., A.D.M. and J.P.; supervision, I.A. and M.J.C.-A. All authors have read and agreed to the published version of the manuscript.

Funding: This research has been funded by a knowledge transfer project with file number 6400/2 financed by the Junta de Castilla y León, and the city council of Ávila.

Institutional Review Board Statement: Not applicable.

Informed Consent Statement: Not applicable.

Data Availability Statement: No data were used for the research described in the article.

Conflicts of Interest: The authors declare no conflicts of interest.

References

1. Romero, J.; Albertos, I.; Díez-Méndez, A.; Poveda, J. Control of postharvest diseases in berries through edible coatings and bacterial probiotics. *Sci. Hortic.* **2022**, *304*, 111326. [CrossRef]
2. Bayram, B.; Ozkan, G.; Kostka, T.; Capanoglu, E.; Esatbeyoglu, T. Valorization and application of fruit and vegetable wastes and by-products for food packaging materials. *Molecules* **2021**, *26*, 4031. [CrossRef] [PubMed]
3. Silva, V.; Silva, A.; Ribeiro, J.; Aires, A.; Carvalho, R.; Amaral, J.S.; Barros, L.; Igrejas, G.; Poeta, P. Screening of chemical composition, antimicrobial and antioxidant activities in pomegranate, quince, and persimmon leaf, peel, and seed: Valorization of autumn fruits by-products for a one health perspective. *Antibiotics* **2023**, *12*, 1086. [CrossRef] [PubMed]
4. Plazzotta, S.; Cottes, M.; Simeoni, P.; Manzocco, L. Evaluating the environmental and economic impact of fruit and vegetable waste valorisation: The lettuce waste study-case. *J. Clean. Prod.* **2020**, *262*, 121435. [CrossRef]
5. Lau, K.Q.; Sabran, M.R.; Shafie, S.R. Utilization of vegetable and fruit by-products as functional ingredient and food. *Front. Nutr.* **2021**, *8*, 661693. [CrossRef]
6. Batool, F.; Kurniawan, T.A.; Mohyuddin, A.; Othman, M.H.; Aziz, F.; Al-Hazmi, H.; Goh, H.H.; Anouzla, A. Environmental impacts of food waste management technologies: A critical review of life cycle assessment (LCA) studies. *Trends Food Sci. Technol.* **2023**, *143*, 104287. [CrossRef]
7. Chandan, A.; John, M.; Potdar, V. Achieving UN SDGs in food supply chain using blockchain technology. *Sustainability* **2023**, *15*, 2109. [CrossRef]
8. Kroll, C.; Warchold, A.; Pradhan, P. Sustainable Development Goals (SDGs): Are we successful in turning trade-offs into synergies? *Palgrave Commun.* **2019**, *5*, 140. [CrossRef]
9. Milho, C.; Silva, J.; Guimarães, R.; Ferreira, I.C.; Barros, L.; Alves, M.J. Antimicrobials from medicinal plants: An emergent strategy to control oral biofilms. *Appl. Sci.* **2021**, *11*, 4020. [CrossRef]
10. Okonogi, S.; Duangrat, C.; Anuchpreeda, S.; Tachakittirungrod, S.; Chowwanapoonpohn, S. Comparison of antioxidant capacities and cytotoxicities of certain fruit peels. *Food Chem.* **2007**, *103*, 839–846. [CrossRef]
11. Mutahar, S.S.; Mutlag, M.A.O.; Najeeb, S.A.Z. Antioxidant activity of pomegranate (*Punica granatum* L.) fruit peels. *Food Nutr. Sci.* **2012**, *3*, 20499. [CrossRef]
12. Chen, J.; Liao, C.; Ouyang, X.; Kahramanoğlu, I.; Gan, Y.; Li, M. Antimicrobial activity of pomegranate peel and its applications on food preservation. *J. Food Qual.* **2020**, *2020*, 8850339. [CrossRef]
13. Abu-Niaaj, L.F.; Al-Daghistani, H.I.; Katampe, I.; Abu-Irmaileh, B.; Bustanji, Y.K. Pomegranate peel: Bioactivities as antimicrobial and cytotoxic agents. *Food Sci. Nutr.* **2024**, *12*, 2818–2832. [CrossRef] [PubMed]
14. Lai, C.S.; Li, S.; Miyauchi, Y.; Suzawa, M.; Ho, C.T.; Pan, M.H. Potent anti-cancer effects of citrus peel flavonoids in human prostate xenograft tumors. *Food Funct.* **2013**, *4*, 944–949. [CrossRef] [PubMed]
15. Rodríguez-González, S.; Gutiérrez-Ruíz, I.M.; Pérez-Ramírez, I.F.; Mora, O.; Ramos-Gomez, M.; Reynoso-Camacho, R. Mechanisms related to the anti-diabetic properties of mango (*Mangifera indica* L.) juice by-product. *J. Funct. Foods* **2017**, *37*, 190–199. [CrossRef]
16. Caballero, B.; Trugo, L.; Finglas, P. *Encyclopedia of Food Sciences and Nutrition*, 2nd ed.; Elsevier Science B.V: Amsterdam, The Netherlands, 2003; p. 6601.
17. Bayav, A.; Şahin, M. Global economic importance of quince: Current situation, forecasting, and competitiveness analysis. *Erwerbs-Obstbau* **2023**, *65*, 509–520. [CrossRef]
18. Ňorbová, M.; Vollmannová, A.; Fedorková, S.; Musilová, J.; Lidiková, J. The forgotten fruit (*Cydonia oblonga* Mill.) and its chemicals composition: A review. *Eur. Food Res. Technol.* **2024**, *250*, 2093–2102. [CrossRef]
19. Food and Agriculture Organization of the United Nations. *World Food and Agriculture-Statistical Yearbook 2020*; Food and Agriculture Organization of the United Nations: Rome, Italy, 2020.
20. Codina, M.C.; González, E.J.; Molina, A.; Carmona, M.; Berruga, M.I. Bio-based films from quince by-products: A sustainable alternative for biodegradable food packaging. *Food Hydrocoll.* **2024**, *157*, 110395. [CrossRef]
21. Rather, J.A.; Yousuf, S.; Ashraf, Q.S.; Mir, S.A.; Makroo, H.A.; Majid, D.; Barba, F.J.; Dar, B.N. Nutritional and bioactive composition, nutraceutical potential, food and packaging applications of *Cydonia oblonga* and its byproducts: A review. *J. Food Compos. Anal.* **2023**, *115*, 105000. [CrossRef]
22. Sun, L.; Liu, H.; Chi, B.; Bian, Y.; Zhou, H.; Tian, Z. Analysis and bioactivity of volatile oil from *Cydonia oblonga* fruit by GC–MS. *Food Anal. Methods* **2022**, *15*, 2725–2733. [CrossRef]

23. Dranca, F.; Oroian, M. Extraction, purification and characterization of pectin from alternative sources with potential technological applications. *Food Res. Int.* **2018**, *113*, 327–350. [CrossRef] [PubMed]
24. Blanda, G.; Rodriguez-Roque, M.J.; Comandini, P.; Flores-Cordova, M.A.; Salas-Salazar, N.A.; Oscar, C.A.; Soto-Caballero, M.C. Phenolic profile and physicochemical characterization of quince (*Cydonia oblonga* Mill) fruits at different maturity index. *Not. Bot. Horti Agrobot. Cluj-Napoca* **2020**, *48*, 2306–2315. [CrossRef]
25. Ashraf, M.U.; Muhammad, G.; Hussain, M.A.; Bukhari, S.N. *Cydonia oblonga* M., a medicinal plant rich in phytonutrients for pharmaceuticals. *Front. Pharmacol.* **2016**, *7*, 163. [CrossRef] [PubMed]
26. Al-Zughbi, I.; Krayem, M. Quince fruit *Cydonia oblonga* Mill nutritional composition, antioxidative properties, health benefits and consumers preferences towards some industrial quince products: A review. *Food Chem.* **2022**, *393*, 133362. [CrossRef] [PubMed]
27. De Bellis, R.; Chiarantini, L.; Potenza, L.; Gorassini, A.; Verardo, G.; De Marco, R.; Benayada, L.; Stocchi, V.; Albertini, M.C.; Fraternale, D. High production of secondary metabolites and biological activities of *Cydonia oblonga* Mill. pulp fruit callus. *J. Funct. Foods* **2022**, *94*, 105133. [CrossRef]
28. Cunniff, P.; Washington, D. Official methods of analysis of AOAC International. *J. AOAC Int.* **1997**, *80*, 127A.
29. AOAC. Official Method 2003.05. In *Official Methods of Analysis of AOAC International*, 18th ed.; AOAC International: Gaithersburg, MD, USA, 2005.
30. AOAC. Official Method 923.03. In *Official Methods of Analysis of AOAC International*, 18th ed.; AOAC International: Gaithersburg, MD, USA, 2005.
31. AOAC. Methods 968.06 & 992.15. In *Official Methods of Analysis of AOAC International*, 18th ed.; AOAC International: Gaithersburg, MD, USA, 2005.
32. Prosky, L.; Asp, N.G.; Schweizer, T.F.; Devries, J.W.; Furda, I. Determination of insoluble, soluble, and total dietary fiber in foods and food products: Interlaboratory study. *J. Assoc. Off. Anal. Chem.* **1998**, *71*, 1017–1023. [CrossRef]
33. AOAC. Official Method 982.14. In *Official Methods of Analysis of AOAC International*, 15th ed.; AOAC International: Gaithersburg, MD, USA, 1990.
34. Santos, H.M.; de Higuera, J.M.; de Araujo Nogueira, A.R. In vitro bioaccessibility of essential minerals from raw and cooked Tilapia fillet: Method validation and analysis by synchronous vertical dual view ICP OES. *Food Chem. Adv.* **2022**, *1*, 100080. [CrossRef]
35. Slinkard, K.; Singleton, V.L. Total phenol analyses: Automation and comparison with manual methods. *Am. J. Enol. Vitic.* **1977**, *28*, 49–55. [CrossRef]
36. Re, R.; Pellegrini, N.; Proteggente, A.; Pannala, A.; Yang, M.; Rice-Evans, C. Antioxidant activity applying an improved ABTS radical cation decolorization assay. *Free Radic. Biol. Med.* **1999**, *26*, 1231–1237. [CrossRef]
37. Brand-Williams, W.; Cuvelier, M.E.; Berset, C.L.W.T. Use of a free radical method to evaluate antioxidant activity. *LWT-Food Sci. Technol.* **1995**, *28*, 25–30. [CrossRef]
38. Zicari, G.; Carraro, E.; Bonetta, S. The Regulation (EC) N. 1924/2006 of the European Parliament and of the Council of the 20 december 2006 on nutrition and health claims made no foods. *Prog. Nutr.* **2007**, *9*, 264–273.
39. Othman, S.; Añibarro-Ortega, M.; Dias, M.I.; Ćirić, A.; Mandim, F.; Soković, M.; Ferreira, I.C.; Pinela, J.; Barros, L. Valorization of quince peel into functional food ingredients: A path towards "zero waste" and sustainable food systems. *Heliyon* **2022**, *8*, e11042. [CrossRef] [PubMed]
40. Coimbra, P.; Marona, B.; Henriques, M.H.; Campos, L.; Gomes, D.M.; Vitorino, C.; Sousa, J.J.; Braga, M.E.; Gaspar, M.C. Edible films based on potato and quince peels with potential for the preservation of cured cheese. *Food Packag. Shelf Life* **2023**, *40*, 101176. [CrossRef]
41. Dimitriu, L.; Preda, D.; Constantinescu-Aruxandei, D.; Oancea, F. Quince Pomace: A Source of Fiber Products and Polyphenols. *Chem. Proc.* **2023**, *13*, 6. [CrossRef]
42. Vasileva, A.L.; Durakova, A.G.; Goranova, Z.T.; Kalaydzhiev, H.R. El potencial del producto alimenticio regional membrillo en forma de polvo como una buena fuente de compuestos bioactivos. *Preprints* **2024**, 2024011888. [CrossRef]
43. Byczkiewicz, S.; Szwajgier, D.; Kobus-Cisowska, J.; Szczepaniak, O.; Szulc, P. Comparative examination of bioactive phytochemicals in quince (*Chaenomeles*) fruits and their in vitro antioxidant activity. *Emir. J. Food Agric.* **2021**, *33*, 293–302. [CrossRef]
44. Serna, J.; Bergwitz, C. Importance of dietary phosphorus for bone metabolism and healthy aging. *Nutrients* **2020**, *12*, 3001. [CrossRef]
45. Krzepilko, A.; Prazak, R. Quince seeds as a potential source of mineral and biological active compounds. *J. Elem.* **2023**, *28*, 107–122. [CrossRef]
46. Ibrahim Anber, M.M.A.; Asadi-Gharneh, H.A. Evaluation of Yield Components, Bioactive Compounds, Antioxidative Activity and Mineral Composition in Quince Genotypes. *Appl. Fruit Sci.* **2024**, *66*, 465–473. [CrossRef]
47. Theuwissen, E.; Mensink, R.P. Water-soluble dietary fibers and cardiovascular disease. *Physiol. Behav.* **2008**, *94*, 285–292. [CrossRef] [PubMed]
48. Bakr, A.F.; Farag, M.A. Soluble dietary fibers as antihyperlipidemic agents: A comprehensive review to maximize their health benefits. *ACS Omega* **2023**, *8*, 24680–24694. [CrossRef] [PubMed]
49. Dai, F.J.; Chau, C.F. Classification and regulatory perspectives of dietary fiber. *J. Food Drug Anal.* **2017**, *25*, 37–42. [CrossRef] [PubMed]

50. Nie, Y.; Luo, F. Dietary fiber: An opportunity for a global control of hyperlipidemia. *Oxid. Med. Cell. Longev.* **2021**, *2021*, 5542342. [CrossRef]
51. Khan, E.; Ahmad, I.Z. An insight into the prophylactic and therapeutic activities of golden apple (*Cydonia oblonga* Mill.) for the future cancer care and prevention: A review. *Ann. Phytomed* **2021**, *10*, 22–35. [CrossRef]
52. Rasheed, M.; Hussain, I.; Rafiq, S.; Hayat, I.; Qayyum, A.; Ishaq, S.; Awan, M.S. Chemical composition and antioxidant activity of quince fruit pulp collected from different locations. *Int. J. Food Prop.* **2018**, *21*, 2320–2327. [CrossRef]
53. Hegedűs, A.; Papp, N.; Stefanovits-Bányai, É. Review of nutritional value and putative health-effects of quince (*Cydonia oblonga* Mill.) fruit. *Int. J. Hortic. Sci.* **2013**, *19*, 29–32. [CrossRef]
54. Benahmed Djilali, A.; Mehraz, R.; Bouacem, K.; Benseddik, A.; Moualek, I.; Nabiev, M.; Benzara, A. Bioactive substances of *Cydonia oblonga* fruit: Insecticidal effect of tannins on *Tribuliumm confusum*. *Int. J. Fruit Sci.* **2021**, *21*, 721–731. [CrossRef]
55. Szychowski, P.J.; Lech, K.; Sendra-Nadal, E.; Hernández, F.; Figiel, A.; Wojdyło, A.; Carbonell-Barrachina, Á.A. Kinetics, biocompounds, antioxidant activity, and sensory attributes of quinces as affected by drying method. *Food Chem.* **2018**, *255*, 157–164. [CrossRef]
56. Aguayo-Rojas, J.; Mora-Rochín, S.; Tovar-Jiménez, X.; Navarro-Cortez, R.O.; Valdez-Morales, M.; Ayala-Lujan, J.L. Phenolic compounds, antioxidant capacity and antihypertensive activity of quince (*Cydonia oblonga* Miller) harvested in Zacatecas. *Polibotánica* **2024**, *57*, 199–212. [CrossRef]
57. Tarighi, S.; Nejad, M.S. Ecofriendly fabrication of silver nanoparticles using quince petal extract and its antibacterial properties against fire blight disease. *J. Nat. Pestic. Res.* **2023**, *4*, 100026. [CrossRef]
58. Lykholat, Y.V.; Didur, O.O.; Drehval, O.A.; Khromykh, N.O.; Sklyar, T.V.; Lykholat, T.Y.; Liashenko, O.V.; Kovalenko, I.M. Endophytic community of *Chaenomeles speciosa* fruits: Screening for biodiversity and antifungal activity. *Regul. Mech. Biosyst.* **2020**, *13*, 130–136. [CrossRef]
59. Altuntas, S.; Korukluoglu, M. Biological activity of optimized phenolic extracts of quince (*Cydonia oblonga* Miller) parts before and after simulated in vitro gastrointestinal digestion. *Food Chem.* **2024**, *437*, 137846. [CrossRef] [PubMed]
60. Maghsoudlou, Y.; Asghari Ghajari, M.; Tavasoli, S. Effects of heat treatment on the phenolic compounds and antioxidant capacity of quince fruit and its tisane's sensory properties. *J. Food Sci. Technol.* **2019**, *56*, 2365–2372. [CrossRef]
61. Anna, B.; Vizma, N.; Dmitry, B. Anti-Helicobacter activity of certain food plant extracts and juices and their composition in vitro. *Food Nutr. Sci.* **2011**, *2*, 7889.
62. Lima, M.D.C.; De Sousa, C.P.; Fernandez-Prada, C.; Harel, J.; Dubreuil, J.D.; De Souza, E.L. A review of the current evidence of fruit phenolic compounds as potential antimicrobials against pathogenic bacteria. *Microb. Pathog.* **2011**, *130*, 259–270. [CrossRef]

sustainability

MDPI

Article

Nudging Householders to Reduce Avoidable Food Waste: The OzHarvest Use It Up Tape

Mark Boulet * and **Nita Lauren**

BehaviourWorks Australia, Monash University, Clayton, VIC 3800, Australia
* Correspondence: mark.boulet@monash.edu

Abstract: Targeting households with food waste reduction interventions represents a critical opportunity to meet global targets to halve food loss and waste. While the evidence base on the effectiveness of food waste interventions is growing generally, less is known about the outcomes of household-focused interventions. This mixed methods study explores how households experienced a behaviourally orientated nudge (the OzHarvest Use it Up Tape^{TM}) and examines its impact on food waste and behaviour change. The "Tape" served multiple functions for households—including as a visual prompt, a labelling device, a planning tool, and a communication tool—and was more effective for large families and for individuals who were disorganised when shopping and cooking. Significant reductions were also identified in participants' fresh vegetable and fruit waste and in the total food amounts they wasted. This study demonstrates the effectiveness of behaviourally orientated nudges, like the Tape, in reducing food waste.

Keywords: food waste; behaviour change; interventions; nudges; households

1. Introduction

The turn of the century has seen a substantial increase in policy and research attention to the global food waste challenge and the urgent need to reduce waste and loss across the entire food system (see, for example, [1–4]). The diverse and severe social, economic, and environmental impacts from food waste have led to the inclusion of target (12.3), to halve global food loss and waste by 2030, in the United Nation's Sustainable Development Goals [5]. While waste and loss occur at all stages (from production to consumption) of the food system, over 900 million tonnes are wasted each year worldwide in the consumption stage (retail, hospitality, and households) alone, with household food wastage representing around 60% of this amount in most countries [3]. The consumption stage, with households as a key target, therefore represents a critical area for effective food waste reduction policies and programs.

A 2019 review [6] of food waste reduction or prevention interventions at the consumption stage by Reynolds and colleagues identified only 13 studies with quantified waste reduction outcomes. Of these, just six tested the impact of interventions targeting the household. Reynolds et al. [6] argued that this small number of studies represents a significant gap in the evidence base relating to food waste reduction interventions, and it was therefore difficult for policy and program managers to "make evidence-based decisions to prevent or reduce consumption stage food waste in a cost-effective manner" [6] (p. 1). They concluded their review with a plea for more "well-designed [food waste] interventions ... [that are] tested using carefully selected methods to understand the outcomes of the intervention and how it works (or not)" [6] (p. 20).

Pleasingly, in the years since Reynolds et al. [6] published their review, there has been a small explosion of published studies exploring and measuring the impacts of food waste reduction interventions at the consumption stage. Studies such as [7–9] have measured the outcomes of broad-based multi-faceted campaigns that combine different consumer engagement approaches, while others [10–12] have measured the outcomes of more targeted interventions, such as food waste reduction tool kits, flexible recipes, and targeted messaging to consumers.

This greatly expanded evidence base has led to further systematic reviews that collate different food waste reduction interventions at the consumption stage and synthesise their outcomes [13–15]. Simões et al. [13] for example, reviewed 96 studies and identified 18 which specifically engaged consumers in reducing food waste from their households. Most of the interventions in these studies were focused on raising consumers' awareness of food waste impacts and providing information on how to reduce or avoid waste. While [13] acknowledged that providing consumers with this type of information was important, they argued that it is often not enough to encourage consumer behaviour change and it needs to be complemented with other intervention types, such as the provision of tools or changes to the home or food retail environments.

A more recent meta-analysis [15] looked at the effectiveness of nudge-based interventions in addressing food waste at the consumption phase. Defining a nudge as "a voluntary, non-intrusive intervention that induces behavioural changes without economic incentives or mandates" [15] (p. 1), they compared the effectiveness of cognitively orientated (changing attitudes, awareness, and knowledge) and behaviourally orientated (modifying the environment in which behaviours occur) nudges. Their meta-analysis revealed that while nudges overall can have a significant effect on reducing food, behaviourally orientated ones are generally more effective in reducing food waste than those that are cognitively orientated (despite the latter being more commonly used in policy and practice). This review [15] highlights the value of nudges in reducing food waste, and echoes a point made by [13] that the evidence base on the effectiveness of nudges and other interventions to reduce food waste in households is still relatively limited (compared to that in other settings, such as hospitality or education institutions).

Our paper aims to strengthen the current evidence base by exploring how households experienced a behaviourally orientated food waste reduction nudge trialled in Australia, and examining its impact on both food waste reduction and behaviour change.

Intervention Development

Founded in 2004, OzHarvest (Website: https://www.ozharvest.org/, accessed on 1 June 2024) is one of Australia's leading national food rescue charities. It has recently complemented its food rescue and education focus with campaigns that engage Australian consumers and households in reducing avoidable food waste. Drawing on previously commissioned research that identified and prioritised food waste reduction behaviours for the Australian context [16], OzHarvest targeted a key behaviour that was considered to be the most impactful (in terms of reducing food waste) and relatively easy for households to apply, namely: householders preparing a regular 'use it up' meal that combines any food in the refrigerator or pantry that needs to be used up (this includes leftover meals and ingredients, as well as items nearing their use-by-date).

To engage Australian households in this behaviour, and in food waste avoidance more generally, OzHarvest developed the Use It Up^{TM} campaign (Website: https://www.ozharvest.org/use-it-up/, accessed on 1 June 2024). As part of the campaign, the Use It Up Tape^{TM} (Website: https://events.ozharvest.org/shop/viewitem/use-it-up-tape, accessed on 1 June 2024) (see Figure 1) was created as a tool to make it easy for householders to adopt the target behaviour. The initial idea was that the brightly coloured product could be used to mark out a space in the refrigerator or pantry to create a 'Use It Up shelf' on which items that needed to be used up could be placed [16]. First made available to the Australian public in October 2021, over 95,000 units of the Use it Up Tape ("the Tape") have been delivered nationally since that time. Under a licensing agreement, an equivalent version was created for The Netherlands ('Eerst Op Tape') and distributed to 40,000 Dutch households. OzHarvest is currently in conversation with other countries about developing additional local versions.

Figure 1. OzHarvest Use It Up Tape^{TM}.

The Tape (and the shelf space it created) was expected to function as a *visual prompt* to remind householders (when they opened the refrigerator or looked in the pantry) of food items or leftover meals that needed to be eaten before they spoiled or before new ones were purchased. Visual prompts are behaviourally orientated nudges that make desired behaviours salient in the minds of individuals [17,18]. It is well established (see, for example, [19,20]) that there are a number of cognitive constraints that prevent individuals from paying active attention to every behaviour (and its implications) that they engage with throughout any given period. This limited attention has been suggested as one of the reasons for the gap between an individual's behavioural intentions (such as using up food items in a meal) and their final actions; commonly known as the intention–action gap [17,21]. Visual prompts such as the Tape and the messages contained on it (e.g., "Eat me", "Cook me up", etc) can help to close this gap by 'nudging' the individual to first recall their intention to use up particular food items before they spoil and to then action this intention by making a meal from these items.

This paper presents a mixed methods intervention impact study of the Tape as a behaviourally orientated nudge to reduce food waste. It describes the design and outcomes of two complementary studies conducted in Australia which explored how households used, and experienced, the Tape, and examined its impact on food waste and on relevant household food provisioning behaviours. The remaining sections of this paper present the methods and results for each study, discuss their findings and implications for food waste policy and programs, and suggest future research opportunities.

2. Materials and Methods

The use and impact of the Tape was explored through two complementary studies:

1. A qualitative observational study that utilised participant-made video diaries to explore how the Tape was used and experienced by different households.
2. A quantitative pre–post study in which participants completed a validated survey-based tool to report their food waste amounts before and after a two-week period of using the Tape.

Each study is described in greater detail below. Taken together, they give insights not just into the food reduction impacts of the Tape, but also its suitability for different households and its different functions to support household food provisioning and reduce waste.

2.1. Study 1

This study is a qualitative observational one, with video diaries used as the primary data collection method. The diaries visually capture participants' use of the Tape and prompt detailed reflection of their experience with it. Video diaries bring the researcher one step closer to the reality of participants' lives and capture what actually happens in situ rather than the more general and abstracted insights that can come from surveys and interviews [22,23].

Research ethics approval was given by the Monash University Ethics Committee (project ID # 28967).

2.1.1. Research Participants

A specialist market research company was engaged to assemble a panel of participants who were financially incentivised (a small stipend) to take part in this research. Panel members are trained in smartphone video recording techniques and upload videos for researcher access to an online platform maintained by the company.

The participants ($n = 9$) in this study were low to middle-income earners from a mix of regional and metropolitan locations along the Australian west coast, primarily around the Western Australian city of Perth (see Table 1). They were the main person in their household responsible for cooking and shopping, their ages ranged from 25–60 years, and they came from households with and without children.

Table 1. Description of qualitative study participants ($n = 9$).

Demographic		Number of Participants
Gender	Female	5
	Male	4
Age	21–30 years 31–40 years 41–60 years	1 4 4
Number of children in household	No children 1–2 children 3–4 children	6 1 2

2.1.2. Research Protocol

Participants were sent a roll of the Tape and provided with basic instructions for its use (mainly that its goal was to help remind them of which food needed to be used up, and that it could be used to mark out a shelf in their refrigerator or pantry). During a two-week period of using the Tape, they were asked to record up to five short (five to ten minute) videos, including:

1. A set-up video showing how the Tape was first used, their initial reflections and their intentions for the next two weeks.
2. Up to three check-in videos (spaced two to four days apart) that showed how the Tape was being used, changes in its use since set up, its general performance and influence on their shopping or cooking practices, and how other household members have interacted with it.
3. A final reflection video on their overall impressions of the Tape, its influence, impacts and outcomes, what changes they might recommend and whether they intend to keep using it.

While the Tape's aim of supporting households to use up leftover meals and ingredients was openly and clearly communicated to participants, the focus on food waste reduction or avoidance was made less explicit. This was to avoid potentially biasing the responses due to perceived social norms against being wasteful with food.

2.1.3. Data Analysis

We received four to six videos from each participant, with a total of 43 videos across the entire sample. Two types of data emerged from this research design: (i) the video recordings of the participants, their refrigerators, pantries, and food, and (ii) their audio transcripts when responding to the prompts above. We were primarily interested in participants' reflections, perspectives and self-reports while recording their video diaries. The audio transcripts were therefore the focus of our analysis, while also noting what we could see of the Tape and its use when it was shown by participants.

We conducted a mainly deductive thematic analysis of the audio transcripts, in which common themes within an individual's video diaries, and across different participants diaries, were noted [24,25].

2.2. Study 2

The second study aimed to quantitatively measure the food waste reduction potential of the Tape and to understand which household food-related behaviours it supported. Following a pre–post trial approach, a validated food waste self-measurement survey was completed by participants before and after they used the Tape for two weeks, and food waste outcomes at each stage were compared.

This study was given ethics approval by the Monash University Ethics Committee (project ID # 32508)

2.2.1. Research Participants

OzHarvest promoted the Tape mainly through its website and extensive social media network (Facebook, X (formally Twitter), and Instagram). This promotion highlighted the potential food waste reduction benefits of the Tape and directed people to the OzHarvest website to order it. While ordering, anyone living within Australia was then invited to participate in the study. No incentives were provided for participation, although at the time of the study, the Tape was made available for free (only postage needed to be paid).

361 participants completed the pre-use survey, and 144 participants completed the post-use survey, with 76 of these participants having fully completed both pre- and post-use surveys ('matched pairs'). A demographic summary is presented in Table 2, which shows that the sample tended to be around 50 years old and most identified as female.

Table 2. Demographic summary of both the entire sample and matched pairs for the quantitative study.

	Entire Sample	**Matched Pairs**
Age (average)	49.74 years	52.10 years
Gender	88.9% identified as female	93.2% identified as female
Household size	31.9% had 4 people	31.5% had 2 people
Number of kids in household	46.5% had no children	49.3% had no children
Education	29.9% had undergraduate degree	28.8% had undergraduate degree
Employment	38.9% were employed full time	26% were employed full time

2.2.2. Research Protocol and Survey Measures

When ordering their Tape from the OzHarvest website, those willing to participate in the study provided their email address and were emailed a link to the pre-survey to complete immediately. At the same time, the Tape was posted to their homes. Within three weeks of the Tape being sent to a participant, they were emailed the link to the post-use survey. With potential delays in postal delivery and participants using the Tape, it was assumed that this period would enable participants to use the Tape for about two weeks overall before completing the post-use survey.

When completing both surveys, participants were asked to provide the four-digit postcode of the Australian suburb where they usually live and their birthdate (i.e., DDMMYY). These two number sets were combined in the analysis to give each participant a unique code, which was used to match their pre- and post-use survey responses (if both were completed).

Both pre- and post-use surveys measured participant food waste based on the Household Food Waste Questionnaire (HFWQ) approach developed—and validated for European countries—by [26], and later refined by [27] for the United States. This asks participants to first identify the different types of food that they discarded and then estimate how much of each type was discarded. Both pre- and post-use surveys took about 10 min each for participants to complete online.

For the first part of each survey, participants reviewed a list of 24 food and drink categories and ticked each category that they had discarded over the past seven days. They were asked to include any edible food and drink they bought online, at the supermarket, as takeaway, or grew themselves, as well as meal leftovers or products that were spoiled or past their expiration date. Bones, peels, pits, or cores, or food and drink thrown away when eating out of the home were not included. Participants were asked to include any food or drink regardless of how it was disposed (i.e., in the bin, compost, or given to pets).

The second phase of the survey asked participants to estimate how much of each food and drink type they discarded, but only for those categories that they had identified previously. We followed [27] and used 'cups' as the more appropriate unit for an Australian audience to estimate how much was thrown out. We also used 'portions', 'pieces' or 'glasses' as other relevant units for different categories (e.g., for fruit, snacks, or beverages, respectively). As per [27], participants were given further guidance to estimate how much of each category they discarded (e.g., "A cup of rice equals 153 g" or "A portion of meat (150 g) refers to one chicken breast, one steak etc"). To reduce survey time and ensure response quality, we did not follow [26] or [27] by asking participants to nominate the 'status' of discarded food (i.e., "completely un-used", "partly used", "meal leftovers"), as this was not critical to the aim of quantifying food waste amounts.

Finally, participants indicated (on a five-point ordinal response scale from "never" to "always") the frequency with which they performed certain household food provisioning behaviours (e.g., checking food stocks before shopping or making a meal with food that needs to be used up). Behaviours included here were not just the target behaviour for the Use It Up campaign described previously, but also included behaviours that were mentioned by participants in Study 1 as something they did because of the Tape.

2.3. Data Analysis

The total quantity of self-reported food waste for each participant was calculated by translating the waste unit for each category (e.g., "cups", "portions") into grams and summing across all categories. Here, we followed the table of average weights for each food and drink category used by [27], which in turn are based on the US Department of Agriculture's estimates for different food categories. Assumptions of normality were tested prior to analysis, where the skew and kurtosis of each food waste category and food provisioning behaviour were checked. The data was considered normal when skew values were between −2 or +2 and kurtosis was between −7 and +7 [28].

Two sets of analyses were conducted to analyse the impact of the Tape on participant food waste outcomes and the frequency of food-related behaviours. The first set focused on comparing the differences between the entire pre- and post-use survey samples. To account for the variations in normality in the food waste data, two types of tests were used to assess differences in food waste between pre- and post-use survey periods; the Welch's Analysis of Variance (Welch's ANOVA), a parametric test that does not assume equal variances, so it can be used for assessing unequal sample sizes (unlike a standard ANOVA), and the Mann–Whitney U test, a non-parametric test that assess differences between independent groups involving non-normal data. A third type of test, ordinal logistic regression (a parametric test that assess relationships between groups on an ordinal response variable), was used to estimate differences in food provisioning behaviours between these two periods.

The second set of analyses focused on those who had completed both pre- and post-use surveys (i.e., matched pairs). Wilcoxon signed-rank tests (a non-parametric test that does not assume normality in data) were used to assess differences in food waste between pre-

and post-use survey periods, and a generalised estimating equations (GEE) analysis was used to assess differences in food provisioning behaviours between these two periods (GEE can be used to assess non-normal repeated measures data). All statistical analyses were performed using SPSS version 28 (IBM SPSS Statistics).

3. Results

3.1. Study 1

Four main themes were identified when coding the video recordings and audio transcripts; (i) how the Tape was used, (ii) the food provisioning functions of the Tape, (iii) the behaviours it supported, and (iv) differences in the usefulness of the Tape based on household type/characteristics. These themes are further described below with illustrative quotes.

3.1.1. Tape Use

Participants used the tape in two different (and often complementary) ways (see Figure 2). The first was to demarcate space in their refrigerator or pantry to place any food to be used up. This was expected, as it was part of the instructions given to participants. Some used it in their pantry and refrigerator, while others only in their refrigerator. No participant used the Tape in their freezers.

The second way saw some participants tear off smaller sections of the Tape as labels on specific items they wanted to use up. This was not expected, as it was not communicated to the participants as an option and revealed a more flexible use of the Tape. Some participants even used specific "Eat me", "Pick me", "Cook me up" sections of the Tape to signify different intentions and uses for food (i.e., "Cook me up" to remind themselves of what needed to be included in the next meal, or "Eat me" as a visual prompt for other household members to eat that item when hungry).

(a) (b)

Figure 2. Example stills taken from participants' diaries showing how the Tape was used. (**a**) To mark individual food items to use up. (**b**) To mark out a shelf in the refrigerator (or pantry) to place items that needed to be used up.

3.1.2. Tape Functions

The Tape provided several different food provisioning functions for participants, who either used it for one main function or for several complementary functions. As was expected during the initial design of the Tape, a prominent function was to act as a visual prompt that reminded participants of foods that needed to be used up when they looked in the refrigerator or pantry—either through the Tape or the shelf it designated.

> *Marking containers with those items with the tape, makes it easy to identify ... [yesterday] my wife spotted one of the containers with some leftover pork noodles in it, and thought, "I'll take that to work, rather than getting lunch at work".*

Participants often had similar items in their refrigerators or pantries which were bought at different times. The Tape was therefore used by some as a labelling device to help distinguish between older and newer versions of certain product.

This is really good for the eggs that we buy. We have eggs here that we bought more recently, but these are the older ones. And sometimes it can get confusing knowing which eggs we bought first. So these ones [labelled with the Tape] we'll use up first.

An intention for the Tape was for it to support householders to make weekly meals with food that needed to be used up. Several participants mentioned how they deliberately used the Tape as a meal planning tool that helped to identify, when they looked in the refrigerator or pantry, the next meals that they would be preparing for their households.

And it's been really useful to just have food I need to cook in one area so that, at a glance maybe in the morning before going to work, I've been able to look at what I need to use up. And come up with an idea of what to cook. An improvement is I've cooked a couple of different dishes . . . new dishes that I haven't tried before, just because [I'm] trying to use up the food.

A completely unexpected use of the Tape was as a communication tool between members of larger households as to what food in the refrigerator or pantry could be eaten if they were hungry between meals or what leftovers should be taken to work for lunch. Other household members could see, by looking for labelled items in the refrigerator or pantry, which food the main person responsible for cooking or shopping wanted them to eat as a snack or take to work for lunch.

It's been really helpful for my husband because he knows what leftovers he needs to eat. . . .. [and] in terms of communicating with the family anything that's on here they can generally eat. That's helpful, rather than having them having to ask me first.

3.1.3. Behaviours Supported

Participants mentioned several specific waste reduction behaviours that the Tape supported. These included making meals that included food that needed to be used up; checking on existing food stocks before shopping and then making a shopping list; buying food types and amounts that they actually need when shopping; and avoiding takeaway/eating out options because they knew that there was food that needed to be used up.

It's been helpful to me for shopping lists . . . because when I look at what I've got and I think of recipes according to what I've got and then make the shopping list based on ingredients for what I want to make during the week. Rather than just think of all new ingredients without thinking first about what we need to use up.

3.1.4. Differences in Usefulness Based on Household Characteristics

The Tape seemed to be more useful for larger (often family-based) households with large volumes of food use and/or disorganised individuals who do not usually plan out their food shopping, storage, and cooking.

. . .if you were already an organised person with good budgeting skills I don't think it would have a huge effect. But because I'm quite disorganised, I found it quite helpful.

But the biggest success was with our daughter, who now can have a look at the food that's in the fridge and choose something [with the Tape on it]. She has a look, she can actually pull the container out and go, yeah, I want some of that.

The Tape was less useful for smaller households (often singles or couples) who buy smaller amounts of food and/or for highly organised individuals who carefully plan their different food provisioning practices.

For myself... I think it's a bit hard because there's not an awful lot of food in the house, I guess, for one person. And you tend to know what you've bought because you're only buying a set amount of food and what's going to be coming out of date as it is.

I use everything I buy regularly and shop for what I'm going to consume. I don't prepare food and let it sit in the fridge. I shop for what I need, I prepare it, if I don't eat it that day... it's consumed the next day or the day after. So, this little simple system [the Tape] isn't effective for me.

3.2. Study 2

This section shows the results of analyses conducted to identify differences in food waste and behavioural frequency outcomes for the entire pre- and post-use survey samples (n = 361 and n = 144, respectively), and for those that completed both pre- and post-use surveys (n = 76 matched pairs). Two participants in the matched pairs did not complete all the questions, and as such were excluded from the analyses assessing differences between matched pairs (resulting in a final sample for matched pairs of n = 74).

3.2.1. Differences between Entire Pre- and Post-Use Samples

Preliminary analyses revealed that a small number of food waste categories contained non-normal data. However, all food provisioning behaviours contained normal data. To ensure a robust analysis, non-parametric tests (i.e., Mann–Whitney U test) were to complement the interpretation of the parametric tests (Welch's ANOVA) of differences in food waste between pre- and post-use samples.

Results from both Welch's ANOVA and Mann–Whitney's U test (see Table 3) indicate that the post-use sample had significantly less fresh vegetable waste and total food waste than the pre-use sample. A significant reduction in fresh fruit waste and bread from pre- to post-use sample was also found but only detected by Mann–Whitney U tests. Unexpectedly, the post-use sample experienced significant increases in fish waste. This effect could have been due to the small number of responses for this category of food waste influencing the accuracy of estimation in the analysis, or the use of non-random sampling and potentially unaccounted for individual differences that may have influenced the magnitude of this effect.

Table 3. Means, sample sizes, Welch statistic, and Mann–Whitney U result of differences in pre- and post-use samples of perceived food waste.

Measure	Pre Mean (g)	Pre Sample Size	Post Mean (g)	Post Sample Size	Welch	df1	df2	p	Mann–Whitney U	p
Fresh veg	**376.56**	**296**	**291.25**	**83**	**8.88**	**1**	**162.40**	**0.003**	**10,211.50**	**0.01**
Non-fresh veg	143.79	33	169	10	0.21	1	12.07	0.65	169	0.92
Fresh fruit	**395.19**	**219**	**331.09**	**48**	**3.59**	**1**	**68.81**	**0.06**	**4339**	**0.04**
Non-fresh fruit	80	14	93.33	3	0.03	1	2.28	0.88	15	0.51
Potatoes	302.40	64	332.80	18	0.13	1	20.64	0.72	516	0.47
Potato products	132.92	13	243.20	5	1.54	1	4.37	0.28	41.50	0.39
Pasta	207.13 [a]	55	229.65	17	0.21	1	25.84	0.65	500	0.64
Rice	210.45	77	213.33	18	0.003	1	21.55	0.96	627	0.49
Beans	224.89	18	92	3	[b]	[b]	[b]	[b]	10.50	0.10
Meat	242.58	96	192.39	23	1.44	1	41.04	0.24	970.50	0.35
Meat alts	171.09	16	278.57	7	3.34	1	14.24	0.09	81	0.10
Fish	**150**	**18**	**337.50** [a]	**6**	**13.28**	**1**	**14.42**	**0.003**	**89**	**0.01**
Sandwich	56.96	92	49.57	23	0.71	1	34.44	0.41	926.50	0.33
Bread	**289.36** [a]	**200**	**217.84** [a]	**63**	**2.48**	**1**	**107.18**	**0.12**	**5141.50**	**0.01**
Cereal	210.67	24	347.43	7	1.14	1	7.81	0.32	114	0.17
Yoghurt	357.27 [a]	88	307.83	23	0.79	1	57.37	0.38	1022	0.94
Cheese	115.69 [a]	52	155.43	14	1.04	1	16.84	0.32	421	0.27
Eggs	180	26	135	6	0.70	1	7	0.43	55.50	0.29
Stews	402.11	57	618.46	13	3.31	1	14.45	0.09	480	0.06
Condiment	249.50 [a]	42	284 [a]	14	0.13	1	22.38	0.72	327	0.51
Candy	61.67	18	58	2	0.04	1	6.59	0.85	44	0.97

Table 3. *Cont.*

Measure	Pre Mean (g)	Pre Sample Size	Post Mean (g)	Post Sample Size	Welch	df1	df2	p	Mann–Whitney U	p
Salty snacks	66.67	18	70	4	0.01	1	3.58	0.92	34	0.86
Nonalcohol	443.71 [a]	62	540	13	0.39	1	15.72	0.54	429	0.70
Alcohol	480.83 [a]	12	396.67	3	0.29	1	12.13	0.60	19	0.87
Total food waste	**1287.03** [a]	**361**	**784.97** [a]	**145**	**22.52**	**1**	**296.08**	**<0.001**	**16,815.50**	**<0.001**

Note: Significant effects are in bold. [a] Food waste category contains non-normal data. [b] Welch's ANOVA could not be performed for Beans category because at least one group had 0 variance.

The ordinal logistic regression results (see Table 4) found that the pre-use survey sample had approximately 48% lower odds of adopting the 'use it up meal' and approximately 85% lower odds of adopting the 'use it up shelf', relative to the post-use survey sample.

Table 4. Ordinal logistic regression results assessing differences in behavioural frequency between the entire pre- and post-use survey samples.

	Model Fitting Information		Goodness-of-Fit			Pseudo R-Square	Parameter Estimates		
	−2 Log Likelihood (Intercept, Final)	χ^2, df	Pearson	Deviance	df	McFadden	Coefficient	95% CI (Lower, Upper)	OR
Make a shopping list	35.83, 34.59	1.25, 1	0.17	0.17	3	0.001	−0.22	−0.62, 0.17	0.80
Check food at home before making a shopping list	34.24, 34.13	0.11, 1	1.61	1.89	3	0.00	−0.06	−0.43, 0.31	0.94
When shopping, only buy what is on shopping list	42.10, 41.91	0.20, 1	1.28	1.28	3	0.00	−0.08	−0.44, 0.28	0.92
Make a use it up meal	**49.76, 37.54**	**12.23 ***, 1**	**1.38**	**1.44**	**3**	**0.01**	**−0.65 *****	**−1.02, −0.28**	**0.52**
Order takeaway	42.71, 39.98	2.73, 1	1.44	1.54	3	0.002	0.30	−0.06, 0.66	1.35
Have a use it up shelf in fridge/pantry	**104.44, 22.85**	**81.58 ***, 1**	**2.10**	**2.14**	**1**	**0.10**	**−1.88 *****	**−2.29, −1.47**	**0.15**

Note: *** $p < 0.001$; χ^2 = chi-square; df = degrees of freedom; CI = confidence interval; OR = odds ratio. Note: Significant effects are in bold.

3.2.2. Differences between Matched Pairs

Results from the Wilcoxon signed-rank tests indicated significant reductions in food waste (see Table 5). Specifically, there was less total food waste, fresh vegetable waste, fresh fruit waste, and meat waste after the introduction of the Tape.

Table 5. Means, standard deviations, and Wilcoxon signed-rank results of food waste and food provision behaviour before and after introduction of the Tape.

Measure	Pre Mean (SD)	Post Mean (SD)	Wilcoxon Z	p	Effect Size
Fresh veg	**294.93 (288.06)**	**133.88 (181.95)**	**−4.98**	**<0.001**	**0.40**
Non-fresh veg	11.97 (57.86)	12.83 (60.99)	−0.05	0.96	0.00
Fresh fruit	**223.22 (253.88)**	**91.09 (174.58)**	**−3.59**	**0.001**	**0.29**
Non-fresh fruit	3.95 (16.34)	0.53 (3.22)	−1.90	0.06	0.15
Potatoes	32.34 (132.88)	45.47 (170.66)	−0.40	0.69	0.03
Potato products	8.42 (38.18)	2.53 (22.02)	−1.63	0.10	0.13
Pasta	26.11 (62.86)	18.53 (88.57)	−1.63	0.10	0.13
Rice	32.34 (72.33)	28.30 (119.53)	−1.08	0.28	0.09

Table 5. *Cont.*

Measure	Pre Mean (SD)	Post Mean (SD)	Wilcoxon Z	*p*	Effect Size
Beans	8.47 (45.51)	1.21 (10.55)	−1.60	0.11	0.13
Meat	**59.21 (138.46)**	**22.70 (77.96)**	**−2.18**	**0.03**	**0.18**
Meat alts	6.91 (46.12)	8.88 (48.94)	−0.28	0.78	0.02
Fish	2.47 (9.36)	6.91 (46.12)	−0.11	0.91	0.01
Sandwich	14.61 (33.76)	11.97 (30.02)	−0.63	0.53	0.05
Bread	112.74 (234.29)	100.18 (269.70)	−0.67	0.50	0.05
Cereal	5.05 (25.09)	16.84 (106.77)	−0.41	0.68	0.03
Yoghurt	82.11 (217.52)	44.21 (131.22)	−1.66	0.10	0.13
Cheese	8.42 (28.32)	8.42 (33.61)	−0.28	0.78	0.02
Eggs	9.47 (42.20)	5.13 (25.17)	−0.72	0.47	0.06
Stews	44.21 (139.73)	48.95 (176.56)	−0.42	0.68	0.03
Condiment	40.25 (162.33)	26.71 (85.62)	−0.32	0.75	0.03
Candy	2.11 (14.08)	0.26 (2.29)	−1.34	0.18	0.11
Salty snacks	4.61 (21.81)	3.68 (19.52)	−0.09	0.93	0.01
Nonalcohol	19.74 (65.28)	10.26 (47.38)	−0.86	0.39	0.07
Alcohol	34.41 (225.19)	9.41 (60.72)	−0.68	0.50	0.06
Total food waste	**1088.04 (931.35)**	**658.87 (759.61)**	**−4.87**	**<0.001**	**0.40**

Note: Significant effects in bold. SD = standard deviation.

The results of the generalised estimating equations are presented in Table 6. The findings showed no significant differences in the adoption of food provisioning behaviours between pre- and post-use survey periods.

Table 6. Generalised Estimating Equations analysis results for adoption of food provisioning behaviours by matched pairs.

Behaviour	Pre/Post Coefficient (SE)	Wald $\chi 2$ (df = 1, n = 74)	*p*
Make a shopping list	0.34 (0.35)	0.92	0.34
Check food at home before making a shopping list	0.03 (0.33)	0.01	0.94
When shopping, only buy what is on shopping list	0.48 (0.30)	2.52	0.11
Make a use-it up meal	0.29 (0.31)	0.88	0.35
Order takeaway	−0.37 (0.29)	1.65	0.20
Have a use-it up shelf in fridge/pantry	0.36 (0.36)	1.01	0.32

Note: SE = standard error; $\chi 2$ = chi-square; df = degrees of freedom; n = sample size; p = significance value.

4. Discussion

4.1. Summary of Principal Findings

The combined aim of the two studies presented in this paper was to explore how households used, and experienced, the OzHarvest Use it Up Tape ("the Tape"), to examine its impact on food waste and to identify any changes in household food provisioning behaviours that may have occurred.

The qualitative video diary study found that households used the Tape to mark out a designated space in their refrigerators or pantries to place food that needed to be used up, or stuck smaller sections of the Tape on specific items that needed to be used up (or did both). The Tape served multiple functions for participating households; it functioned as a visual prompt, a labelling device, a planning tool, and a communication tool, and sometimes had multiple functions for the same household. Participants also indicated that the Tape helped them to engage in different food-related behaviours that reduced food waste, such as making meals with food that needed to be used up (the initial target behaviour) or sticking to their shopping list when shopping. The Tape seemed to be more useful and effective for large family homes and for individuals who were disorganised when shopping and/or cooking.

The quantitative study measured and compared self-reported food waste outcomes and behavioural frequencies before and after a two-week period of Tape use and found that there was a significant reduction in fresh vegetables, fresh fruit, and total food amounts wasted when comparing the entire pre- and post-use samples. The analysis of matched pairs (those who completed both pre- and post-use surveys) also found a significant reduction in fresh vegetables, fresh fruit and total food amounts wasted, as well as for meat. There also seemed to be a significant increase in the frequency across the entire pre- and post-use samples of participants who reported making a use-it-up meal before and after the two-week period, and an increase in designating a use-it-up shelf in fridges/pantries. However, increased engagement in these behaviours was not shown in the matched pairs analyses.

4.2. *Implications for Food Waste Reduction Policies and Programs*

While visual prompt-type nudges have been previously explored with regard to their influence on food waste recycling behaviours [28], to the best of our knowledge this is one of the few studies that looks at their influence on reducing or preventing food waste [13,15]. In their recent study, [11] included reminder stickers and other 'salience tools' in the suite of food waste reduction tools they provided to participants. However, they mainly measured the collective impact of all the tools, rather than looking at their individual impact. Ref. [10] did look more specifically at the influence of different salience tools—a basket for collecting food items that need to be used up, a whiteboard for noting these items, and clips to attach to items—and found that they did not have an added impact on food waste levels.

Our study supports arguments by [15] on the potential of behaviourally orientated nudges such as visual prompts to reduce food waste in households. The most wasted food items in Australian households are fresh vegetables, fresh fruit, and bread/baked goods [29], and the use of the Tape seemed to lead to a reduction in fruit and vegetables wastage in participating households (but no significant reduction for bread/baked goods). This shows that not only did the Tape support an overall reduction in food waste for participants, but was particularly effective in tackling some of the more commonly wasted items in the Australian context. We speculate the wastage of bread and other baked goods may not have been reduced because the Tape may not have been used in the areas in the kitchen where these items are commonly stored (i.e., in bread storage bins and/or freezers).

While the results were not consistent between the analysis of the entire pre- and post-use samples and matched pairs, the quantitative study did suggest that the Tape supported an increase in the target behaviour for the campaign, namely making a regular meal that combines food that needs to be used up. This leads us to conclude that it was the increased frequency of this behaviour which may have led to the food waste reduction outcomes that were measured. This finding is important not just because it highlights the effectiveness of the Tape as a behaviourally orientated nudge, but also because it supports that decision to include this target behaviour in the overall Use It Up campaign as an effective way to reduce food waste in Australia [16].

In their study of the effectiveness of stickers attached to weekly home food delivery plastic bags to prompt consumers to return them for re-use, [17] described these types of reminders as 'action-close' or 'point-of-decision' prompts. They argued that these types of nudges are particularly effective as they "catch decision makers attention in the situation and at the time of the desired behaviour change" (p. 2). Their study showed that the reminder's proximity to the action was critical and that "reminders issued at the time and in the situation of the taking action can bridge limited attention more effectively than conventual, action-distant reminders" (p. 2). For policy and program managers seeking to reduce food waste, this positions the effectiveness of behaviourally orientated nudges like the Tape against more conventional campaigns providing information and raising awareness [11,13]. While an increased consumer awareness of global and personal food waste implications, and what could be done to address these, is important, individuals also need to be prompted of the necessary behaviours at the time and place when they are most relevant, namely when they are standing at their refrigerator or pantry looking for something to eat or cook.

Of final relevance to policy and program managers is the varying effectiveness of the Tape based on the characteristics of the household. The qualitative study showed that family-based households with young children, and those households in which the primary person responsible for shopping or cooking (the dietary gatekeeper) was particularly disorganised, responded the best to the Tape and found it effective. As with any nudge-based (or indeed any other type of) behaviour change intervention, a targeted implementation of the Tape that engages with particular household-types is needed, as is the creation of a broader range of complementary food waste reduction tools that might be relevant for other households [11].

4.3. Limitations and Future Research

While the size of the sample of Study 1 was adequate for a qualitative exploration of the experience of participants in using the Tape, we do note that higher income households were not well-represented. A slightly expanded sample that also investigated how the Tape was used in these types of households would have allowed for more comprehensive conclusions to be reached in Study 1.

The variations noted in the outcomes for the entire pre- and post-use samples and those for the matched pairs might have been due to several factors, namely the quasi-experimental nature of the intervention and the lack of a true control condition. Not only does this make it difficult to establish the Tape as the causal mechanism behind the observed food waste reduction, but the nature of the intervention also means that participants self-nominated their participation, and this might introduce a type of selection bias to the study. Namely, participants were potentially already motivated to reduce food waste from the onset, particularly those that were organised and engaged enough to complete both the pre- and post-use survey. These matched pair participants, in particular, may have been already engaging in certain food provisioning behaviours that reduce food waste, with little opportunity to do them more in the two-week period allotted to their participation.

Related to measurement issues, [26] acknowledge that participants using their food waste measurement survey typically underestimate the amount of food that they waste, a problem which has been highlighted for other survey-based food waste self-reporting tools [30]. However, they still point to the value this tool provides in comparing food waste amounts between households or across time for the same households. The results presented in this study should therefore not be seen as accurate measures of Australian food waste, but rather as a way to test the effectiveness of the Tape by comparing between participating households.

The relatively small sample sizes within the quantitative study may have influenced our ability to detect the effects of the Tape. A follow-up study with a larger sample size might provide more robust findings with regard to food waste and behavioural outcomes, and the sample of the qualitative study could also be expanded to be more nationally representative for Australia. Follow-up quantitative studies with larger sample sizes might also need to consider Odds Ratios (ORs) and the underlying data when interpreting the results of an ordinal logistic regression [31]. ORs can be influenced by other factors in the data, such as the presence of confounding variables [32] or the distribution of the predictor and outcome variables [31]. Given the low sample size, covariates were not included in the analysis and, as such, it is unknown whether the results are influenced by any other factors.

Future research on nudges such as the Tape might include a control condition and repeated measures of outcomes over a longer period to enable more robust conclusions about the long-term use of this type of tool and its impacts. Everitt et al. [7] recently published the outcomes of a broader food waste reduction campaign in Canada that was based on a randomised control trial and repeated measurements from 2017–2020. Quantitative design issues, such as those listed above, highlight the value of including a qualitative element to this research, which has allowed researchers to understand what is going on inside of participants' homes and how they have interacted with the Tape. Future research would benefit from being able to implement a more strictly controlled delivery and the adoption of 'action-close' nudges, such as the Tape, with the inclusion of a true control condition to allow for more stringent comparisons. Additionally, wherever possible, including qualitative measures is recommended to complement the accuracy of self-reported measures and behaviours.

5. Conclusions

When engaging consumers to play their part in tackling the global food waste challenge, it is becoming increasing clear that simply providing information about the challenge and what they need to do is not enough to lead to a meaningful behaviour change. Action-close behaviourally orientated nudges and prompts, such as the OzHarvest Use it Up Tape, play an important role in supporting consumers to turn their intentions to reduce food waste into action by engaging them when they are paying attention to food, i.e., when opening their refrigerator or pantry in order to prepare a meal. There is an opportunity to create multi-faceted interventions that map daily consumer behaviours related to food waste and utilise a mix of both behaviourally and cognitively orientated nudges, to provide the appropriate support for different behaviours and ensure ongoing food waste reduction outcomes.

Author Contributions: Conceptualisation, M.B.; methodology, M.B. and N.L.; formal analysis, M.B. and N.L.; investigation, M.B.; resources, M.B.; data curation, M.B. and N.L.; writing—original draft preparation, M.B. and N.L.; writing—review and editing, M.B. and N.L.; project administration, M.B.; funding acquisition, M.B. All authors have read and agreed to the published version of the manuscript.

Funding: This research project was commissioned by OzHarvest and partially supported by funding from the Australian Government's Environmental Restoration Fund and from the Western Australian Department of Water and Environmental Regulation.

Institutional Review Board Statement: The study was conducted in accordance with the Declaration of Helsinki, and approved by the Institutional Ethics Committee of Monash University (project ID # 28967 and # 32508).

Informed Consent Statement: Informed consent was obtained from all subjects involved in the study.

Data Availability Statement: The datasets presented in this article are not readily available, as they belong to OzHarvest. As the research commissioners, requests to access the datasets should be directed to OzHarvest.

Acknowledgments: We are very grateful to Ronni Kahn AO, Ruby Wake, Monique Llewellyn, Annika Stott and Fiona Nearn at OzHarvest for their collaboration in this research, guidance, and support. We would also like to thank all research participants for their involvement.

Conflicts of Interest: The authors declare no conflicts of interest.

References

1. Candeal, T.; Brüggemann, N.; Bruns, H.; Casonato, C.; Diercxsens, C.; García-Herrero, L.; Gil, J.M.; Haglund, Y.; Kaptan, G.; Kasza, G.; et al. *Tools, Best Practices, and Recommendations to Reduce Consumer Food Waste—A Compendium*; Publications Office of the European Union: Luxembourg, 2023.
2. FAO. *The State of Food and Agriculture 2019. Moving forward on Food Loss and Waste Reduction*; Food & Agriculture Organization of the United Nations: Rome, Italy, 2019.

3. UNEP. *Food Waste Index Report 2021*; United Nations Environmental Program: Nairobi, Kenya, 2021.
4. Willett, W.; Rockström, J.; Loken, B.; Springmann, M.; Lang, T.; Vermeulen, S.; Garnett, T.; Tilman, D.; DeClerck, F.; Wood, A.; et al. Food in the Anthropocene: The EAT-Lancet Commission on healthy diets from sustainable food systems. *Lancet* **2019**, *393*, 447–492. [CrossRef] [PubMed]
5. Transforming Our World: The 2030 Agenda for Sustainable Development. Available online: https://sustainabledevelopment.un.org/post2015/transformingourworld/publication (accessed on 10 May 2024).
6. Reynolds, C.; Goucher, L.; Quested, T.; Bromley, S.; Gillick, S.; Wells, V.K.; Evans, D.; Koh, L.; Carlsson Kanyama, A.; Katzeff, C.; et al. Consumption-stage food waste reduction interventions—What works and how to design better interventions. *Food Policy* **2019**, *83*, 7–27. [CrossRef]
7. Everitt, H.; van der Werf, P.; Seabrook, J.A.; Gilliland, J.A. The proof is in the pudding: Using a randomized controlled trial to evaluate the long-term effectiveness of a household food waste reduction intervention during the COVID-19 pandemic. *Circ. Econ. Sustain.* **2023**, *3*, 881–898. [CrossRef] [PubMed]
8. Shu, Y.; Booker, A.; Karetny, J.; O'Keefe, K.; Rees, K.; Schroder, L.; Roe, B.E. Evaluation of a community-based food waste campaign using a national control group. *Waste Manag.* **2023**, *160*, 101–111. [CrossRef] [PubMed]
9. Soma, T.; Li, B.; Maclaren, V. An evaluation of a consumer food waste awareness campaign using the motivation opportunity ability framework. *Resour. Cons. Recycl.* **2021**, *168*, 105313. [CrossRef]
10. Cooper, A.; Lion, R.; Rodriguez-Sierra, O.E.; Jeffrey, P.; Thomson, D.; Peters, K.; Christopher, L.; Zhu, M.J.H.; Wistrand, L.; van der Werf, P.; et al. Use-up day and flexible recipes: Reducing household food waste by helping families prepare food they already have. *Resour. Cons. Recycl.* **2023**, *194*, 106986. [CrossRef]
11. van Herpen, E.; Wijnen, T.; Quested, T.; Reynolds, C.; Sharda, N. Convenient tools and social norms: Measuring the effectiveness of an intervention to reduce household food waste. *J. Clean Prod.* **2023**, *429*, 139604. [CrossRef]
12. Nisa, C.F.; Bélanger, J.J.; Schumpe, B.M. Assessing the effectiveness of food waste messaging. *Environ. Sci. Policy* **2022**, *132*, 224–236. [CrossRef]
13. Simões, J.; Carvalho, A.; Gaspar de Matos, M. How to influence consumer food waste behavior with interventions? A systematic literature review. *J. Clean Prod.* **2022**, *373*, 133866. [CrossRef]
14. Tian, X.; Xia, Z.; Xie, J.; Zhang, C.; Liu, Y.; Xu, M. A meta-analytical review of intervention experiments to reduce food waste. *Environ. Res. Lett.* **2022**, *17*, 064041. [CrossRef]
15. Zhang, J.; Huang, Y.; Zhu, J.; Zhao, L. A meta-analysis on the effectiveness of food-waste reducing nudges. *Food Policy* **2023**, *120*, 102480. [CrossRef]
16. Boulet, M.; Stott, A.; Kneebone, S. Which behaviours matter? Prioritising food waste reduction behaviours for targeted policy and program approaches. *J. Environ. Manag.* **2023**, *345*, 118668. [CrossRef]
17. Essl, A.; Steffen, A.; Staehle, M. Choose to reuse! The effect of action-close reminders on pro-environmental behavior. *J. Environ. Econ. Manag.* **2021**, *110*, 102539. [CrossRef]
18. Sussman, R.; Gifford, R. Please turn off the lights: The effectiveness of visual prompts. *Appl. Ergon.* **2012**, *43*, 596–603. [CrossRef]
19. Kahneman, D. *Thinking, Fast and Slow*; Farrar, Straus and Giroux: New York, NY, USA, 2011.
20. Thaler, R.H.; Sunstein, C.R. *Nudge: Improving Decisions about Health, Wealth and Happiness*; Penguin Books: London, UK, 2002.
21. Conner, M.; Norman, P. Understanding the intention-behaviour gap: The role of intention strength. *Front. Psychol.* **2022**, *130*, 923464. [CrossRef] [PubMed]
22. Boulet, M.; Hoek, A.C.; Raven, R. The gaze of the gatekeeper: Unpacking the multi-level influences and interactions of household food waste through a video elicitation study. *Resour. Cons. Recycl.* **2021**, *171*, 105625. [CrossRef]
23. Dodds, S.; Bulmer, S.; Murphy, A. Incorporating visual methods in longitudinal transformative service research. *J. Serv. Theory Pract.* **2018**, *28*, 434–457. [CrossRef]
24. Braun, V.; Clarke, V. Using thematic analysis in psychology. *Qual. Res. Psychol.* **2006**, *3*, 77–101. [CrossRef]
25. Cho, J.Y.; Lee, E.-H. Reducing confusion about grounded theory and qualitative content analysis: Similarities and differences. *Qual. Rep.* **2014**, *19*, 1–20. [CrossRef]
26. van Herpen, E.; van Geffen, L.; Nijenhuis-de Vries, M.; Holthuysen, N.; van der Lans, I.; Quested, T. A validated survey to measure household food waste. *MethodsX* **2019**, *6*, 2767–2775. [CrossRef]
27. Shu, Y.; Roe, B.E.; Bender, K. Adapting, refining and expanding a validated questionnaire to measure food waste in U.S. households. *MethodsX* **2021**, *8*, 101377. [CrossRef] [PubMed]
28. Shearer, L.; Gatersleben, B.; Morse, S.; Smyth, M.; Hunt, S. A problem unstuck? Evaluating the effectiveness of sticker prompts for encouraging household food waste recycling behaviour. *Waste Manag.* **2017**, *60*, 164–172. [CrossRef] [PubMed]
29. Karunasena, G.G.; Pearson, D. *Fight Food Waste Cooperative Research Centre Australian Household Food Waste: A Summary of Behaviours, Attitudes, Perceived and Actual Food Waste*; Fight Food Waste Cooperative Research Centre: Urrbrae, Australia, 2021.
30. van Herpen, E.; van der Lans, I.; Holthuysen, N.; Nijenhuis-de Vries, M.; Quested, T. Comparing wasted apples and oranges: An assessment of methods to measure household food waste. *Waste Manag.* **2019**, *88*, 71–84. [CrossRef] [PubMed]

31. Agresti, A. *Categorical Data Analysis*, 3rd ed., John Wiley & Sons: Hoboken, NJ, USA, 2013.
32. Rothman, K.J.; Greenland, S.; Lash, T.L. *Modern Epidemiology*, 3rd ed.; Lippincott Williams & Wilkins: Amsterdam, The Netherlands, 2008.

sustainability

MDPI

Article

Development of Healthy and Clean-Label Crackers Incorporating Apple and Carrot Pomace Flours

Saeed Salari [1], Thomas Castigliego [1], Joana Ferreira [1,*], Ana Lima [2] and Isabel Sousa [1]

1 LEAF—Linking Landscape, Environment, Agriculture and Food—Research Center, Associate Laboratory TERRA, Instituto Superior de Agronomia, Universidade de Lisboa, Tapada da Ajuda, 1349-017 Lisboa, Portugal; saeedsalari@isa.ulisboa.pt (S.S.); thomas.castigliego@studenti.unimi.it (T.C.); isabelsousa@isa.ulisboa.pt (I.S.)

2 Veterinary and Animal Research Centre (CECAV), Faculty of Veterinary Medicine, Lusófona University, 376 Campo Grande, 1749-024 Lisbon, Portugal; agusmaolima@gmail.com

* Correspondence: jpferreira@isa.ulisboa.pt

Abstract: The valorization of fruit and vegetable side-streams from the juice industry is an important contribution to the optimization of food resources and is an environmentally friendly practice in line with the concepts of circular economy and sustainability. The aim of this work is to incorporate them back into the food value chain by adding them as ingredients in staple foods like crackers. This is also important in terms of food fortification, as they are rich in nutrients and bioactive compounds. Crackers are popular snacks with a huge global market value, enjoyed by consumers of all ages. The current study aims to integrate flour from dried apple and carrot pomaces, resulting from juice processing, as natural ingredients with potential health benefits. The incorporation levels ranged from 20 to 40% dry weight in crackers, and their impact on physicochemical and mechanical properties was evaluated, as well as bioactivity (potential impact on health) and sensory acceptance. The addition of pomaces resulted in significant changes in texture and color, as well as enhancing the antioxidant activity of the crackers. Crackers containing pomace flours, except for the cracker with 40% carrot pomace, showed a high overall sensory acceptability and good intentions to buy.

Keywords: valorization; pomace flour incorporation; food fortification; crackers; bioactivity; sustainability

1. Introduction

In recent years, due to knowledge of the benefits that fruit and vegetables (F&V) have on health and the demand for natural foods, F&V has become one of the first choices in a healthy diet and its consumption is strongly advised by the WHO (e.g., the "five a day program") [1–3].

Even with the fruit juice and confectionery industry having been well established for a long time, F&V are still accountable for up to 20% of food waste and losses along the food supply chain [4]. Therefore, the side-streams from these industries can be stabilized by drying and grinding waste into flours rich in fiber and bioactive compounds, to be used like natural food ingredients and to enhance the health benefits and technological functionality of several food products.

Major food trends show that the production of F&V has steadily increased worldwide. For example, the total production rose up to 59 and 68% between 2000 and 2021, reaching 910 and 1150 million tons, respectively [5]. Almost 50% is processed as juice, and millions of tons of waste are being generated that could be a big challenge for the environment, but at the same time this could be considered an interesting side-stream, as this waste is known to be a source of functional compounds such as phenolics and fiber [6–8]. An advantage of that combination (fiber and phenolic compounds) is their bound capacity, as fiber can deliver bioactive compounds and act as a vehicle for their transport along the gastrointestinal tract, allowing their release in the gut after fiber fermentation by the gut microbiota. To this purpose, the valorization of pomaces as food ingredients could be an interesting and efficient way to promote health benefits [9,10].

Fiber refers to a group of carbohydrate polymers with ten or more monomers and a degree of polymerization (DP) higher than 10 that are not readily digestible nor absorbable in the small intestine, but could be fermentable by the human gut microbiota [11,12]. Examples include indigestible oligosaccharides, cellulose, hemicellulose, arabinoxylan, β-glucan, inulin, gum, pectin, and resistant starch [12,13]. Several studies show the link between taking in adequate fiber and a healthy gut and reduced risk of depression, obesity, and chronic diseases such as diabetes type 2 and cardiovascular and coronary heart disease [13–15].

Phenolic compounds are molecules characterized by an aromatic ring and one or more hydroxyl substitutes, and their binding capacity to mono- and polysaccharides increases their structural heterogeneity. For that reason, more than 8000 phenolics can be identified in nature and almost 75% of them are flavonoids in plants [16]. Fruits in particular are considered rich sources of flavonoids and phenolic acids, including gallic, ellagic, and vanillic acids. These compounds are important for their therapeutic potential as they can act as free radical scavengers or antioxidants, participating in the oxidative stress process, which may play a decisive role in the aging process and the development of many neurodegenerative, metabolic, and inflammatory disorders [17,18].

Concurrent with the rising demand for foods with functional and healthier properties, snacking has become a huge trend, with a value estimated at EUR 495.60 billion in 2023 and is expected to grow with a CAGR of 6.29% during the forecast period of 2023–2028 [19]. Consumers all over the world are moving towards preferring food that is easy to carry and readily accessible, making snack foods one of the best options [20]. Food industries are now launching fortified products enriched with vitamins, protein, and nutrients, giving consumers snacks with nutritional support [21].

Food functionality can be enhanced by using F&V pomace due to its functional qualities. Many types of pomace are used in a broad range of baked goods, including cakes, muffins, cookies, rock buns, and crackers [22–27]. To the best of our knowledge, there is a lack of information in the literature on the maximum level of incorporation of pomaces to develop crackers with the highest phenolic content and antioxidant activity. The present work studied the influence of replacing wheat flour with carrot and apple pomace flours (CPF and APF, respectively) in different percentages up to 40% in crackers to be consumed in snacking. The influence on the physicochemical and sensory properties and on the antioxidant capacity of wheat-based crackers was evaluated, as well as the sensory acceptability.

2. Materials and Methods

2.1. Materials

2.1.1. Chemicals

Methanol and ethanol were purchased from Fisher Chemicals (Fair Lawn, NJ, USA). DPPH (2,2-diphenyl-1-picryl-hydrazyl-hydrate), Trolox (6-hydroxy-2,5,7,8-tetramethylchroman-2-carboxylic acid), acetic acid, sodium acetate, TPTZ (2,4,6-tris(2-pyridyl)-s-triazine), hydrochloric acid, iron (III) chloride hexahydrate, Folin–Ciocalteu reagent, sodium carbonate, gallic acid, calcium chloride, and sodium hydroxide were obtained from Sigma-Aldrich (Saint Louis, MO, USA). Ultrapure water was purchased from the Synergy® Water Purification System from Merck Millipore (Burlington, MA, USA). All chemicals used were of analytical or HPLC grade.

2.1.2. Ingredients

AP and CP flours were provided by ALITEC—Alimentos Tecnológicos SA (Nazaré, Portugal). Due to the high moisture level in pomaces and the high risk of microbial contamination and oxidation, the drying procedure was conducted by the company using a drying tunnel (Tecnofruta, Valencia, Spain) at 80–85 °C and 55 Hz of air flow for 110 min, and pomaces were ground and packaged afterwards (Ferneto, Vagos, Portugal). Other ingredients for cracker preparation, including wheat flour T55 (76% carbohydrates, 10% protein, 3.5% fiber, and 1.3% fat), baking powder, fine sea salt, white sugar, and vegetable oil, were purchased from the local market.

2.2. Methods

2.2.1. Proximate Composition

Pomace flours were analyzed in terms of moisture, minerals, ash content, total fat, crude fiber, and crude protein following the international standard methods. Moisture and ash content were determined gravimetrically (AACC method 44-15.02) [28]. Minerals (Na, K, Ca, Mg, P, S, Fe, Cu, Zn, Mn, and B) were estimated using Inductively Coupled Plasma—Optical Emission Spectroscopy (iCap Series-7000 plus series ICP-OES, Thermo Fisher Scientific, Waltham, MA, USA), following the procedure described by Marrero et al., (2013) [29], and according to AACC 40-75.01 [30]. Total fat was determined by the Soxhlet method according to AACC 30-25-01 [31]. Crude fiber was determined by the Weende method (AOAC method 978.10) [32]. Crude protein was determined by using an NDA 701 Dumas nitrogen analyzer and the common conversion factor of 6.25 [33]. Total starch quantification was performed by using the Megazyme Total Starch (AA/AMG) Assay kit and following the Rapid Total Starch (RTS) method, that is, according to AOAC method 996.11 with a slight modification [34].

2.2.2. Crackers Manufacture

Crackers were prepared according to the following formulation, developed in our lab in previous studies [35], with 59% commercial all-purpose wheat flour T55, 1.5% baking powder, 1% salt, 1% sugar, 7.5% vegetable oil, and 30% distilled water (*w*/*w*). Pomace flours, at 20 and 40% (*w*/*w*) incorporation levels (AP20 and CP20, AP40 and CP40, for 20% and 40% of incorporation of apple and carrot pomaces) were added to the same formulation by substituting a corresponding amount of wheat flour. In the case of the 40% crackers, the water content was increased and adjusted to develop a workable dough with suitable consistency. The ingredients were weighed based on a 300 g batch and mixed in a food processor (Bimby, Vorwerk, Germany) to obtain a homogeneous dough. Then, the dough was left to rest for 10 min and then laminated into thin sheets using a pasta roller machine. The laminated dough was divided into pieces using a square mold (75 × 75 mm). Each piece was then slightly perforated. Next, the crackers were baked at 180 °C in a forced-air convention oven (Unox, Cadoneghe, Italy) for approximately 10 min. Then, they were dried for 30 min at 60 °C and cooled for 30 min at room temperature, and then placed in hermetic glass jars for storage. Part of the cracker batches were promptly ground into powder using the food processor (Bimby, Vorwerk, Germany), and then frozen for further biochemical analysis and antioxidant potential evaluation.

2.2.3. Dough Rheology

Viscoelastic Behavior

The small amplitude oscillatory shear (SAOS) rheology measurements were conducted using a rheometer (Haake Mars III—Thermo Scientific, Dreieich, Germany) with a UTC—Peltier system to determine the viscoelastic properties of the dough, with and without APF and CPF, at 20 °C. The stress sweep test at 1 Hz was performed for the determination of the linear viscoelastic region to select the critical stress to be applied during the SAOS measurements. Then, the frequency sweep test allowed the acquisition of the storage (G') and loss (G'') moduli at frequencies ranging from 0.01 Hz to 100 Hz, while maintaining a constant shear stress within the linear viscoelastic region of each sample. Each sample (control; AP20; AP40; CP20 and CP40) was placed in the bottom plate of a 20 mm serrated parallel plate (PP20) with a 1 mm gap. To stop moisture loss during testing, liquid paraffin was applied to the sample edges. Each formulation was tested at least in triplicate.

Mixolab—Mixing and Pasting Curves

The impact of APF and CPF incorporation at 0%, 10%, and 20% *w/w* wheat flour basis on the dough during mixing and pasting was assessed using the Mixolab2 instrument (Chopin Technologies, Paris, France), following the Chopin+ protocol, at a constant water absorption of 55 g/100 g, determined in a previous test. The test settings used were similar to those described by [36]. In the case of the 40% pomace incorporation, the dough was too tough to be tested; it was over the limits of the torque in this equipment.

The Mixolab parameters evaluated were as follows: water absorption (WA% at 14% moisture basis): the amount of water required to achieve a dough of appropriate consistency (target); dough development time (DDT): the time it takes for dough to develop during mixing to reach C1 (maximum torque during mixing to determine water absorption); dough stability (DS): the duration during which the dough maintains its structural integrity around C1—11% [37]; C2 (Nm): minimum torque value when the Mixolab starts heating the dough, reflecting the gluten quality; C3 (Nm): peak torque obtained after C2, expressing starch gelatinization; C4 (Nm): decrease after C3, representing the cooking stability; and C5 (Nm): the torque value obtained by the end of the test, representing starch gelification during the cooling stage [36]. The results are in triplicate for each blend, as well as for the control.

Dimensions

The characteristic dimensions, width (W) and thickness (T), of 10 crackers from each formulation were measured using a digital caliper model 684132 (Lee Tools, Housten, TX, USA). The spread ratio (W/T) was calculated accordingly.

2.2.4. Color

With a CIE standard illuminant D65, a 2-degree field of view, and a d/0° viewing angle, the Minolta CR-400 (Japan) colorimeter was used to measure the color of the cracker samples. The results were expressed in terms of L^*, lightness (values increasing from 0 to 100); a^*, redness to greenness (60 to −60 positive to negative values, respectively); and b^*, yellowness to blueness (60 to −60 positive to negative values, respectively) according to the CIELab system. By applying the formula $\Delta E^* = [(\Delta L^*)^2 + (\Delta a^*)^2 + (\Delta b^*)^2]^{1/2}$, the total color difference between the crackers was calculated using average $L^*a^*b^*$ values. The measurements were performed with a white standard ($L^* = 94.61$, $a^* = -0.53$, and $b^* = 3.62$) at room temperature and under the same light conditions. Measurements were replicated ten times for each formulation (one measurement per cracker).

2.2.5. Moisture Content and Water Activity

These properties were measured with a PMB Humidity Analyzer (AE Adam GmbH, Felde, Germany), after checking it against the reference gravimetric method and LabMaster-aw neo (Novasina AG, Lachen, Switzerland).

2.2.6. Texture

Instrumental texture analysis was carried out in a TA.XTplus (Stable Micro Systems, Godalming, UK) texturometer. "Three-point bending" or "snap" tests were performed using a double clamp set and 3 mm thick knife blade at 1 mm/s probe speed, with a 5 kg load cell, and at a controlled (20 ± 1 °C) room temperature. Three textural parameters of the cracker were evaluated: peak force or hardness (N), first break distance or brittle deformation (mm), and total area of work or total energy at rupture or toughness (J).

2.2.7. Sensory Analysis

Cracker samples with 20 and 40% of pomace flours, as well as the control samples, were tested by an untrained sensory analysis panel (n = 44, age: 18–49). The cracker samples were evaluated in terms of appearance, color, smell, taste, texture, and overall acceptability (six levels, to avoid the center bias, from "very pleasant" to "very unpleasant").

The buying intention was also assessed, from "would certainly buy" to "certainly wouldn't buy" (four levels). In compliance with EN ISO 8589 standard, the assays were carried out in a standardized sensory analysis room [38].

2.2.8. Antioxidant Potential

To prepare sample extracts, 2 g of pomace flours or cracker powders were weighed in a test tube and extracted with 20 mL of ethanol 96% at ambient temperature mixed using an overhead shaker (Heidolph Instruments, Schwabach, Germany) overnight. Then, the extracts were centrifuged at 3220 g for 10 min. The supernatant was recovered and stored at −24 °C until use.

To evaluate the radical scavenger potential, the DPPH assay was performed by mixing 3.9 mL of DPPH radical solution (0.06 mM in methanol, Sigma-Aldrich, St. Louis, MO, USA) and 100 µL of sample extract. The reaction mixtures were vortexed and incubated in darkness at room temperature for 40 min and the absorbance was measured at 515 nm. The antioxidant capacity of the samples was expressed in terms of µmol of Trolox equivalent antioxidant capacity (TEAC) per g of sample (Trolox calibration curve: 0 to 1000 µg.mL^{-1}, R^2 = 0.9958) and corresponding radical scavenging activity (RSA). A control assay without pomace extract was also performed. Analyses were conducted in triplicate.

Another way of looking at the antioxidant potential is by measuring the ferric reducing power through FRAP assay. This was performed by mixing 2.7 mL of FRAP solution, 270 µL of distilled water, and 90 µL of sample extract. The reaction mixtures were vortexed and incubated in a water bath at 37 °C for 30 min and the absorbance was measured at 595 nm. The antioxidant capacity of the samples was expressed in terms of µmol of TEAC per g of sample (Trolox calibration curve: 0 to 800 µg.mL^{-1}, R^2 = 0.9971). Analyses were conducted in triplicate.

2.2.9. Total Phenolic Content

The TPC was determined using the Folin–Ciocalteu method and gallic acid as a standard, as proposed earlier by Singleton and Rossi, 1965. To 150 µL aliquots of each sample, 2.4 mL of deionized water and 140 µL of Folin–Ciocalteu reagent (Sigma-Aldrich, St. Louis, MO, USA) were added and vortexed. After 3 min, 300 µL of sodium carbonate was added and vortexed again and then stored in darkness at room temperature for 2 h. The absorbance of each sample was measured at 725 nm. Results were expressed in gallic acid equivalents (mg GAE g^{-1}) through a calibration curve (gallic acid: 0 to 200 µg.mL^{-1}, R^2 = 0.9998) (Sigma-Aldrich, St. Louis, MO, USA).

2.2.10. Statistical Analysis

Statistical analysis of the experimental data was performed using SPSS (version 29, IBM, Armonk, NY, USA), through variance analysis (one-way ANOVA), and by the Tukey test as the post hoc at a significance level of 95% ($p < 0.05$). All results are presented as average ± standard deviation.

3. Results and Discussion

3.1. Physicochemical Characteristics of the Raw Pomaces and Crackers

The results for the proximate composition of APF, CPF, and crackers are shown in Table 1. It is worth noticing that the moisture content is about 60% higher in CP than in AP, and a value below 14.5% is considered the limit value for stable flours in cereals [39]. As side streams, pomaces are not taken into full consideration and, therefore, after the juice extraction, they are submitted to a dying process at 80/85 °C for 110 min. But since the final moisture content depends on several factors, namely the drying process and conditions, to be able to re-introduce these by-products into the food chain again, appropriate industrial controlled routines must be implemented. In the case of crackers, the lower moisture content of crackers including APF compared to ones including CPF could be due to the difference in the initial moisture content of the pomaces and the cooking time.

Table 1. Proximate composition of the apple and carrot pomace flours and crackers (% dw). Results are expressed as average ± standard deviation (n = 3), followed by an alphabet letter. Different letters mean different significant results (Tukey's HSD; $p \leq 0.05$).

Sample	Moisture	Ash	Fiber	Fat	Nitrogen	Carbohydrates, Including Starch	Starch
APF	10.8 ± 1.08 ^{a}	1.3 ± 0.01 ^{a}	21.9 ± 0.84 ^{a}	1.8 ± 0.13 ^{a}	0.7 ± 0.01 ^{a}	59.7 ± 2.39 ^{a}	13.8 ± 0.49 ^{a}
CPF	18.9 ± 0.31 ^{c}	7.3 ± 0.07 ^{b}	10.9 ± 0.37 ^{b}	1.0 ± 0.12 ^{b}	1.3 ± 0.01 ^{b}	51.3 ± 4.40 ^{b}	5.8 ± 0.23 ^{b}
WF	14.1 ± 0.45 ^{b}	0.5 ± 0.02 ^{c}	0.5 ± 0.01 ^{c}	1.0 ± 0.04 ^{b}	1.3 ± 0.01 ^{b}	76.1 ± 0.44 ^{c}	75.3 ± 0.51 ^{c}
Control	4.2 ± 0.28 ^{d}	2.4 ± 0.08 ^{d}	1.1 ± 0.06 ^{d}	10.4 ± 0.02 ^{c}	1.5 ± 0.02 ^{c}	72.7 ± 0.31 ^{d}	51.1 ± 0.85 ^{d}
AP20	2.3 ± 0.10 ^{e}	2.8 ± 0.15 ^{e}	1.4 ± 0.30 ^{d}	11.0 ± 0.05 ^{d}	1.3 ± 0.00 ^{b}	74.3 ± 0.37 ^{d}	49.5 ± 0.56 ^{de}
AP40	2.5 ± 0.09 ^{e}	2.9 ± 0.05 ^{e}	5.4 ± 0.55 ^{e}	11.3 ± 0.39 ^{d}	1.1 ± 0.01 ^{d}	70.4 ± 1.38 ^{d}	48.8 ± 0.11 ^{e}
CP20	3.2 ± 0.18 ^{de}	3.4 ± 0.03 ^{ef}	1.3 ± 0.04 ^{d}	10.1 ± 0.15 ^{c}	1.4 ± 0.01 ^{e}	73.1 ± 0.10 ^{d}	44.8 ± 0.14 ^{f}
CP40	3.5 ± 0.13 ^{de}	4.1 ± 0.07 ^{f}	2.6 ± 0.17 ^{f}	10.8 ± 0.05 ^{d}	1.3 ± 0.00 ^{b}	70.9 ± 0.14 ^{d}	39.2 ± 0.46 ^{g}

Superscript, lowercase letters indicate the significant differences between different fractions. APF: apple pomace flour; CPF: carrot pomace flour; WF: wheat flour; Control: cracker without pomace; AP20: cracker with 20% apple pomace flour; AP40: cracker with 40% apple pomace flour; CP20: cracker with 20% carrot pomace flour; CP40: cracker with 40% carrot pomace flour.

Regarding ash content, as expected, the value in carrot pomace is 5.6 times higher than in apple pomace, and this is consistent with the detailed results from the individual minerals content (Table S1). In fact, as carrot is a root and apple is a fruit, the proximity to the soil and the function of the root, absorbing water and minerals to feed the plant, might explain this difference [40,41]. Subsequently, crackers incorporating 20 to 40% of CPF have the highest ash content (3.4 and 4.1% dw) compared to the AP20 and AP40 (2.8 and 2.9% dw), respectively.

Results for fiber are much higher for apple (almost 2-fold), probably because the pomace is enriched in apple skin, seeds, and stalks [42]. Accordingly, AP40 has double the content of fiber compared to CP40, as expected. The fat content of both apple and carrot pomaces is considerably low (between 1 and 1.8% dw), and is slightly higher in apple pomace, as the apple skin has some non-polar components at the surface, contributing to the overall fat composition [43].

For protein results, when using the value from the Dumas equipment, without the conversion factor, for nitrogen, the value for carrot, again as a root and involved in taking the nitrogen out from the soil, is two times higher than in apple pomace. This does not mean that carrot has more protein, and this was the reason we decided to keep the values without applying the conversion factor of nitrogen into protein. Since wheat flour has more protein (10% dw) and nitrogen (see Section 2.1.2), the incorporation of pomaces results in the reduction in the nitrogen content of crackers, particularly in the case of AP40, in which the nitrogen content is the lowest (1.1% dw).

Carbohydrates were calculated as the difference for the other compounds after converting nitrogen into protein by a factor of 6.25, and the value was markedly higher in apple pomace. However, for crackers, there is no significant difference in carbohydrates when incorporating APF and CPF, which could be due to the degradation of carbohydrates during the Maillard reaction or caramelization, which could possibly happen when cooking at 180 °C [44–46]. Regarding the starch, CPF has a lower content than APF, which is mainly due to the fact that carrot, as a root, has only negligible amounts of starch and these gradually diminish before harvest. And these results are consistent with the results obtained for crackers. It has also been shown that the increase in pomace substitution with wheat flour leads to a remarkable reduction in starch in the final product. The presence of seeds possibly contributes to the higher content of starch in APF [47].

Analyzing the mineral composition presented in Table S1, both pomaces are enriched in potassium, and this is particularly higher in carrot pomace (almost 5-fold when compared to apple pomace). As a root, carrot accumulates a high concentration of minerals that are diffused from the soil towards the roots. Phosphorus, calcium, and sodium were also identified in considerable amounts.

3.2. Rheology of Cracker Dough

The rheological properties of different formulations were analyzed using the small amplitude oscillatory shear (SAOS). The presence of specific structures that can partially store energy and partially recover upon stress release gives rise to the material's viscoelastic properties. When the applied stress is released, a significant portion of the same energy will be lost irretrievably. Thus, certain materials are known to exhibit both elastic and viscous behaviors. An oscillating rheometer can be used to record variations in the conservative and loss moduli's values based on temperature and frequency [48]. The analysis involves two parameters: G′ and G″. The storage modulus (G′) represents the portion of energy that can be utilized to recover deformation and describes the proportion of elastic properties in the material under study. Conversely, the portion of energy lost or dispersed during sinusoidal deformation is characterized by the loss modulus (G″) [49]. The mechanical behavior at 20 °C is presented in Figure 1 by a frequency sweep from 0.01 up to 100 Hz using a stress value within the viscoelastic region (structure is not damaged), previously determined by a stress sweep at 1 Hz.

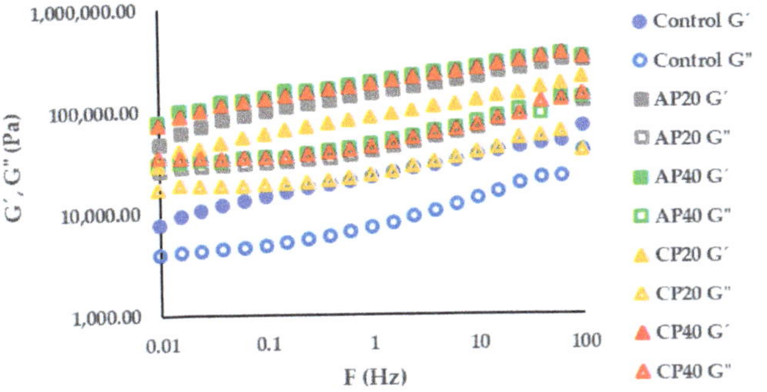

Figure 1. Mechanical spectra (G′ and G″ as a function of frequency) of cracker doughs at different levels of incorporation of apple and carrot pomace flours.

Figure 1 shows the frequency sweep (or mechanical spectrum) of cracker doughs with the addition of apple and carrot pomaces. The values of the G′ and G″ moduli depend on the internal structure of the systems.

It was found that, in each analyzed case (Figure 1), the elastic properties predominated over the viscous ones, with the conservative modulus (G′) values being higher than the loss modulus (G″). The presence of pomaces in the formulation and their growing share increases the values of G′ and G″ moduli and spectra, which are all higher by about tenfold in Pa values, compared with the control. The behavior of cracker doughs is viscoelastic, with G′ being higher than G″ and both values being frequency-dependent, as is the general characteristic behavior of doughs. There is a difference between the addition of 20% apple and 20% carrot pomace flours, with a higher impact for apple pomace flour addition, which can be seen on values for G′ at 1 and 10 Hz (Figure 2); this could be due to the higher level of starch (more than 2-fold) in apple pomace. However, there are no substantial differences between the spectra of 40% apple pomace addition and 40% of carrot, as they all have a considerable amount of carbohydrates, over 50%, increasing the dough consistency but not modifying the spectra trend and balance between elastic and viscous components. In fact, based on G′ values extracted from the frequency sweep at 1 Hz and 10 Hz (Figure 3), one can confirm that the addition of pomace exerted an evident effect, increasing the value of the dynamic viscoelastic properties of the dough.

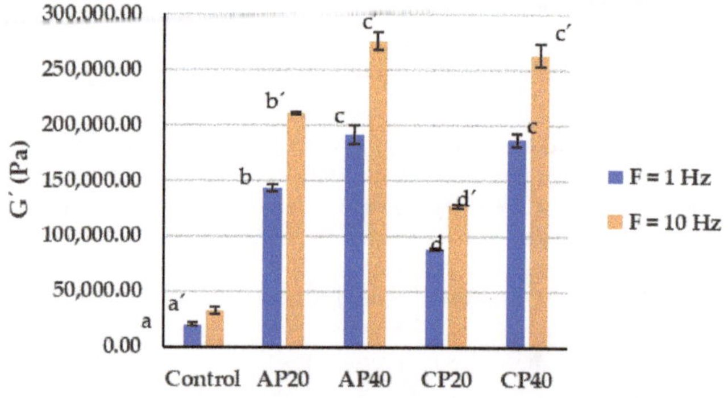

Figure 2. Storage modulus of cracker doughs with apple and carrot pomace at frequencies of 1 and 10 Hz. Results are expressed as average ± standard deviation of triplicates (n = 3), followed by an alphabet letter. Different letters mean significant different results (Tukey's HSD; $p \leq 0.05$).

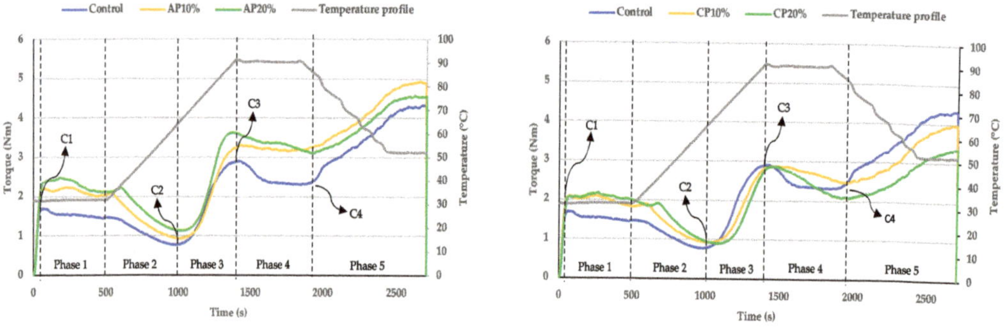

(a) (b)

Figure 3. Mixolab curves of (**a**) mixture of wheat flour with 10 and 20% apple pomace flour and (**b**) mixture of wheat flour with 10 and 20% of carrot pomace flour vs. wheat flour.

3.3. Evaluation of the Mixing and Pasting Characteristics

The Mixolab Chopin+ protocol [50] was performed to evaluate the influence of incorporating apple and carrot pomace flours at different levels on the mixing and pasting behavior of the mixture of wheat flour and pomace flour at a constant water absorption level (55 g/100 g) (Figure 3). This method determines the consistency of dough as the torque exerted by the dough on the kneading pieces, reproducing the overall processing during kneading and the first part of baking to follow the behavior of the protein matrix at a constant temperature (30 °C), followed by the role of starch by applying temperature profile heating at about 90 °C, and subsequently cooling down to 50 °C. The decision to use constant water absorption was taken to allow us to compare the behavior of the different dough systems.

In the first phase of the analysis, dough development time (DDT), the time needed for the gluten network to form (the time needed to reach the first peak in torque—C1), is an essential parameter to evaluate. For wheat flour, this period usually ranges from 0.99 to 7.36 min, and strong flours are characterized by showing a long DDT. The C1 is influenced mainly by the quality of the protein, which is responsible for the gluten matrix, the size of starch granules, and the level of starch degradation [51,52]. C1 and DDT show an increasing trend for doughs with both pomace flours; in accordance with the increase in viscoelastic parameters, the fiber and starch present in pomaces dilutes the gluten and increases the time taken to develop the inner structure of the dough.

The value of stability for different wheat-based flours ranges between 1.43 and 9.13 min [53,54]. The Mixolab method to calculate stability is complex (the time during which the upper frame is bigger than C1–11%). Therefore, stability was determined using the time the dough stayed at a stable value of consistency (Nm). For the additions of apple pomace flour (Figure 3a), the stability of the dough was reduced due to a dilution effect of the wheat gluten. However, the stability was not affected much by the carrot pomace flour additions (Figure 3b), and the high levels of hemicelluloses in roots should be responsible for this. Nevertheless, the torque was always higher with the addition of pomace flours, demanding compensation with a specific amount of added fiber due to the reduction in gluten and starch contents.

In the second stage, when warming starts, the torque decreases to the minimum value (C2) which was attributed to the weakening of the internal network under mechanical shear stress and protein destabilization [51,55,56]. The decrease in torque from C1 to C2, as well as its rate (given by the slope α), was higher after the addition of pomace flours compared to the control, although C2 values were all very similar. This must be due to the dilution of the gluten content and the protein's weaker network.

The third phase of the mixolab, when it reaches its top temperature (about 90 °C), was evaluated via the C3 and slope β parameters. The C3 torque ranged from 2.9 (control) up to 3.6 Nm (AP20) due to the presence of an increased amount of fiber, in accordance with the results from the rheometer. This can be explained by the higher content in starch of the apple pomace, compared to the carrot, and a good synergy remaining with the wheat starch. A similar trend is also noticeable for the slope β, an indicator of starch gelatinization rate, running in parallel, but C3 peaks increase with apple pomace addition and keep the same value as carrot pomaces.

This temperature-regulated testing after phase 1 gives emphasis to the phenomena which occurred mainly with starch and fiber, represented by the C4 decreasing torque. This phase relates to the vulnerability of the gelatinized starch granule to the enzymatic hydrolysis by amylases [51]. The highest hot gel stability, i.e., the value of minimum torque, was observed for the samples containing apple pomace (C4 = 3.2), probably due to the highest content of starch.

The starch gellification phase 5 was different for the apple and the carrot pomaces. For the former, the gel strength was higher than the control, as apples have a lot of starch, especially if they are not too ripe. For carrots, the fibers were not so prone to become organized and form a gel matrix with the wheat starch at the end, and the torque was lower than the control.

3.4. Color and Dimensions

The pomace crackers exhibited different, visually appealing colors, as shown in Figure 4. The color parameters in terms of lightness (L^*), balance between redness and greenness (a^*), and balance between yellowness and blueness (b^*) are represented in Figure 5.

(a) (b)

Figure 4. Cracker control vs (**a**) AP20 and CP20 and (**b**) AP40 and CP40 (from left to right).

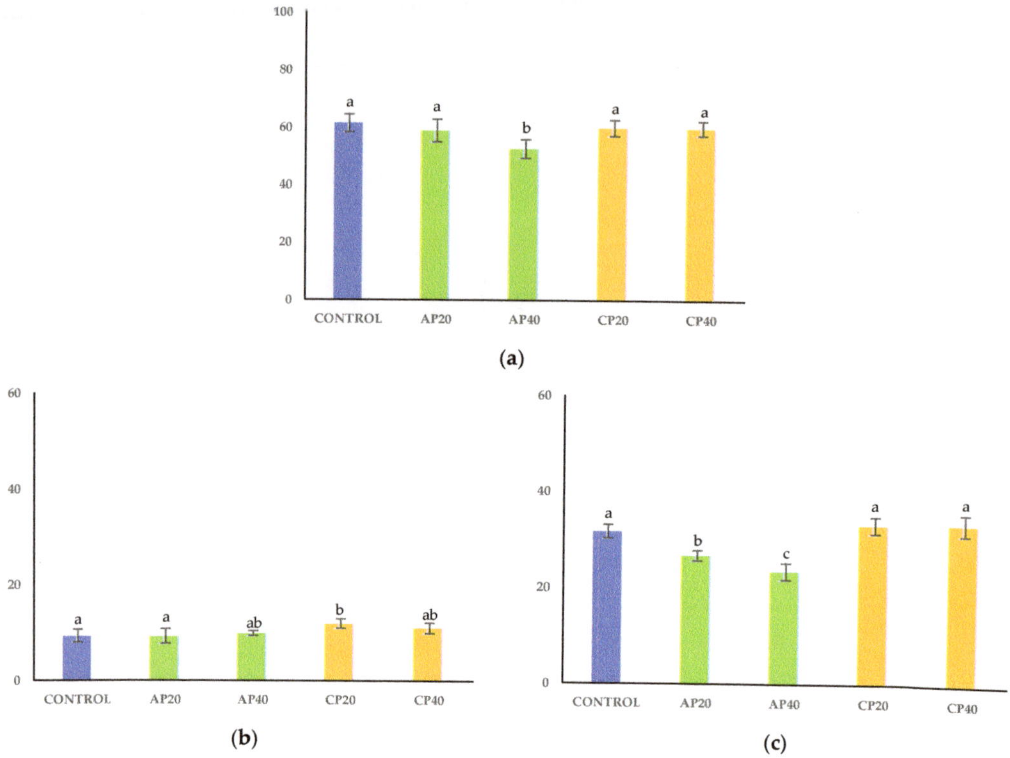

Figure 5. The color parameters of crackers: (**a**) lightness (L^*), (**b**) balance between redness and greenness (a^*), (**c**) balance between yellowness and blueness (b^*). Results are expressed as average ± standard deviation (n = 10), followed by an alphabet letter. Different letters mean significantly different results (Tukey's HSD; $p \leq 0.05$).

The incorporation of pomace flour did not result in any significant changes in relation to the control (L^* 61.5) in terms of lightness, except for crackers with 40% AP that showed a darker color. For crackers with AP, the lightness decreased significantly from 58.9 to 52.7 when increasing the incorporation level from 20 to 40%, but no significant changes were observed between crackers with 20 and 40% CP. The reason could be related to the higher amount of simple reducing sugar contained in apple compared to carrot and wheat flour which, during cooking, undergoes a Maillard reaction, conferring a darker color on the final product.

Regarding a^*, crackers showed positive values in the red domain and there was no difference between the control and with AP, but when CP was incorporated, there was an evident increase. However, concerning b^*, an opposite trend was observed in which CP did not show any significant difference to the control, contrary to AP. The result revealed a decrease in yellowness and redness in crackers with AP and CP, respectively, by increasing the amount of pomace included. The reaction kinetics of pigment degradation, namely β-carotene in carrot and Lutein in apple, upon a high temperature during baking, might be dependent on the initial pigment concentration [35]. Furthermore, the Maillard reaction between proteins and reducing sugars in both sources resulted in the formation of brown-colored compounds like melanoidins, which affect visual color perception and consequently the placement of the sample within the $L^*a^*b^*$ in the tridimensional space. In addition, the volume changes and moisture loss that take place during baking can also have a significant impact on the crackers' appearance. As will be discussed later, crackers which presented lower dimensions and lower water content showed that those parameters can influence the color perception and accelerate surface browning [57].

The results of the total color differences between wheat-flour-based crackers (WF) used as the control and those that included pomace flours from both apple and carrot are shown in Table 2. A noticeable impact on color was seen when 20% and 40% of the apple and carrot pomace flours were added (ΔE* > 5) [53].

Table 2. Total color difference calculated for the crackers with apple and carrot pomace flour at different levels of incorporation in comparison with the WF cracker (control). Data expressed as means.

Sample	ΔE*
Control	-
AP20	10.58
AP40	10.97
CP20	11.10
CP40	11.15

Characteristic dimensions of all crackers are presented in Figure 6. It was observed that there was a significant ($p < 0.05$) reduction in the crackers' thickness (1.8–2.2 mm for AP and 2.0–2.8 mm for CP) when compared to the control (3.3 mm) when increasing the amount of pomace flour. This decreasing trend is related to the incapacity of the dough to expand during baking, as wheat gluten, responsible for this volume increase by building the network that holds the gas produced during leavening, is diluted. In addition, the presence of fiber that interferes with starch gelatinization, competing for water, will further reduce the expansion of the structure, which is a crucial factor for product development. These differences led to higher spread ratios for the pomace crackers, which were statistically significant ($p < 0.05$) for all incorporation levels.

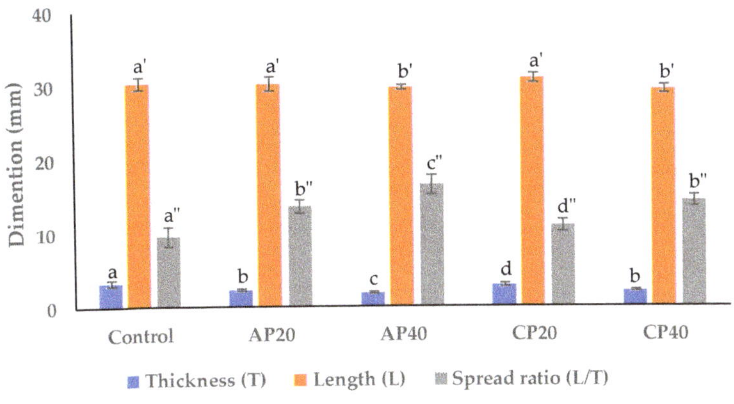

Figure 6. The texture parameters of crackers. Results are expressed as average ± standard deviation (n = 10), followed by an alphabet letter. Different letters mean significantly different results (Tukey's HSD; $p \leq 0.05$).

3.5. *Moisture Content and Water Activity of Crackers*

The quality parameters in low-moisture foods are influenced by water content and/or water activity (aw), which has a significant impact on their crispiness and sensory appeal. The food materials became softer and more stale and lost their crispiness above a critical aw value, typically around 0.5 [58,59].

According to Figure 7, the control cracker presented an initial aw of 0.3 and with the addition of pomaces, the aw decreased significantly with apple but not for carrot, due to its higher level of moisture content (18%). The decrease in aw corresponds to highly crispy products [58], but for the cracker with 40% of carrot pomace the aw increased up to 0.4, determining a softer texture which was not well appreciated by the panel in the sensory analysis (to be discussed later in Section 3.7).

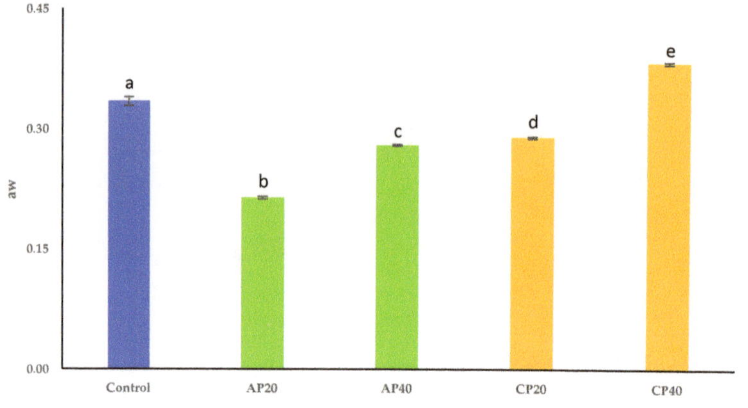

Figure 7. Evaluation of the water activity (aw) of crackers. Results are expressed as average ± standard deviation (n = 10), followed by an alphabet letter. Different letters mean significantly different results (Tukey's HSD; $p \leq 0.05$).

Regarding the water content (Table 1), the same behavior was observed but without overpassing the control level (4.2). Also, in this case, the apple crackers still showed a better crispiness (2.32–2.45), which is the main parameter of texture acceptance for this category of product. It is possible that adding pomace in large quantities will result in a weaker gluten network that is less effective at trapping gas bubbles and water molecules. This result means that the addition of pomaces, particularly for apple pomaces, has a positive impact on the texture of the crackers.

3.6. Texture

Since consumers value a crunchy and crisp texture highly, texture is one of the key factors in cracker appreciation [60]. One of the most suitable instrumental analysis tests to assess the texture of these kinds of brittle food samples is the "three-point bending" or "snap" test, in which the cracker is leaned upon two support beams while a third moves down (parallel) into the middle point of the sample, causing the sample to fracture into two equal pieces.

The crackers with 20% incorporation showed significant ($p > 0.05$) decreases in hardness and toughness in relation to the control crackers, although there seemed to be a tendency for both parameters to increase with the 40% addition of pomace. However, a different behavior was detected with the deformation results. With AP, there was a positive correlation between the amount incorporated and the deformation, and therefore the brittleness reduced, while in CP the reverse was found. The crackers with 20% of AP and CP had a thinner structure (showing a higher spread ratio), which led to lower resistance to breakage, as also described by some other authors for *P. tricornutum* crackers [35,61]. In the case of AP and CP 40%, the addition of water in the recipe to develop a desired cracker resulted in a higher hardness and toughness and less brittleness (Figure 8).

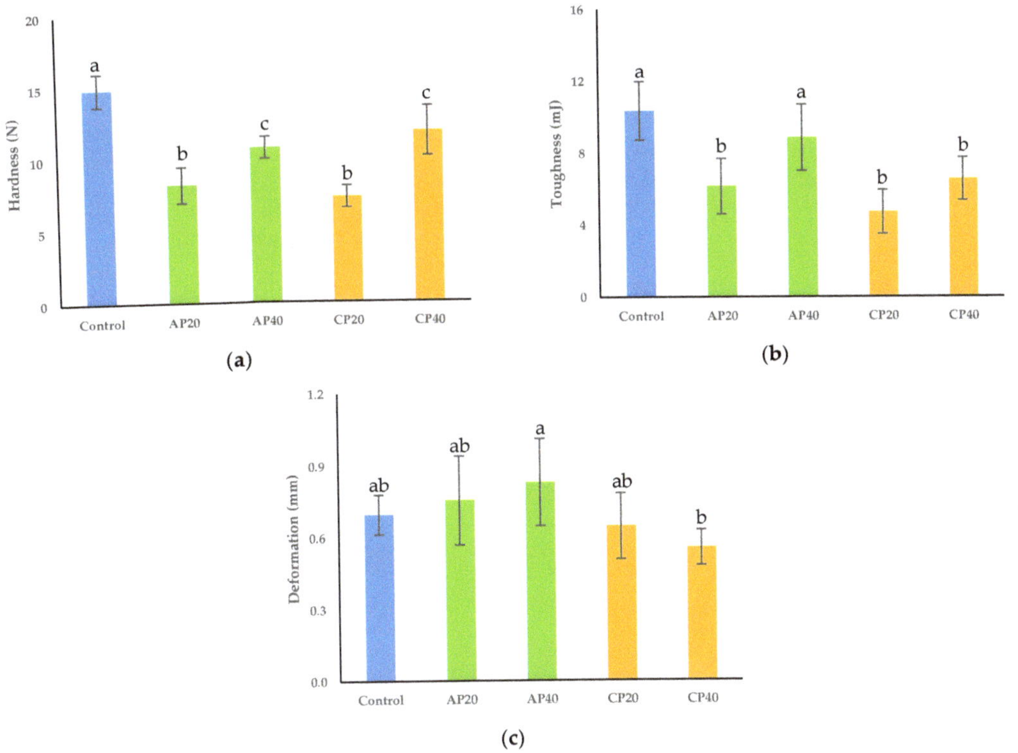

Figure 8. The results of texture parameters of crackers: (**a**) hardness, (**b**) toughness, and (**c**) deformation. Results are expressed as average ± standard deviation (n = 10), followed by an alphabet letter. Different letters mean significantly different results (Tukey's HSD; $p \leq 0.05$).

3.7. Sensory Evaluation

Sensory analysis trials were carried out on samples with 20 and 40% AP and CP pomace flours and the control. Figure 9 represents the average scores of the sensorial parameters, as evaluated by the panel.

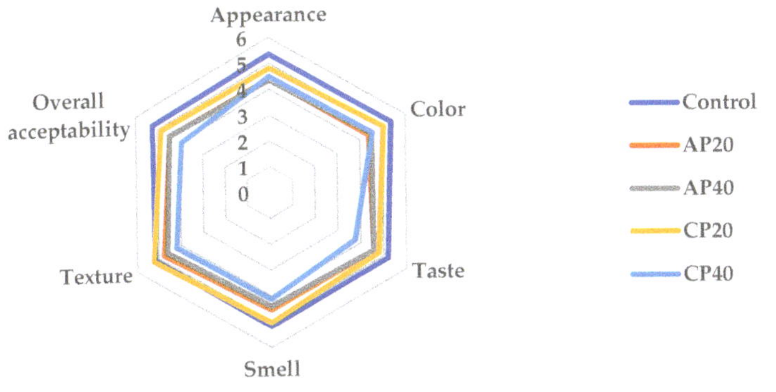

Figure 9. Responses of the sensory analysis panel tasters (n = 44) regarding crackers. Sensory attributes were classified as follows: 1—dislike very much, 2—dislike moderately, 3—dislike slightly, 4—like slightly, 5—like moderately, 6—like very much.

The control sample showed high global sensory scores (>5) and was preferred over the crackers with pomace flours, which was as expected for this amount of pomace flour included. Better sensorial scores (>5) that were even higher than the control were presented by the texture of the crackers with 20% CP, which could be related to the instrumental texture and aw/water content results, indicating a crisp texture [35]. The CP20 was preferred, considering all the parameters evaluated, while crackers with 40% CP showed the lowest scores for taste, smell, texture, and overall acceptability due to their softer texture.

More than 70% of the panelists agreed that crackers with 20% CP received the highest sensory ratings and said that they would probably or definitely purchase this product (Figure 10).

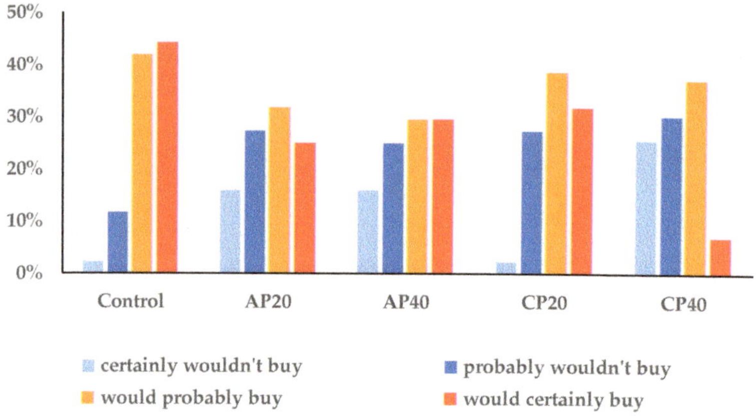

Figure 10. Responses of the sensory analysis panel tasters (n = 44) in terms of buying intention for crackers.

3.8. Total Phenolic Content (TPC) and Antioxidant Potential

The TPC was evaluated on the ethanolic extracts of crackers, as shown in Figure 11. The TPC found in all the crackers' formulation was significantly different ($p > 0.05$) when compared to the control crackers (1.2 mg GAE g^{-1}), except for the CP20 formulation, showing a TPC similar to the one found in the control. This might be due to the lower TPC and antioxidant activity found in the carrot pomace flour used in the preparation [62]. When increasing pomace flour content from 20 to 40% the TPC significantly increased from 3.7 to 4.2 mg GAE g^{-1} ($p < 0.05$) when using apple pomace flour and from 1.9 to 3.5 mg GAE g^{-1} when using carrot pomace flour.

The antioxidant capacity of the crackers was determined using the DPPH and FRAP methods (Figure 12). All the crackers prepared with the two pomace flours showed higher antioxidant potential when compared to the control. Also, apple pomace promoted an improvement in the antioxidant potential when compared with the carrot pomace. It is also evident that upon increasing the amount of pomace flour that was incorporated in the crackers from 20% to 40% an improvement in the antioxidant potential of the crackers was achieved.

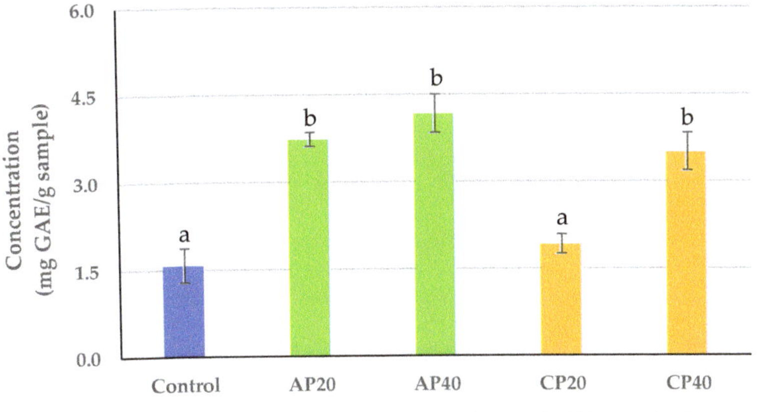

Figure 11. Total phenolic content of crackers, expressed as mg GAE/g sample. Results are expressed as average ± standard deviation (n = 3), followed by an alphabet letter. Different letters mean significantly different results (Tukey's HSD; $p \leq 0.05$).

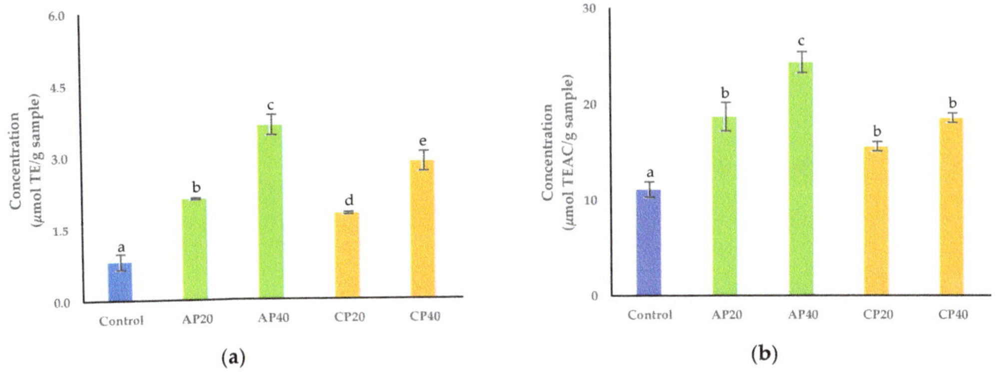

Figure 12. AAT of crackers, measured by DPPH (**a**) and FRAP (**b**), expressed as µmol TE/g sample. Results are expressed as average ± standard deviation (n = 3), followed by an alphabet letter. Different letters mean significantly different results (Tukey's HSD; $p \leq 0.05$).

4. Conclusions

The study's results indicate that adding apple and carrot pomaces to wheat crackers enhanced their total phenolic content and antioxidant activity. The fiber content, especially in AP40 and CP40 variants, improved the dough's rheological properties and the crackers' texture, increasing hardness and brittleness. The sensory analysis results revealed that these texture changes, except for CP40, were well received by the panelists. Among the variants, CP20 emerged as the most preferred cracker, resembling the control in taste and buying intention. Conversely, CP40 had the lowest buying intention, largely due to its bitterness, likely resulting from the carrot residues post-juice extraction.

The incorporation of apple and carrot pomaces in staple foods like crackers is a sustainable way to introduce these residues back into the food chain, adding them as ingredients for food fortification with potential health benefits.

Supplementary Materials: The following supporting information can be downloaded at: https://www.mdpi.com/article/10.3390/su16145995/s1. Here is the caption of S1. Table S1: Minerals composition determined by ICP-EOS for apple and carrot pomace flours. The results are the mean of triplicates shown as mean ± standard deviation (n = 3) and were reported as mg of mineral per 100 g of pomace flour.

Author Contributions: Conceptualization, J.F., A.L. and I.S.; methodology, J.F., A.L. and I.S.; validation, J.F., A.L. and I.S.; formal analysis, S.S., T.C., J.F., A.L. and I.S.; investigation, S.S. and T.C.; data curation, S.S., T.C. and J.F.; writing—original draft preparation, S.S., T.C. and J.F.; writing—review and editing, J.F., A.L. and I.S.; visualization, J.F., A.L. and I.S.; supervision, J.F., A.L. and I.S.; project administration, J.F. and I.S.; funding acquisition, I.S. All authors have read and agreed to the published version of the manuscript.

Funding: This research was funded by the (i) Project VIIAFOOD—Plataforma de Valorização, Industrialização e Inovação comercial para o AgroAlimentar, ref C64492956-00000040 (Project n 37), Agendas Mobilizadoras para a Inovação Empresarial Aviso N° 02/C05-i01/2022, através do Plano de Recuperação e Resiliência (PRR) e Fundos Europes NextGeneration EU; (ii) LEAF-Linking Landscape, Environment, Agriculture and Food Research Center, Instituto Superior de Agronomia, Universidade de Lisboa, Portugal (FCT UIDB/04129/2020). S.S. was supported by an FCT PhD grant (2022.13785.BD).

Data Availability Statement: The original contributions presented in the study are included in the article/Supplementary Material, further inquiries can be directed to the corresponding author/s.

Conflicts of Interest: The authors declare no conflicts of interest.

References

1. Hurtado-Barroso, S.; Trius-Soler, M.; Lamuela-Raventós, R.M.; Zamora-Ros, R. Vegetable and Fruit Consumption and Prognosis Among Cancer Survivors: A Systematic Review and Meta-Analysis of Cohort Studies. *Adv. Nutr.* **2020**, *11*, 1569–1582. [CrossRef]
2. Milajerdi, A.; Ebrahimi-Daryani, N.; Dieleman, L.A.; Larijani, B.; Esmaillzadeh, A. Association of Dietary Fiber, Fruit, and Vegetable Consumption with Risk of Inflammatory Bowel Disease: A Systematic Review and Meta-Analysis. *Adv. Nutr.* **2021**, *12*, 735–743. [CrossRef]
3. Głąbska, D.; Guzek, D.; Groele, B.; Gutkowska, K. Fruit and Vegetable Intake and Mental Health in Adults: A Systematic Review. *Nutrients* **2020**, *12*, 115. [CrossRef] [PubMed]
4. Santos, D.; Pintado, M.; Lopes da Silva, J.A. Potential Nutritional and Functional Improvement of Extruded Breakfast Cereals Based on Incorporation of Fruit and Vegetable By-Products—A Review. *Trends Food Sci. Technol.* **2022**, *125*, 136–153. [CrossRef]
5. FAO; United Nations. *Agricultural Production Statistics (2000–2021)*; Food and Agriculture Organization: Québec City, QC, Canada, 2021.
6. Kandemir, K.; Piskin, E.; Xiao, J.; Tomas, M.; Capanoglu, E. Fruit Juice Industry Wastes as a Source of Bioactives. *J. Agric. Food Chem.* **2022**, *70*, 6805–6832. [CrossRef] [PubMed]
7. Farvid, M.S.; Spence, N.D.; Holmes, M.D.; Barnett, J.B. Fiber Consumption and Breast Cancer Incidence: A Systematic Review and Meta-Analysis of Prospective Studies. *Cancer* **2020**, *126*, 3061–3075. [CrossRef] [PubMed]
8. Xu, X.; Zhang, J.; Zhang, Y.; Qi, H.; Wang, P. Associations between Dietary Fiber Intake and Mortality from All Causes, Cardiovascular Disease and Cancer: A Prospective Study. *J. Transl. Med.* **2022**, *20*, 344. [CrossRef] [PubMed]
9. Rocchetti, G.; Gregorio, R.P.; Lorenzo, J.M.; Barba, F.J.; Oliveira, P.G.; Prieto, M.A.; Simal-Gandara, J.; Mosele, J.I.; Motilva, M.-J.; Tomas, M.; et al. Functional Implications of Bound Phenolic Compounds and Phenolics–Food Interaction: A Review. *Compr. Rev. Food Sci. Food Saf.* **2022**, *21*, 811–842. [CrossRef] [PubMed]
10. Dong, Y.; Gao, M.; Qiu, W.; Song, Z. Uptake of Microplastics by Carrots in Presence of As (III): Combined Toxic Effects. *J. Hazard. Mater.* **2021**, *411*, 125055. [CrossRef]
11. Korczak, R.; Slavin, J.L. Definitions, Regulations, and New Frontiers for Dietary Fiber and Whole Grains. *Nutr. Rev.* **2020**, *78* (Suppl. 1), 6–12. [CrossRef]
12. Fu, J.; Zheng, Y.; Gao, Y.; Xu, W. Dietary Fiber Intake and Gut Microbiota in Human Health. *Microorganisms* **2022**, *10*, 2507. [CrossRef] [PubMed]
13. Swann, O.G.; Kilpatrick, M.; Breslin, M.; Oddy, W.H. Dietary Fiber and Its Associations with Depression and Inflammation. *Nutr. Rev.* **2020**, *78*, 394–411. [CrossRef]
14. Snauwaert, E.; Paglialonga, F.; Vande Walle, J.; Wan, M.; Desloovere, A.; Polderman, N.; Renken-Terhaerdt, J.; Shaw, V.; Shroff, R. The Benefits of Dietary Fiber: The Gastrointestinal Tract and Beyond. *Pediatr. Nephrol.* **2023**, *38*, 2929–2938. [CrossRef]
15. Waddell, I.S.; Orfila, C. Dietary Fiber in the Prevention of Obesity and Obesity-Related Chronic Diseases: From Epidemiological Evidence to Potential Molecular Mechanisms. *Crit. Rev. Food Sci. Nutr.* **2023**, *63*, 8752–8767. [CrossRef] [PubMed]
16. Vuolo, M.M.; Lima, V.S.; Maróstica Junior, M.R. Chapter 2—Phenolic Compounds: Structure, Classification, and Antioxidant Power. In *Bioactive Compounds*; Campos, M.R.S., Ed.; Woodhead Publishing: Cambridge, UK, 2019; pp. 33–50. [CrossRef]
17. Mutha, R.E.; Tatiya, A.U.; Surana, S.J. Flavonoids as Natural Phenolic Compounds and Their Role in Therapeutics: An Overview. *Future J. Pharm. Sci.* **2021**, *7*, 25. [CrossRef] [PubMed]
18. Rudrapal, M.; Khairnar, S.J.; Khan, J.; Dukhyil, A.B.; Ansari, M.A.; Alomary, M.N.; Alshabrmi, F.M.; Palai, S.; Deb, P.K.; Devi, R. Dietary Polyphenols and Their Role in Oxidative Stress-Induced Human Diseases: Insights Into Protective Effects, Antioxidant Potentials and Mechanism(s) of Action. *Front. Pharmacol.* **2022**, *13*, 806470. [CrossRef]

19. Snack Food—Worldwide | Statista Market Forecast. Statista. Available online: https://www.statista.com/outlook/cmo/food/confectionery-snacks/snack-food/worldwide (accessed on 29 December 2023).
20. Contini, C.; Boncinelli, F.; Marone, E.; Scozzafava, G.; Casini, L. Drivers of Plant-Based Convenience Foods Consumption: Results of a Multicomponent Extension of the Theory of Planned Behaviour. *Food Qual. Prefer.* **2020**, *84*, 103931. [CrossRef]
21. Alongi, M.; Anese, M. Re-Thinking Functional Food Development through a Holistic Approach. *J. Funct. Foods* **2021**, *81*, 104466. [CrossRef]
22. Nakov, G.; Brandolini, A.; Estivi, L.; Bertuglia, K.; Ivanova, N.; Jukić, M.; Komlenić, D.K.; Lukinac, J.; Hidalgo, A. Effect of Tomato Pomace Addition on Chemical, Technological, Nutritional, and Sensorial Properties of Cream Crackers. *Antioxidants* **2022**, *11*, 2087. [CrossRef]
23. Imeneo, V.; Romeo, R.; Gattuso, A.; De Bruno, A.; Piscopo, A. Functionalized Biscuits with Bioactive Ingredients Obtained by Citrus Lemon Pomace. *Foods* **2021**, *10*, 2460. [CrossRef]
24. Usman, M.; Ahmed, S.; Mehmood, A.; Bilal, M.; Patil, P.J.; Akram, K.; Farooq, U. Effect of Apple Pomace on Nutrition, Rheology of Dough and Cookies Quality. *J. Food Sci. Technol.* **2020**, *57*, 3244–3251. [CrossRef]
25. Alongi, M.; Melchior, S.; Anese, M. Reducing the Glycemic Index of Short Dough Biscuits by Using Apple Pomace as a Functional Ingredient. *LWT* **2019**, *100*, 300–305. [CrossRef]
26. Badjona, A.; Adubofuor, J.; Amoah, I.; Diako, C. Valorisation of Carrot and Pineapple Pomaces for Rock Buns Development. *Sci. Afr.* **2019**, *6*, e00160. [CrossRef]
27. Baldán, Y.; Riveros, M.; Fabani, M.P.; Rodriguez, R. Grape Pomace Powder Valorization: A Novel Ingredient to Improve the Nutritional Quality of Gluten-Free Muffins. *Biomass Convers. Biorefinery* **2023**, *13*, 9997–10009. [CrossRef]
28. Technical, A. Moisture--Air-Oven Methods. In *AACC International Approved Methods*; AACC International: St. Paul, MN, USA, 2009. [CrossRef]
29. Marrero, J.; Rebagliati, R.J.; Leiva, E.; Londonio, A.; Smichowski, P. Inductively Coupled Plasma Optical Emission Spectrometric Determination of Fifteen Elements in Dietary Supplements: Are the Concentrations Declared in the Labels Accurate? *Microchem. J.* **2013**, *108*, 81–86. [CrossRef]
30. Technical, A. Determination of Minerals by Inductively Coupled Plasma Spectroscopy. In *AACC International Approved Methods*; AACC International: St. Paul, MN, USA, 2009. [CrossRef]
31. Technical, A. Crude Fat in Wheat, Corn, and Soy Flour, Feeds, and Mixed Feeds. In *AACC International Approved Methods*; AACC International: St. Paul, MN, USA, 2009. [CrossRef]
32. Wendt Thiex, N.J.; Latimer, G.W., Jr. (Eds.) Fiber. In *Official Methods of Analysis of AOAC INTERNATIONAL*; Oxford University Press: Oxford, UK, 2023. [CrossRef]
33. *ISO 16634-2:2016*; Food Products—Determination of the Total Nitrogen Content by Combustion According to the Dumas Principle and Calculation of the Crude Protein Content. ISO: Geneva, Switzerland, 2016. Available online: https://www.iso.org/standard/66661.html (accessed on 15 December 2023).
34. *AOAC 996.11-2005*; Starch (Total) in Cereal Products. Amyloglucos—$14.60. AOAC Official Method. AOAC: Rockville, MD, USA, 2015. Available online: http://www.aoacofficialmethod.org/index.php?main_page=product_info&products_id=1546 (accessed on 11 March 2024).
35. Batista, A.P.; Niccolai, A.; Bursic, I.; Sousa, I.; Raymundo, A.; Rodolfi, L.; Biondi, N.; Tredici, M.R. Microalgae as Functional Ingredients in Savory Food Products: Application to Wheat Crackers. *Foods* **2019**, *8*, 611. [CrossRef]
36. Švec, I.; Hrušková, M. The Mixolab Parameters of Composite Wheat/Hemp Flour and Their Relation to Quality Features. *LWT—Food Sci. Technol.* **2015**, *60*, 623–629. [CrossRef]
37. Kim, H.R.; Kim, M.-R.; Ryu, A.-R.; Bae, J.-E.; Choi, Y.-S.; Lee, G.B.; Choi, H.-D.; Hong, J.S. Comparison of Rheological Properties between Mixolab-Driven Dough and Bread-Making Dough under Various Salt Levels. *Food Sci. Biotechnol.* **2023**, *32*, 193–202. [CrossRef] [PubMed]
38. *ISO 8589:2007*; Sensory Analysis—General Guidance for the Design of Test Rooms. ISO: Geneva, Switzerland, 2007. Available online: https://www.iso.org/standard/36385.html (accessed on 24 December 2023).
39. Commission Regulation (EC) No 687/2008 of 18 July 2008 Establishing Procedures for the Taking-Over of Cereals by Intervention Agencies or Paying Agencies and Laying down Methods of Analysis for Determining the Quality of Cereals (Codified Version). 2008. Volume 192. Available online: http://data.europa.eu/eli/reg/2008/687/oj/eng (accessed on 29 December 2023).
40. White, P.J.; Broadley, M.R. Biofortification of Crops with Seven Mineral Elements Often Lacking in Human Diets—Iron, Zinc, Copper, Calcium, Magnesium, Selenium and Iodine. *New Phytol.* **2009**, *182*, 49–84. [CrossRef]
41. Karley, A.J.; White, P.J. Moving Cationic Minerals to Edible Tissues: Potassium, Magnesium, Calcium. *Curr. Opin. Plant Biol.* **2009**, *12*, 291–298. [CrossRef]
42. Skinner, R.C.; Gigliotti, J.C.; Ku, K.-M.; Tou, J.C. A Comprehensive Analysis of the Composition, Health Benefits, and Safety of Apple Pomace. *Nutr. Rev.* **2018**, *76*, 893–909. [CrossRef] [PubMed]
43. Verardo, G.; Pagani, E.; Geatti, P.; Martinuzzi, P. A Thorough Study of the Surface Wax of Apple Fruits. *Anal. Bioanal. Chem.* **2003**, *376*, 659–667. [CrossRef] [PubMed]
44. Smuda, M.; Glomb, M.A. Fragmentation Pathways during Maillard-Induced Carbohydrate Degradation. *J. Agric. Food Chem.* **2013**, *61*, 10198–10208. [CrossRef] [PubMed]

45. Hollnagel, A.; Kroh, L.W. Degradation of Oligosaccharides in Nonenzymatic Browning by Formation of α-Dicarbonyl Compounds via a "Peeling Off" Mechanism. *J. Agric. Food Chem.* **2000**, *48*, 6219–6226. [CrossRef] [PubMed]
46. Tamanna, N.; Mahmood, N. Food Processing and Maillard Reaction Products: Effect on Human Health and Nutrition. *Int. J. Food Sci.* **2015**, *2015*, 526762. [CrossRef] [PubMed]
47. Zhang, F.; Wang, T.; Wang, X.; Lü, X. Apple Pomace as a Potential Valuable Resource for Full-Components Utilization: A Review. *J. Clean. Prod.* **2021**, *329*, 129676. [CrossRef]
48. Raymundo, A.; Batista, A.; Sousa, I. Rheology Applied to Food Product Design. 2017, pp. 63–118. Available online: https://www.researchgate.net/publication/324260187_Rheology_applied_to_food_product_design (accessed on 28 April 2024).
49. Pycia, K.; Juszczak, L. The Influence of the Addition of Nuts on the Thermal and Rheological Properties of Wheat Flour. *Molecules* **2021**, *26*, 3969. [CrossRef]
50. Kahraman, K.; Sakıyan, O.; Ozturk, S.; Koksel, H.; Sumnu, G.; Dubat, A. Utilization of Mixolab® to Predict the Suitability of Flours in Terms of Cake Quality. *Eur. Food Res. Technol.* **2008**, *227*, 565–570. [CrossRef]
51. Tran, K.D.; Konvalina, P.; Capouchova, I.; Janovska, D.; Lacko-Bartosova, M.; Kopecky, M.; Tran, P.X.T. Comparative Study on Protein Quality and Rheological Behavior of Different Wheat Species. *Agronomy* **2020**, *10*, 1763. [CrossRef]
52. Bánfalvi, Á.; Németh, R.; Bagdi, A.; Gergely, S.; Rakszegi, M.; Bedő, Z.; Láng, L.; Vida, G.; Tömösközi, S. A Novel Approach to the Characterization of Old Wheat (*Triticum Aestivum* L.) Varieties by Complex Rheological Analysis. *J. Sci. Food Agric.* **2020**, *100*, 4409–4417. [CrossRef]
53. Khemiri, S.; Khelifi, N.; Nunes, M.C.; Ferreira, A.; Gouveia, L.; Smaali, I.; Raymundo, A. Microalgae Biomass as an Additional Ingredient of Gluten-Free Bread: Dough Rheology, Texture Quality and Nutritional Properties. *Algal Res.* **2020**, *50*, 101998. [CrossRef]
54. Baljeet, S.Y.; Yogesh, S.; Ritika, B.Y. Physicochemical and Rheological Properties of Indian Wheat Varieties of *Triticum aestivum*. *Qual. Assur. Saf. Crops Foods* **2017**, *9*, 369–382. [CrossRef]
55. Rosell, C.M.; Collar, C.; Haros, M. Assessment of Hydrocolloid Effects on the Thermo-Mechanical Properties of Wheat Using the Mixolab. *Food Hydrocoll.* **2007**, *21*, 452–462. [CrossRef]
56. Haros, M.; Ferrer, A.; Rosell, C. Rheological Behaviour of Whole Wheat Flour. In Proceedings of the 13th World Congress of Food Science & Technology, Nantes, France, 17–21 September 2006; p. 681.
57. Davidson, I. *Biscuit Baking Technology: Processing and Engineering Manual*; Elsevier: Amsterdam, The Netherlands, 2023.
58. Arimi, J.M.; Duggan, E.; O'Sullivan, M.; Lyng, J.G.; O'Riordan, E.D. Effect of Water Activity on the Crispiness of a Biscuit (Crackerbread): Mechanical and Acoustic Evaluation. *Food Res. Int.* **2010**, *43*, 1650–1655. [CrossRef]
59. Peleg, M. A Mathematical Model of Crunchiness/Crispness Loss in Breakfast Cereals1. *J. Texture Stud.* **1994**, *25*, 403–410. [CrossRef]
60. Tunick, M.H.; Onwulata, C.I.; Thomas, A.E.; Phillips, J.G.; Mukhopadhyay, S.; Sheen, S.; Liu, C.-K.; Latona, N.; Pimentel, M.R.; Cooke, P.H. Critical Evaluation of Crispy and Crunchy Textures: A Review. *Int. J. Food Prop.* **2013**, *16*, 949–963. [CrossRef]
61. Nowak, D.; Jakubczyk, E. The Freeze-Drying of Foods—The Characteristic of the Process Course and the Effect of Its Parameters on the Physical Properties of Food Materials. *Foods* **2020**, *9*, 1488. [CrossRef]
62. Salari, S.; Ferreira, J.; Lima, A.; Sousa, I. Effects of Particle Size on Physicochemical and Nutritional Properties and Antioxidant Activity of Apple and Carrot Pomaces. *Foods* **2024**, *13*, 710. [CrossRef]

sustainability

MDPI

Review

Valorization of Peach By-Products: Utilizing Them as Valuable Resources in a Circular Economy Model

Nikoletta Solomakou, Aikaterini M. Drosaki, Kyriakos Kaderides, Ioannis Mourtzinos and Athanasia M. Goula *

Department of Food Science and Technology, School of Agriculture, Forestry and Natural Environment, Aristotle University, 54124 Thessaloniki, Greece; nsolomak@agro.auth.gr (N.S.); adrosaki@agro.auth.gr (A.M.D.); kaderidisk@gmail.com (K.K.); mourtzinos@agro.auth.gr (I.M.)
* Correspondence: athgou@agro.auth.gr

Abstract: Peach processing generates significant amounts of by-products including peels, pomace, and seeds that are often discarded as waste, despite their rich content of bioactive components. Various methods, such as solvent extraction, ultrasound-assisted extraction, and alkaline and acid hydrolysis, have been employed to recover valuable components from peach by-products. These compounds have shown potential applications in the food, pharmaceutical, and cosmetic industries due to their antioxidant, antimicrobial, and anti-inflammatory properties. Furthermore, these wastes can also be used to produce functional ingredients, natural colorants, and dietary supplements. Alternative uses include animal feed, composting materials, and biofuels. This comprehensive review provides an overview of the valorization of peach by-products, focusing on the isolation of valuable compounds, the techniques used, and the potential applications of the obtained compounds.

Keywords: bioactive compounds; extraction; peach; peel; pomace; seeds

1. Introduction

Peach (*Prunus persica*) is a widely valued and economically important fruit sought after for its sweet and juicy flavor. Global peach production and processing generate large amounts of waste, including peels, pits, and pomace, which are often overlooked and treated as waste. However, as the sustainability paradigm gains traction across industries, there is growing interest in exploring innovative ways to unlock the hidden potential of these peach by-products.

The valorization of peach by-products represents a compelling opportunity to transform what was once considered waste into valuable resources. By harnessing the rich array of bioactive compounds, essential nutrients, and functional components present in peach peels, seeds, and pomace, numerous industries can embrace more sustainable practices and foster a circular economy. This review paper aims to provide an extensive overview of the current research and applications surrounding the valorization of peach by-products.

Key objectives of this review include:

Uncovering the nutritional and bioactive potential: The biochemical composition of peach by-products will be explored, shedding light on their diverse nutritional content, such as antioxidant compounds and dietary fibers. By understanding the richness of these by-products, novel avenues for their utilization in the food and nutraceutical sectors can be identified.

Innovative and green extraction and processing techniques: The review will delve into various conventional and alternative extraction methods used to isolate and preserve the valuable compounds of peach by-products. Special attention will be given to environmentally friendly approaches that minimize energy consumption, solvent usage, and waste generation.

Environmental and economic impact: An essential aspect of valorization is its potential to contribute to sustainability goals and environmental conservation. We will assess the environmental benefits of diverting peach by-products from landfills and the potential for generating revenue streams, thus making these practices economically viable.

In conclusion, the objective of this review is to provide a comprehensive understanding of the valorization of peach by-products, accentuating their potential as valuable resources within a circular economy. By shedding light on innovative extraction techniques, industrial applications, and environmental impacts, this review aims to stimulate further research and encourage industries to embrace sustainable practices that harness the untapped potential of peach by-products.

2. Peach and Its Products and By-Products

Peaches (*Prunus persica*) are classified as "stone fruits" because their seeds are protected by a tough, stone-like endocarp. Peaches and nectarines, belonging to the same *Prunus persica* species and *Rosaceae* family, yielded a combined production of over 24.5 million tons in 2020 [1]. Apart from their economic significance, peaches offer substantial nutritional advantages due to their abundant content of organic acids, sugars, vitamins, and minerals [2].

The peach fruit is composed of three distinct parts. The first, constituting approximately 75.2% of the fruit's weight, is the succulent and yellow-hued pulp, or mesocarp. With a wide-ranging taste that alternates between acidic and sweet, its average pH typically fluctuates from 3.5 to 4.0 [3]. The peel, or exocarp, forms the second part, accounting for around 22.5% of the fruit. The final part comprises the endocarp, better known as the stone, enclosing the seed within a sturdy shell. Depending on the peach species, the seed makes up 5.0 to 12.5% of the fruit's weight [4]. On average, the stone comprises 6% seed and 94% seed shell [5].

Industrial peach processing depends on the final product. Among processed peach products, canned peaches in syrup account for 93%, peach jam for 6%, and peach juice for 1% [6]. The most popular products are peach syrup (canned or in glass jars) and peach puree concentrate, the latter being used as an ingredient in recipes such as baby food, juices, jams, pulp, and yogurt. Figure 1 shows the peach-processing flow chart for these products and their by-products (seeds, peels, and pomace).

Essentially, the procedure of processing peaches into syrup includes harvesting, selection, chemical peeling (a 1.5% to 2% sodium hydroxide solution, near boiling temperature), pitting, and steam blanching. The syrup is then poured into glass or can packaging for final pasteurization. Peach concentrate is obtained by washing/sorting, removing leaves and kernels (formed by seeds and seed coats), crushing, heating (90–95 °C), evaporation (60–75 °C), degassing, bottling, and sterilization (105–120 °C).

According to Plazzotta et al. [7], the global annual processing of peaches to produce juices amounts to approximately 15 million metric tons, resulting in an estimated 10% discarded materials depending on the fruit's ripeness. This implies that, considering the worldwide yearly peach production and accounting for 10% of residues or by-products [7], 2.4 million tons of peach wastes are generated on a global scale annually. The growth of fruit-harvesting and -processing activities, coupled with inadequate handling techniques, serves to augment the by-product output, which currently remains vastly underutilized across the globe. However, many research works underscore that peach by-products are rich in valuable compounds, such as oils (in seeds), phenolic compounds (in peel and pomace), pectin (in peel), and proteins (in seeds) [8,9].

Traditionally, peach by-products were discarded, contributing to environmental waste and loss of valuable resources. In recent years, a shift towards sustainable practices and circular economy concepts has prompted a reevaluation of these peach by-products. Researchers and industries alike are recognizing the potential value of these discarded components and are exploring innovative methods to harness their inherent nutritional and functional properties. Through innovative extraction and processing techniques, these by-products are being transformed into a range of value-added products, such as functional ingredients, natural antioxidants, essential oils, dietary fibers, fertilizers, and even biofuels (Figure 2). The valorization of these by-products not only minimizes waste but also contributes to the diversification of products and revenue streams within the peach industry.

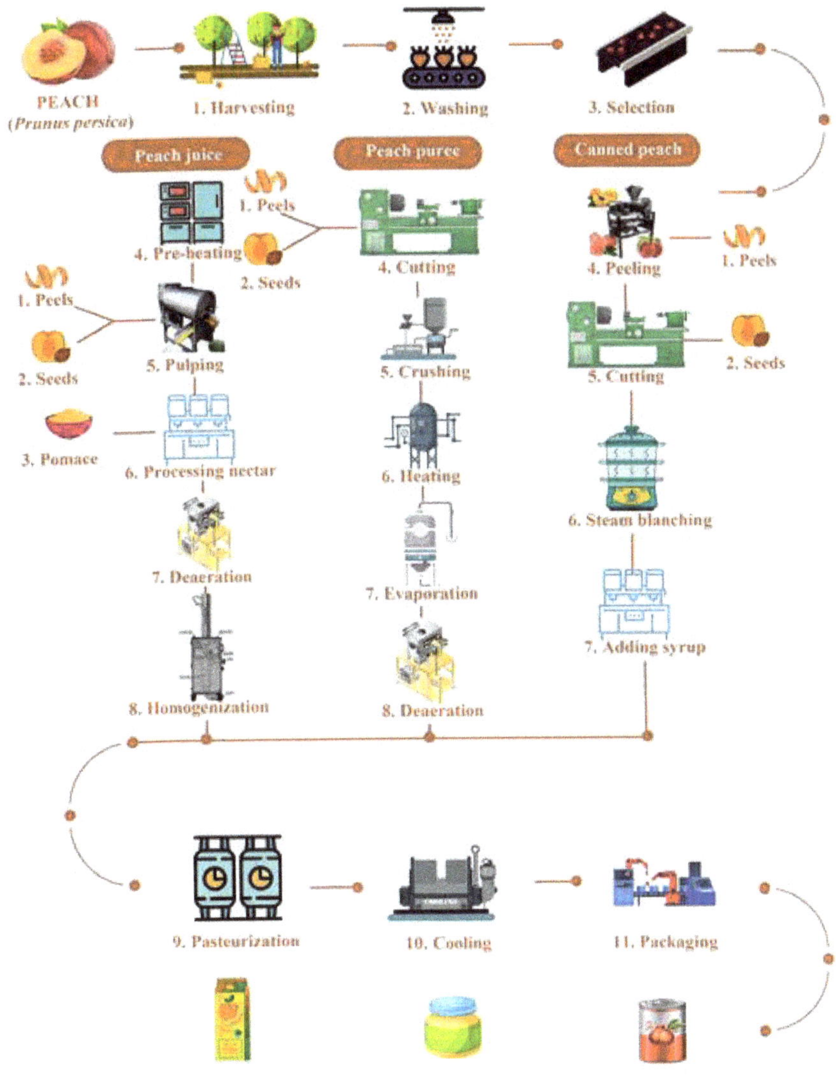

Figure 1. Processing of peach and production of by-products.

The increasing interest in valorization and potential applications of peach by-products is obvious in the increasing trend of publications (Figure 3), where up until August 2023, 211 documents regarding peach by-products have been published. Among these publications, 63% are associated with peach seeds, 27% belong to studies on peach peel, and 10% revolve around peach pomace. Twenty-four of these papers correspond to the extraction of seed oil; twenty-one deal with the valorization of peach by-products by the extraction of phenolic compounds; nine refer to isolation of carotenoids; whereas the extraction of pectin is studied in eight works.

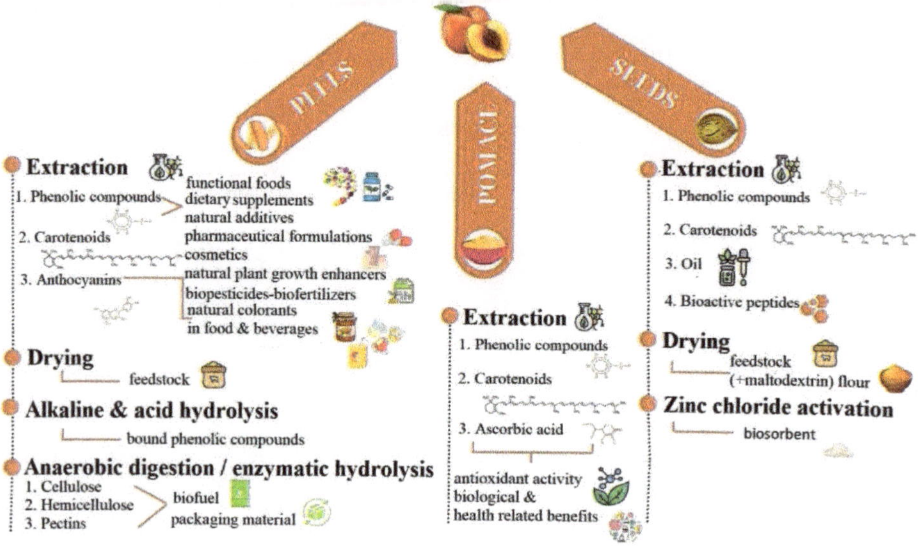

Figure 2. Valorization of peach by-products.

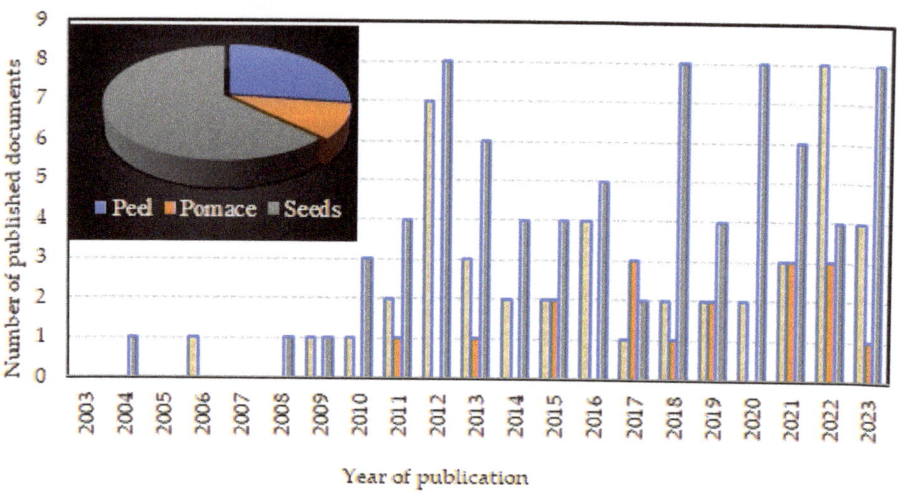

Figure 3. Scopus database results for published articles on peach by-product (last accessed 30 October 2023, query string entered: "peach peel", "peach pomace", "peach seeds").

3. Peach Peels

3.1. Peach Peel Identification

The composition of peach peels has been extensively studied, revealing the presence of a great variety of value-added components. According to the international literature, peach peels are a rich source of valuable compounds such as phenolic compounds, in addition to dietary fibers, pectin, and minerals. The chemical composition of peach peels is presented in Table 1 and varies depending on parameters such as peach variety, growing conditions, harvest maturity stage, and analytical methods used [10].

Table 1. Chemical composition of peach peels.

Component	Content	Reference
Moisture	88.04 ± 0.30%	[11]
Sugars	7.58 ± 0.25%, 9.29–18.96 mg/g dw, 5.42–10.2 g/100 g fw	[11–13]
Total dietary fibers	1.31 ± 0.20%	[11]
Protein	1.14 ± 0.15%	[11]
Fat	0.08%	[11]
Total ash	0.49 ± 0.01%	[11]

dw: dry weight, fw: fresh weight.

Many studies mentioned that fruit peels present several beneficial organic components, which are highly accumulated in the peel compared to the pulp [14]. Additionally, Saidani et al. [12] measured various phenolic compounds in peach peels, including hydroxycinnamic acids (e.g., chlorogenic acid), flavonoids (e.g., catechins, quercetin), and anthocyanins (e.g., cyanidin-3-glucoside) (Table 2). Peach peels exhibit a higher concentration of total phenolic components, ranging from two to three times more than the levels found in the flesh and whole extracts [15]. These compounds have antioxidant properties and are associated with potential health benefits, such as reduced oxidative stress, inflammation, and protection against certain diseases. Furthermore, peach peels are a great source of carotenoids, including β-carotene, zeaxanthin, and lutein [16].

Table 2. Phenolic composition of peach peels.

Phenolic Compound	Content	Reference
	Flavonols	
Kaempferol-3-rhamnoside	65.64–129.32 mg/100 g fw	[17]
Dihydroquercetin-3-glucoside	38.96–130.8 mg/100 g fw	[17]
Dihydroquercetin-3-galactoside	23.09–160.55 mg/100 g fw	[17]
Kaempferol-3-galactoside	14.73–211.08 mg/100 g fw	[17]
3′-Methylquercetin	6.98–12.58 mg/100 g fw	[17]
Quercetin	5.34–12.81 mg/100 g fw	[17]
Isorhamnetin-3-rutinoside	4.61–22.66 mg/100 g fw	[17]
Kaempferol-3-glucuronide	3.76–21.32 mg/100 g fw	[17]
Dihydromyricetin-3-glucoside	3.17–12.68 mg/100 g fw	[17]
Quercetin-3-rutinoside	0.53–50.61 mg/100 g fw	[15,17–19]
Quercetin-3-glucoside	0.52–9.69 mg/100 g fw	[15]
Dihydromyricetin	0.31–1.45 mg/100 g fw	[17]
Dihydrokaempferol	0.28–0.45 mg/100 g fw	[17]
Kaempferol	0.18–2.98 mg/100 g fw	[17]
Quercetin-3-galactoside	0.16–79.11 mg/100 g fw	[12,15,17,18]
Kaempferol-3-glucoside	0.08–78.89 mg/100 g fw	[12,17]
	Anthocyanins	
Cyanidin-3-rutinoside	0.18–6.35 mg/100 g fw	[15,18,19]
Cyanidin-3-glucoside	0.07–32.51 mg/100 g fw	[12,15,18,19]
	Hydroxycinnamic acids	
trans-p-coumaric acid	8.3–18.5 mg/100 g dw	[13]
Coumaric acid	2.9–3.2 mg/100 g fw	[20]
trans-ferulic acid	2.6–13.7 mg/100 g dw	[13]
trans-caffeic acid	2.6–12.8 mg/100 g dw	[13]
trans-sinapic acid	2.2–7.2 mg/100 g dw	[13]
Ferulic acid	1.2 mg/100 g fw	[20]
Caffeoylquinic acid derivative	0.25–0.98 mg/100 g fw	[12]
Chlorogenic acid	0.1–47.05 mg/100 g fw 84.2–355.9 mg/100 g dw	[12,13,15,18–20]
4-caffeoylquinic acid	0.08–0.70 mg/100 g fw	[12]
Neochlorogenic acid	0.03–34.6 mg/100 g fw	[12,15,18–20]
p-coumaroylquinic acid	0.03–0.12 mg/100 g fw	[12]

Table 2. *Cont.*

Phenolic Compound	Content	Reference
	Hydroxybenzoic acids	
Gallic acid	4.47–8.48 mg/100 g fw	[18]
Protocatechuic acid	3.30–27.3 mg/100 g dw	[13]
	Flavan-3-ols	
Catechin 3′,5-diglucoside	2.25–4.32 mg/100 g fw	[17]
Epicatechin	0.64–20.6 mg/100 g fw	[15,17–20]
Catechin	0.12–31.89 mg/100 g fw	[12,15,17–20]
	Flavanones	
Eriodictyol-7-rutinoside	5.20–29.83 mg/100 g fw	[17]
Naringenin-7-glucuronide	3.79–13.01 mg/100 g fw	[17]
Hesperetin	3.43–13.75 mg/100 g fw	[17]
Naringenin-7-glucoside	3.20–32.46 mg/100 g fw	[17]
Hesperetin-7-rutinoside	2.63–55.79 mg/100 g fw	[17]
Naringenin-7-rutinoside	1.32–9.43 mg/100 g fw	[17]
Naringenin	0.56–2.13 mg/100 g fw	[17]
Eriodictyol-7-glucoside	0.45–1.97 mg/100 g fw	[17]
Eriodictyol	0.36–0.51 mg/100 g fw	[17]
Eriodictyol-7-neohesperidoside	0.33–16.95 mg/100 g fw	[17]
	Flavones	
Luteolin-7-glucuronide	8.45–156.89 mg/100 g fw	[17]
Luteolin-7-rutinoside	0.64–20.3 mg/100 g fw	[17]
Luteolin	0.05–2.97 mg/100 g fw	[17]
	Proanthocyanidins	
PAC-B type dimer	119.13–1762.13 mg/100 g fw	[17]
PAC-A type dimer	2.87–7.24 mg/100 g fw	[17]
PAC-B type tetramer	0.44–3.53 mg/100 g fw	[17]
PAC-A type trimer	0.07–0.29 mg/100 g fw	[17]
Procyanidin B1	0.04–49.23 mg/100 g fw	[12,15,19]
Procyanidin B2	0.02–0.10 mg/100 g fw	[12]

dw: dry weight, fw: fresh weight.

Mannino et al. [17] utilized untargeted analysis methodologies (HPLC-DAD-ESI-MS) to identify 37 different phytochemicals in hydroalcoholic extracts of the peel from two peach varieties, as presented in Table 2. In research conducted by Patra and Baek [21], the composition of peach peels was examined, confirming the presence of several components, including chlorogenic acid, epicatechin, cyanidin-3-glycoside, catechin, and rutin. Additionally, in their study, Saidani et al. [12] measured the phenolic content of peach peel and identified several compounds, such as quercetin-3-galactoside, a combination of quercetin-3-O-glucoside and quercetin-3-o-rutinoside, and kaempferol-3-O-glucoside. Hydroxycinnamic acids, such as p-coumaroylquinic acid, 4-caffeoylquinic acid, chlorogenic acid, neochlorogenic acid, and a derivative of caffeoylquinic acid, were also found. Notably, anthocyanin cyanidin-3-O-glucoside was also detected in the analyzed peach peel samples.

According to those findings, the flavonoid content in peach peels varied from 39–245 mg equivalent of catechins per 100 g of fresh weight (fw), while for peach pomace, the range was 8–112 mg equivalent of catechins per 100 g fw, considering nine different peach cultivars. Saidani et al. [12] also identified chlorogenic acid as the predominant hydroxycinnamic acid in peach peel, ranging from 6.74 to 31.2 mg/100 g fw, followed by neochlorogenic acid (1.02–7.98 mg/100 g fw) and anthocyanins (0.24 to 17.6 mg cyanidin-3-glycoside per 100 g fw). These components were found to be more abundant in peach peels compared to peach pulp. Anthocyanins are primarily found in the peel of peaches, similar to flavonols. However, in some cultivars, a small amount of anthocyanin pigment can also be detected in the flesh, specifically in the area surrounding the stone. The concentration of anthocyanins in the peel is generally consistent with the percentage of red color observed on the epidermis [15]. Furthermore, Redondo et al. [20]

mentioned that peach peel has a higher concentration of quercetin (7.1 mg 100/g fw) compared to other fruits such as apricot (6.4 mg 100/g fw) and plum (5.8 mg 100/g fw).

Dabbou et al. [18] examined the effect of peach variety (Early May Crest, Sweet Cap, and O'Henry) and harvesting stage (commercial ripening and full ripening) on the phenolic profile and antioxidant activity of peach by-products. Specifically, according to this research, regardless of the harvesting stage, the peach peel contained higher concentrations of phenolics, including total hydroxycinnamic acids, total anthocyanins, and total flavonols, compared to the peach pulp. Regarding the examined peach varieties, the O'Henry variety had the highest carotenoid content, despite a decrease in the peel during ripening. On the other hand, Sweet Cap exhibited the highest phenol content, which further increased in the peel as the fruit ripened.

Regarding carbohydrates, peach peels accumulate different types of soluble sugars and polyols, such as sucrose, glucose, fructose, and sorbitol [12,13,22], and a percentage of dietary fibers, including cellulose, hemicellulose, and pectins. The study by Lu et al. [23] revealed that carbohydrates accounted for 52.20% of the total content of peach peel flour, with soluble dietary fiber (pectin) comprising 27.30% of the carbohydrate fraction. Dietary fibers are tightly correlated with digestive health and regulation of blood sugar levels, among other health benefits [24].

According to Mihaylova et al. [13], three tricarboxylic acids were quantified in both peach pomace and peel tissue, malic, quinic, and citric acids. These acids contribute to the characteristic tangy flavor of peaches and contribute to the shelf life of the fruits [25].

3.2. *Valorization of Peach Peels*

Over the last few years, peach peels have gained attention due to their potential for valorization. A wide range of techniques can be employed to effectively utilize peach peels, extracting their valuable components, such as phenolic compounds, flavonoids, and other antioxidants, and turning them into valuable products [12]. Extraction techniques like maceration, ultrasound-assisted extraction, and alkaline and acid hydrolysis can be employed to isolate these compounds from the peach peels (Table 3). The extracted compounds can be used as natural antioxidants, food additives, or even as raw materials of nutraceuticals.

Table 3. Extraction of antioxidants from peach peels.

Conditions	Yield	Reference
	Ultrasound-assisted extraction	
80% MeOH, 60 kHz, 30 W, 30 min	TPC: 4.58–12.68 mg GAE/g dw Neochlorogenic acid: 5.77–342.75 mg/kg dw Chlorogenic acid: 52.2–1631.25 mg/kg dw Procyanidin B1: 54.76–539.22 mg/kg dw Catechin: 60.14–1030.06 mg/kg dw Cyanidin-3-glucoside: 9.33–670.59 mg/kg dw Quercetin-3-galactoside: 8.45–396.49 mg/kg dw Quercetin-3-glucoside: 2.45–581.21 mg/kg dw Quercetin-3-rutinoside: 59.15–193.25 mg/kg dw Kaempferol-3-rutinoside: 16.91–110.86 mg/kg dw	[26]
50% EtOH, 42 kHz, 30 min, room temperature	TPC: 8.38–18.81 mg GAE/g dw	[27]
80% EtOH, 50 °C, 30 min (free) 2 M NaOH, 18 h, 30 °C, pH 1.5–2.0, ethyl acetate (bound-alkaline) MeOH/H_2SO_4 (90:10), 70 °C, 24 h, sonication, pH 12.0, ethyl acetate (bound-acid)	TPC: 6.82–13.12 mg GAE/g dw (free) 7–31% of total phenolics (bound) TF: 164.14–515.83 µg QE/g dw (free) TAC: 327.84–1246.77 µg Cy-gluE/g dw (free) 0–49% of total anthocyanins (bound)	[13]

Table 3. *Cont.*

Conditions	Yield	Reference
	Maceration Extraction	
0.05% HCl in methanol, dark	TAC: 1–8 mg Cy-gluE/100 g fw	[28]
80% MeOH, 1 min blending	TPC: 877–1896 mg GAE/kg	[29]
80% MeOH, 8 h, room temperature	TPC: 1209.3–1354.5 mg GAE/100 g dw TF: 599.7–785.5 mg CE/100 g dw	[2]
MeOH/H_2O/formic acid (60:38:2)	TPC: 88.9–277.0 mg GAE/100 g fw TF: 39.3–245 mg CE/100 g fw TAC: 0.55–17.6 mg Cy-gluE/100 g fw	[12]
1 M NaOH, vacuum, 25 °C, 18 h, pH < 2.0	TPC: 0.61–0.91 g/100 g dw	[24]
1% HCl/EtOH, pH 3.0, 60 °C, 1 h (anthocyanins) Acetone + BHT, 24 h, 4 °C (carotenoids, lycopene)	TAC: 0–3.58 g/kg fw Chlorophyll a: 2.34–81.36 g/kg fw Chlorophyll b: 2.94–31.13 g/kg fw Carotenoid: 1.78–19.83 g/kg fw Lycopene: 0.73–1.49 mg/kg fw b-carotenoid: 0.31–10.63 mg/kg fw	[30]
50% EtOH, pH 2.0, 1 h, shaking, 70% acetone, shaking (free) MeOH/H_2SO_4 (90:10), 85 °C, 20 h (bound)	TPC: 79.14–167.10 mg GAE/100 g fw (free) 52.93–84.02 mg GAE/100 g fw (bound)	[31]
Hexane, 20 min, shaking 180 rpm, 0.1% methanolic KOH, 6 °C, 45 min (carotenoids) MeOH/H_2O/formic acid (47.5:47.5:5), 20 min (phenolic compounds)	Cyanidin-3-glucoside: 74–178 mg/100 g dw Chlorogenic acid: 52–136 mg/100 g dw Procyanidin B1: 84–148 mg/100 g dw Procyanidin B3: 80–128 mg/100 g dw Procyanidin B2: 12–41 mg/100 g dw Catechin: 69–106 mg/100 g dw Quercetin-3-glucoside: 8–19 mg/100 g dw Quercetin-3-rutinoside: 8–13 mg/100 g dw Neoxanthin: 10.3–13.6 μg/g dw Zeaxanthin: 10.1–18.7 μg/g dw Lutein: 9.6–15.1 μg/g dw Lutein epoxide: 8.2–20.6 μg/g dw β-carotene: 7.5–16.4 μg/g dw	[16]

dw: dry weight, fw: fresh weight, TPC: total phenolic content, TAC: total anthocyanin content, TF: total flavonoids, GAE: gallic acid equivalents, Cy-gluE: cyanidin-3-glucoside equivalents, CE: catechin equivalents, QE: quercetin equivalents.

3.2.1. Maceration Extraction of Phenolic Compounds

Maceration extraction is a widely used technique for the recovery of phenolic components and carotenoids from peach peels. Regarding phenolic components, a suitable solvent, typically water, alcohol, or a hydroalcoholic mixture, is used to extract the bioactive components present in the peel. The solvent is chosen based on its ability to effectively dissolve and extract phenolics from the plant matrix. Liu et al. [28] determined the anthocyanin content of two peach varieties (Hujingmilu and Yulu) using the pH differential method. Additionally, Chang et al. [29] studied the phenolic content and antioxidant activity of eight clingstone peach cultivars (Andross, Bolinha, Corona, Halford, Kakamas, Ross, Walgant, and breeding line 18-8-23). Specifically, phenolic components of peach peels were extracted using 80% aqueous methanol as solvent, and, consequently, peach extracts were analyzed by HPLC and anthocyanins, hydroxycinnamates, flavonols, and flavan-3-ols were detected (Table 3). Regarding the extraction of peach peel free phenolics, Liu et al. [31] also estimated the total phenolic content of peach peels of four different Chinese commercial cultivars (Hujingmilu, Dahonghua, Fenghuayulu, and Wulingyulu) using two different solvents (ethanol and acetone). The phenolic content in peach peel tissue was found to be

45.5–64.8% higher compared to the flesh, indicating that removing the peel could result in significant nutrient loss.

In their extensive study, Saidani et al. [12] examined nine commercial peach cultivars and qualified hydroxycinnamates, flavanols, and anthocyanins using UPLC. For extracting the aforementioned components, the researchers used a mixture of water, methanol, and formic acid. In particular, a total of 12 phenolic components were identified in the peach peels, which included five hydroxycinnamic acids, three flavan-3-ols, three flavonols, and one anthocyanin (specifically, cyanidin-3-O-glucoside), presented in Table 3. Within the peach cultivars examined, the total polyphenol composition in the peel tissue ranged from 88.9 to 277.0 mg gallic acid equivalents/100 g fw. Similarly, the total flavonoid content varied from 39.3 to 245 mg catechin equivalents/100 g fw, while the level of total anthocyanins (compounds responsible for the red color of peach skin) reached quantities up to 17.6 mg cyanidin-3-glucoside equivalents/100 g fw. On the contrary, de Escalada Pla et al. [24] reported lower concentrations of total polyphenols (0.61–0.91 g/100 g dw).

3.2.2. Ultrasound-Assisted Extraction of Phenolic Compounds

Ultrasound-assisted extraction (UAE) has gained attention as an efficient technique for the recovery of polyphenols from various plant materials, including peach peels. This method utilizes high-frequency sound waves to enhance the extraction process by promoting mass transfer and disrupting cell structures, leading to improved extraction efficiency [32].

Several studies have demonstrated the effectiveness of UAE in extracting phenolic components from peach peels. For example, Zhao et al. [26] found that an extraction time of 30 min at 60 kHz, 30 W using 80% MeOH as solvent resulted in the extraction of total phenolics from peach peels of 17 different Chinese peach cultivars at a concentration up to 12.68 mg gallic acid equivalents/g dw. Furthermore, various antioxidants, such as chlorogenic acid, procyanidin B1, catechin, neochlorogenic acid, cyanidin-3-O-glucoside, quercetin-3-O-glucoside, quercetin-3-O-rutinoside, quercetin-3-O-galactoside, and kaempferol-3-O-rutinoside, were identified, quantified, and are presented in Table 3. Additionally, the researchers compared the phenolic profiles of peach pulp and peels. In general, both tissues contained predominantly chlorogenic acid and catechins. However, the peel tissue exhibited higher levels of phenolic compounds compared to the pulp, whereas flavonols and anthocyanins were primarily detected in peach peels. Mihaylova et al. [13] investigated the recovery of free phenolics from peach peels using UAE of 30 min at 50 °C and 80% EtOH as solvent and quantified total phenolics (6.82–13.12 mg gallic acid equivalents/g dw), anthocyanins (327.84–1246.77 μg cyanidin-3-glucoside/g dw), and flavonoids (164.14–515.83 μg quercetin equivalents/g dw) of peach peels from eight different Bulgarian cultivars.

3.2.3. Alkaline and Acid Hydrolysis for Extraction of Bound Phenolic Compounds

Regarding bound phenolics, alkaline and acid hydrolysis are two commonly used techniques for the recovery of these compounds from different plant matrices. These hydrolysis techniques involve the use of alkaline or acid solutions to break down linkages between phenolic compounds and different macronutrients, facilitating their extraction and recovery. According to the literature, alkaline hydrolysis involves the use of potassium hydroxide (KOH), sodium hydroxide (NaOH), or ammonium hydroxide (NH_4OH) to break down the ester bond linking of phenolic acids to the cell walls and thus is an effective way to release phenolic components from polysaccharides [33]. In acid hydrolysis, acid solutions, such as hydrochloric acid (HCl) or sulfuric acid (H_2SO_4), hydrolyze glycosidic bonds and solubilize sugars and leave ester bonds intact [34]. Chen et al. [35] further support the notion that alkaline hydrolysis is more effective than acid hydrolysis in releasing phenolic components. Their findings align with the idea that alkaline conditions facilitate a more efficient extraction process, resulting in higher phenolic compound yields. In contrast, acid hydrolysis may lead to a higher loss of phenolic components during the extraction process due to the elevated temperatures used, whereas alkaline treatment is performed at room temperature [33].

In the case of peach peels, a distinct number of studies have been conducted regarding the extraction of bound phenolics. For instance, Liu et al. [31] used acid hydrolysis for the recovery of bound phenolics from peach peels. Specifically, an extraction with a mixture of MeOH and H_2SO_4 at a ratio of 90:10, at 85 °C, for 20 h resulted in an extracted phenolic content up to 84.02 mg gallic acid equivalents/100 g fw, a value quite lower than the reported concentration of free phenolics (Table 3). In the case of Mihaylova et al. [13], they applied both acid and alkaline hydrolysis for the extraction of bound phenolics. The findings of the study showed that alkaline hydrolysis was a more effective method for extracting phenolic compounds from peach peels. Moreover, the results indicated that the studied peach peel varieties predominantly contained free phenolics, as the proportion of bound phenolics in the total phenolic content ranged from 7 to 31%.

3.2.4. Exploitation of Peach Peel Extract

Extracts derived from peach peels can be utilized in the development of functional foods, dietary supplements, and natural additives. Incorporating peach peel extracts into food products can enhance their nutritional value and provide additional health benefits [10]. Furthermore, the utilization of peach peels as a source of antioxidants promotes sustainability by reducing waste and maximizing the potential value of this by-product [36]. Phenols in foods have generally demonstrated greater effectiveness in preventing lipid peroxidation compared to many vitamins [27]. These natural antioxidants have been documented to exhibit greater potency, efficiency, and safety compared to synthetic antioxidants [37].

Specifically, the antioxidants extracted from peach peels can be used as natural food additives and preservatives. They can help prolong the shelf life of food products by preventing oxidative degradation and microbial growth. Additionally, the extract can be incorporated into functional foods and beverages to enhance their nutritional value and provide health benefits [14,19,38,39].

Furthermore, peach peels contain carotenoids and anthocyanins, the natural pigments which are responsible for the vibrant colors of the fruit. These pigments can be extracted by conventional methods, such as maceration, or novel techniques, such as ultrasound-assisted extraction, and used as natural colorants in the food and beverage industry. They can also be employed in the production of a variety of food products such as jams, jellies, beverages, and other products, providing an alternative to synthetic colorants [40,41]. It should be noted that according to Kultys and Kurek [40], industrial production of certain carotenoids, including beta-carotene, lutein, lycopene, and zeaxanthin, is carried out on a large scale for their utilization as ingredients in food and supplements. The market for carotenoids is anticipated to witness substantial growth, increasing from USD 1.5 billion in 2019 to USD 2.0 billion by 2026. This growth can be attributed to the rising demand for natural carotenoids as food colorants, along with advancements in carotenoid recovery techniques. It is important to consider the low stability of carotenoids in the presence of oxygen, light, and high temperatures and to take care of all processing conditions. When handling carotenoids in an industrial setting or during food processing, these factors must be carefully managed to ensure the stability of these valuable compounds.

The antioxidant and anti-inflammatory properties of peach peel extracts make them valuable for the development of pharmaceutical and cosmetic formulations. These extracts can be utilized in the production of nutraceuticals, dietary supplements, and pharmaceutical drugs and can also be incorporated into skincare products [19,40,42]. Wadhwa et al. [42] discussed the importance of peach antioxidants as substitutes for synthetic food antioxidants, such as butylated hydroxyanizole (BHA) and butylated hydroxytoluene (BHT), whereas the presence of these phytochemicals exhibits antioxidative, antimicrobial, and immune-modulatory effects. Schilderman et al. [43] reported that high doses of BHT may have toxic effects.

Additionally, antioxidants derived from peach peels can also be employed in agricultural and horticultural applications. Specifically, they can potentially be used as natural plant growth enhancers, biopesticides, and biofertilizers. The extract's antioxidants and antimicrobial properties can help protect plants from oxidative stress and diseases, promoting healthier plant growth and increased crop yield [19,42]. According to Bento et al. [19], many compounds of plant tissues have been proven to exert antimicrobial activity, such as acids, aliphatic alcohols, aldehydes, isoflavonoids, ketones, and terpenes. Inhibition of bacteria has been scientifically proven against *Staphylococcus*, *Bacillus*, *Klebsiella*, and *Escherichia* strains [44,45]. Antifungal activity has also been reported for plant defensins, which are plant-derived proteins, with a small size and a high concentration of cysteine. For example, defensin PpDFN1 has been identified in peaches and has shown antifungal activity against fungi species that commonly affect plant tissues, such as *Monilinia*, *Penicillium*, and *Botrytis* species [46]. Another protective aspect of peach peel extracts is their antiparasitic effect. In particular, such extracts have been tested against helminths and other nematodes that affect humans and the poultry industry [47]. Indeed, the antiparasitic effect was confirmed and compared to commercial drugs that are commonly used, presenting similar results.

3.2.5. Other Uses of Peach Peels

Additionally, peach by-products and especially peach peels are considered to be a great source of dietary fibers, such as cellulose, hemicellulose, and pectins [24,42]. These fibers can be isolated and processed to produce functional dietary fiber ingredients that can be used to enhance the nutritional value and functional properties of various food products, such as baked goods, beverages, and dairy products [48].

Typically, the food industry's by-products can be valued as a feedstock. Peach peels can be dried, ground, and incorporated into animal feed formulations, providing a source of dietary fibers, vitamins, and minerals, while reducing waste and providing a sustainable feed option [40,42].

Another method of peach peel management is its utilization for biofuel production. Through different biological processes (e.g., anaerobic digestion or enzymatic hydrolysis), the carbohydrates present in the peels can be converted into biofuels like biogas or bioethanol. This can contribute to renewable energy generation and reduce the environmental impact of waste disposal. According to Wadhwa et al. [42], the untreated peach peels can be used directly as a substrate for microbial growth or they can undergo enzymatic treatments to enhance their potential for bioenergy production. Specifically, high concentrations of cellulose, pectin, and hemicellulose in these plant tissues can function as an appropriate substrate for fermentation using *S. cerevisae* with encouraging results.

The use of peach peel as packaging material has gained significant attention in recent years due to its potential to address environmental concerns associated with traditional packaging methods. The composition of peach peel, which consists of cellulose, hemicellulose, lignin, and other bioactive compounds, makes it suitable for packaging applications. These natural components provide mechanical strength, barrier properties, and antimicrobial activity, which are essential for preserving and protecting different types of products. A distinct number of studies have been conducted regarding the formation of films from peach peels and their usage as a potential packaging material. Specifically, according to Lu et al. [23], the formed yellow peach skin film shows significant potential for being utilized in the field of oil packaging, exhibits outstanding mechanical properties, and possesses the ability to effectively inhibit oil oxidation, minimizing the peroxide value from 60.32 meq/kg (control sample) to 50.75 meq/kg (film formed with a combination of peach peel, sodium alginate, and glycerol).

4. Peach Seeds

4.1. Peach Seed Identification

The peach fruit yields various valuable by-products, with the endocarp being another one. The endocarp consists of a seed covered by a hard shell, known as the seed shell or kernel shell. On average, the seed constitutes about 6% of the endocarp, while the kernel shell makes up about 94% [5]. The seed's weight accounts for 5 to 12.5% of the entire fruit, depending on the peach species [4,49]. The chemical composition of peach seeds is presented in Table 4.

Table 4. Chemical composition of peach seeds.

Component	Content	Reference
Moisture	4.1–6.9%	[4,24,49,50]
Sugars	12.91–47.44%	
Total dietary fibers	1.8–4.0%	
Protein	2.67–26.77	
Fat	37.69–48.41%	
Total ash	3.36–3.82%	

This by-product is rich in pectin and also in bioactive components such as phenolics and vitamins, posing a challenge for the scientific community to recover and utilize these substances in various industries such as pharmaceuticals, food, and cosmetics [10]. Efforts are being made to enhance the circular economy and food sustainability by recovering bioactive compounds from peach seeds. Notably, the seeds are abundant in phenolic compounds, carotenoids, fatty acids, and protein [13,51]. Researchers identified 18 phenolic compounds using liquid chromatography photodiode array quadrupole time-of-flight mass spectrometry (LC-MS/QTof) analysis. These compounds can be categorized into flavons, flavonols, flavan-3-ols (monomers, dimers, and polymeric procyanidins), and phenolic acids (hydroxycinnamic and hydroxybenzoic acids) (Table 5). Regarding the flavan-3-ols, six compounds were identified, with catechin being the primary monomeric flavan-3-ol, along with minor amounts of epicatechin and its derivatives like epicatechin gallate, epigallocatechin, gallocatechin, and epigallocatechin gallate [52]. Procyanidins were found to be the major class of phenolic compounds, with procyanidin B1 and other minor dimer procyanidins also identified [53]. Polymeric procyanidins in both dimer and trimer forms were detected in peach seeds [4].

Hydroxycinnamic acids comprised the second main group of polyphenolic compounds found in peach seeds, with five identified compounds, primarily caffeoylquinic derivatives, especially chlorogenic acid [4,54]. Other hydroxycinnamic acid derivatives, such as neochlorogenic acid, coumaroylquinic acids, and phenylpropanoid o-diphenol phaselic acid, were also detected [55,56]. Peach kernels also contain protocatechuic acid 4-O-glucoside, 2-hydroxybenzoic acid, 2,3-dihydroxybenzoic acid, ellagic acid acetyl-xyloside, and 3-O-methylgallic acid [55]. Hydroxyphenylpropanoic acids, including 3-hydroxy-(3-hydroxyphenyl) propionic acid, dihydrocaffeic acid 3-O-glucuronide, and 3-hydroxy-3-(3-hydroxyphenyl) propionic acid, were also identified [55]. Flavonols and flavonoids were another group of polyphenols identified in peach seeds, with compounds like quercetin, quercetin 3-galactoside, 3-rutinoside, 3-glucoside, isorhamnetin 3-rutinoside, kaempferol 3-galactoside, hesperidin, and luteolin [53,57]. Some of these compounds were also found in other fruits, while hesperidin-7-rutinoside and luteolin-7-glucoside were not previously detected in peach but are found in other plants [58].

Peach seeds are a source of carotenoids, including β-carotene and xanthophylls like zeaxanthin, violaxanthin, and β-cryptoxanthin. The concentrations of these compounds vary based on factors like peach variety, cultivation region, fruit maturity, and climate [59,60]. Zeaxanthin was the most prevalent carotenoid found in peach seeds, followed by β-carotene, while other compounds were present in trace amounts [4,61].

Table 5. Phenolic composition of peach seeds.

Phenolic Compound	Content	Reference
	Flavonols	
Quercetin-3-O-glucoside	2.87 mg/100 g dw	[4]
Kaempferol-3-O-glucoside	1.98–63.14 mg/100 g dw	[4]
Luteolin-7-glucoside	1.61 mg/100 g dw	[4]
Kaempferol-7-neohesperidoside	0.62 mg/100 g dw	[4]
Hesperidin-7-rutinoside	0.55 mg/100 g dw	[4]
Isorhamnetin-3-O-glucoside	0.53–66.67 mg/100 g dw	[4]
	Hydroxycinnamic Acids	
Neochlorogenic acid	130.07 mg/100 g dw	[4]
Chlorogenic acid	72.92–1727.05 mg/100 g dw	[4]
cis-5-p-coumaroyloquinic acid	21.93–190.8 mg/100 g dw	[4]
2-O-caffeoyl-L-malate	17–130.52 mg/100 g dw	[4]
3-O-p-coumaroyloquinic acid	9.6–70.22 mg/100 g dw	[4]
Gallic acid	2.98 mg/100 g dw	[55]
Caffeic acid	0.98 mg/100 g dw	[55]
	Hydroxybenzoic Acids	
p-hydroxybenzoic acid	18.64 mg/100 g dw	[55]
Ellagic acid	0.77–9.42 mg/100 g dw	[4]
	Flavan-3-ols	
Procyanidin B1	150.65 mg/100 g dw	[4]
Procyanidin B2	28.12 mg/100 g dw	[4]
Epicatechin	18.62–33.74 mg/100 g dw	[4]

dw: dry weight.

Peach stones have a significant lignocellulosic composition, with a protective network mainly comprising lignin, followed by cellulose and hemicellulose [62,63]. The kernel and seed of peach consist of 46% cellulose, 14% hemicellulose, and 33% lignin [5].

Shukla and Kant [64], using conventional Soxhlet extraction with different solvents, found 7.48% crude fat in dry peach seeds. The peach seed oil is of considerable importance in medicine due to its high content of unsaturated fatty acids [65] (Table 6). Oleic acid constitutes 55.2% of the total fatty acids, followed by linoleic acid at 30.8%, while palmitic acid, stearic acid, and α-linolenic acid were also identified [65,66].

Table 6. Fatty acid composition of peach seed oil.

Fatty Acids	Content	Reference
	Unsaturated Fatty Acids	
Oleic acid	55–74%	[67]
Linoleic acid	12–31%	[67]
	Saturated Fatty Acids	
Stearic acid	23.70%	[65]
Palmitic acid	7.97%	[65]
α-linolenic acid	0.11%	[65]

Furthermore, peach seeds are rich in proteins, comprising approximately 40% of the seed's content [8,68]. Protein content in peach seeds was found to be about 29.4% [64]. The proteins in peach seeds include superoxide dismutase, an antioxidant enzyme, as well as 14 bioactive peptides [68].

In conclusion, peach seeds offer a wealth of valuable substances, such as phenolic components, carotenoids, lignocellulosic compounds, proteins, and fatty acids, making them a potential source for various applications in different industries. The recovery and utilization of these compounds present exciting opportunities for enhancing the circular economy and promoting food sustainability.

4.2. Valorization of Peach Seeds

4.2.1. Extraction of Oil

Peach seeds are rich in oil, comprising approximately 48.4% of their composition, which offers significant health and nutritional benefits, primarily due to its high content of oleic and linoleic acids [66,69–71]. Extracting oil from peach seeds can be achieved through both conventional and alternative methods. Conventional methods include hydrodistillation, Soxhlet extraction, and maceration, while an alternative method is supercritical fluid extraction, which is also utilized for extracting oil from apricot kernels and walnuts (Table 7) [71–75].

Table 7. Extraction of oil from peach seeds.

Extraction Method	Conditions	Yield *	Fatty Acid Composition	Reference
Soxhlet	Hexane, 70/80/90 °C	38%	Oleic acid: 74% Linoleic acid: 15%	[76]
Maceration	Hexane/ethanol	22%/17%	Oleic acid: 74% Linoleic acid: 15%	[76]
Supercritical fluid extraction	5% ethanol at 50 °C/300 bar	24%	Oleic acid: 60–65% Linoleic acid: 15–20%	[10,77]
Maceration	130 mL petroleum ether at 65 °C for 2.5 h	30–50%	Oleic acid: 55.2% Linoleic acid: 30.8% Palmitic acid: 7.97% Stearic acid: 2.37% α-linoleic acid: 0.11%	[65]
Soxhlet	n-hexane	46.4 ± 1.3%	Oleic acid: 74.55% Linoleic acid: 16.85%	[67]

* g oil/100 g dry weight.

Soxhlet extraction involves the use of different solvents, dichloromethane, ethanol, n-hexane, and ethyl acetate, with ethanol or dichloromethane yielding the highest extraction rates due to the polarity of the extracted compounds. The solubility is enhanced as these solvents easily penetrate the solid matrix [71,78].

Maceration, another conventional method, yields lower results, likely because of its lower temperature compared to Soxhlet extraction. The high viscosity at lower temperatures hinders solvent penetration into the matrix, leading to decreased extraction efficiency [71,76,79]. Even lower extraction yields are observed with hydrodistillation, a method relying on the use of a polar solvent like water. The high viscosity and surface tension of water limit oil extraction [79].

In contrast, supercritical carbon dioxide offers comparative advantages over conventional methods. This alternative method involves low temperatures and energy consumption, solvent recycling, and the possibility of adjusting solvents, making it a preferred choice [75,80–82]. When extracting peach seed oil using this method, a yield of 23.5% dry basis is achieved with pure CO_2 at 50 °C/300 bar [10]. Researchers have studied various combinations of temperature and pressure to optimize extraction. An increase in temperature at low pressure decreases the extraction yield, likely due to reduced solvent density. Conversely, an increase in pressure leads to higher yields as the vapor pressure is enhanced, outweighing the reduction in solvent density [71]. Additionally, increasing pressure at a constant temperature increases the density of CO_2 and, consequently, the extraction efficiency [81,83,84].

4.2.2. Exploitation of Peach Seed Oil

The peach kernel and the oil extracted from the peach seed are equally important and find applications in various industries, including food, cosmetics, pharmaceuticals, and energy production [71,85–87].

Peach seed oil is an innovative source of bioactive components, such as essential fatty acids, carotenoids, and phenolic compounds [88,89]. Notably, it contains high quantities of linoleic acid, which plays a vital role in cell membrane synthesis and tissue regeneration, making peach seed oil valuable in pharmaceutical and cosmetic industries [87,90]. The abundance of vitamin E, particularly γ-tocopherol, provides the oil with strong antioxidant properties, further increasing its attractiveness to the pharmaceutical and cosmetic sectors [91]. Furthermore, Sodeifian and Sajadian [67] analyzed the total phenolic compounds in peach seed oil and quantified them at 334.5 mg GAE/100 g oil. Using gas chromatography–mass spectrometry (GC–MS), they determined that unsaturated fatty acids comprised 86% of the total content, with oleic and linoleic acids being the primary representatives at 55–74 and 12–31%, respectively [67]. Additionally, the saturated fatty acids palmitic, stearic, and α-linolenic acid were estimated at 7.97, 2.37, and 0.11%, respectively [10,65].

In addition to its applications in pharmaceuticals and cosmetics, peach seed oil has been explored for medical purposes. Studies have shown that the oil enhances blood circulation, reduces blood stasis, and decreases abnormal blood lipid levels, thereby slowing down the progression of atherosclerosis [65,92]. This effect is attributed to the reduction of tissue factor protein levels, limiting the formation of atherosclerotic plaque and the anti-inflammatory and antioxidative activities of the oil [65,93]. Studies have even indicated that peach seed oil has potential benefits in mitigating the effects of cerebral ischemia [94]. The antioxidant properties, derived from the presence of phenolic compounds, further contribute to the potential protection against various human diseases [95].

Furthermore, peach seed oil is utilized as a food supplement due to its bioactive compound content, which provides protection against oxidation [96]. It has been found to be effective in preventing enzymatic browning in fruits and vegetables and inhibiting lipid oxidation and fungal growth [97,98]. The valuable source of polyphenols in peach seed oil contributes to its antioxidant activity and inhibition of enzyme activity [10]. In addition, the high content of unsaturated acids contributes to the oil's antioxidant activity, making it a preferred choice for the food industry [67].

4.2.3. Recovery of Bioactive Components

Both conventional and alternative extraction techniques are employed to obtain bioactive components from peach seeds. However, there is a growing preference for alternative methods due to their lower energy consumption, reduced solvent usage, environmental friendliness, and ability to produce final products of higher quality [10,99]. Various studies have explored different extraction methods to obtain the antioxidants present in peach seeds. Hong et al. [55] used high-pressure liquid chromatography (HPLC) in combination with a photodiode array detector (PDA) and found that peach seeds exhibited a total phenolics content (TPC) of 0.47 (mg gallic acid equivalents (GAE)/g), a total flavonoids content (TFC) of 0.18 (mg quercetin equivalents (QE)/g), and a total tannins content (TTC) of 0.07 (mg catechin equivalents (CE)/g). Additionally, the antioxidant activity measured by the 2,2′-diphenyl-1-picrylhydrazyl antioxidant assay (DPPH) was 0.98 (mg ascorbic acid equivalents (AAE)/g), and the total antioxidant capacity (TAC) was 0.27 (mg AAE/g) [13]. Peach seed was found to have the highest radical-scavenging capacity among all stone fruits [55].

Similarly, Nowicka and Wojdyło [4] confirmed the high antioxidant capacity of peach seeds by examining 20 different peach varieties using untargeted analysis (LC-QTOF-MS/MS). They identified and quantified the phenolic content, with total polyphenols ranging from 3.8 to 12.7 g/100 dry matter and cyanogenic glycoside content varying between 17.4 and 245.7 mg/100 dry matter [10]. The flavan-3-ol dimers, procyanidin B1 and procyanidin B2, were estimated at approximately 150.65 and 28.12 mg/100 g dry matter, respectively. The subsequent group, comprising hydroxycinnamic acids and hydroxybenzoic acids, ranged from 130.94 to 2275.95 mg/100 g [4]. Notable compounds of this group include chlorogenic acid, neochlorogenic acid, and ferulic acid [20]. The high polyphenol content was confirmed by the FRAP method, which indicated the highest antioxidant capacity of the peach seeds (3.3 mmol Trolox/100 fw) as compared to the peach peel (2.2 mmol Trolox/100 fw) and pulp (0.2 mmol Trolox/100 fw) [10].

Peach seeds were found to contain significant amounts of carotenoids, including β-carotene, xanthophylls (mono- or dihydroxylated carotenoids), zeaxanthin, violaxanthin, and β-cryptoxanthin [100]. The total carotenoids were measured at 109.3 mg β-carotene equivalents/100 g, with β-carotene and β-cryptoxanthin estimated at 7.8 μg/g and 1.01 μg/g, respectively [4]. This high carotenoid content contributes to the pharmaceutical industry's interest in peach seeds due to their potent antioxidant activity [100].

Finally, peach seeds were found to contain a protein concentration of 29.36% (dry basis) [64]. Combinatorial peptide ligand library (CPLL) technology was used to detect 97 unique genetic products from peach seeds, with 1 identified protein specifically related to peach seeds and 14 bioactive peptides [8,68,101]. The presence of antioxidant and antihypertensive peptides in peach seeds makes them a valuable source of bioactive compounds for food applications [68].

4.2.4. Other Uses of Peach Seeds

The peach kernel offers versatile utilization both as a single by-product and separately as the kernel and seed. In the former case, the entire endocarp is either naturally decomposed in landfills or subjected to drying, combustion, pyrolysis, or gasification to produce energy [102–104]. However, peach seeds can serve various beneficial purposes. For instance, Redondo et al. [20] highlighted the potential use of peach seeds as animal feed due to their antioxidant activity and polyphenol content.

Another application was proposed by Qiu et al. [105], who studied the sugar yields of peach seeds with the use of deep eutectic solvent (DES) as a biomass pretreatment method. The DES treatment significantly increased glucose yields by approximately 90% and facilitated the extraction of lignin, with 70.2% of lignin being obtained from peach seeds [10]. This biomass could be utilized in the conversion of biofuels and chemicals, offering a sustainable approach.

Uysal et al. [5] demonstrated the production of biosorbent from peach seeds through the creation of activated carbon with zinc chloride activation. The process involved bio-oil production at different temperatures (300 and 400 °C) followed by activation through precarbonization and zinc chloride impregnation at temperatures ranging from 500 to 700 °C. The resulting activated carbon exhibited excellent adsorption capacities, particularly for phenol and methylene blue, with values ranging from 51.6 to 64.9 mg/g and 104.2 to 121.9 mg/g, respectively. Moreover, the peach seed powder was proved to be a potential adsorbent for removing Acid Blue 25 (AB25), a common basic dye, from aqueous solutions, with an adsorption time of 120 min [106].

Finally, peach seeds can serve as a source of nutrients due to their centesimal composition and specific characteristics [107]. Efforts have been made to produce flour from peach seeds. Pelentir et al. [108] worked on the addition of maltodextrin in the drying process of flour, resulting in a final product with high contents of oleic and linoleic acids (around 50% each), which are relatively scarce in vegetable oils. Additionally, the flour contains varying proportions of starch and protein [107,108]. Although this flour holds promise as an innovative product, further toxicological studies are required before considering its integration into the human diet [107].

5. Peach Pomace

5.1. Peach Pomace Identification

Peach pomace represents a large portion of the by-product generated during peach juice processing (ca. 24% of fruit weight). The chemical composition of peach pulp is presented in Table 8. It contains a variety of phytochemicals, the concentration of which depends on many factors, such as peach maturity, horticultural practices, genotype, postharvest storage conditions, geographic origin, and processing procedure [109]. It is rich in various bioactive components, such as polyphenols, carotenoids, vitamins, minerals, and amino acids, which are linked with promotion of health (Table 9).

Table 8. Chemical composition of peach pomace.

Component	Content	Reference
Moisture	65.84–84.76%	[110,111]
Sugars	12.14–26.38% 10.8–15.7 g/100 g fw	[12,110,111]
Total dietary fibers	1.78%	[110]
Protein	0.68%	[110]
Fat	0.21%	[110]
Total ash	0.43–0.56%	[110,111]

fw: fresh weight.

The extraction technique and the kind of peach have an immediate impact on the phenolic concentration. According to Vizzotto et al.'s [112] research, red-fleshed peaches had a higher phenolic content than light-fleshed peaches. The ethanolic pomace extract was found to be richer in phenolics than the methanolic extract [113], whereas Loizzo et al. [110] concluded that peach pomace is characterized by a higher phenolic concentration than peach peels and seeds. However, Liu et al. [109] and Saidani et al. [12] reported that the phenolic content of peach peels is substantially higher than that of peach pomace. According to Vizzotto et al. [112], the peach cultivars with the highest phenolic content have a bitter flavor. During fruit development and ripening, the phenolic composition of pomace was found to be reduced [54]. Additionally, Liu et al. [109] found that both the pulp and peel of late-maturing varieties had higher total phenolic contents than early-maturing types.

Table 9. Antioxidant compounds in peach pomace.

Component	Content	Reference
Total phenolics	105.1 ± 1.21 mg GAE/g extract 3.62–19.4 mg GAE/100 g fw 24.83–86.33 mg of GAE/100 g fw 3.5–4.5 mg/g dw 711.7–881.3 mg GAE/100 g dw	[113] [12] [109] [54] [2]
Phenols	921.8 ± 2.5 mg CGA/100 g fw 461 ± 308 mg CGA/100 g fw	[110,112]

Table 9. *Cont.*

Component	Content	Reference
Flavonoids	726.5 ± 8.2 mg QCT/100 g fw 17.76 ± 130.17 mg RE/100 g fw 301.3–499.7 mg CE/100 g	[110] [109] [2]
Anthocyanins	148.7 ± 83 mg C3G/100 g fw	[112]
Flavan-3-ols	116–214 mg/100 g 0.05–1.89 mg/g dw	[54,114]
Hydroxycinnamic acids	103–303 mg/kg	[114]
Chlorogenic acid	15.029 ± 1.3 mg/kg extract 0.12–1.82 mg/g dw 3.58–14.22 mg/100 g fw	[109] [54] [109]
Neochlorogenic acid	2.13–12.14 mg/100 g fw	[109]
Total carotenoids	13.79 ± 2.45 μg/g fw 61.9 ± 1.8 mg β-catotene/100 g fw 2.8 ± 0.9 mg β-catotene/100 g fw	[115] [110] [112]
β-carotene	5.07–28.9 μg/g dw	[115]
β-cryptoxanthin	2.19–88.05 μg/g dw	[115]
Zeaxanthin	1.33–19.08 μg/g dw	[115]
Lutein	0.83–10.8 μg/g dw	[115]
(E/Z)-phytoene	0.41–8.8 μg/g dw	[115]
Ascorbic acid	4.15–14.2 mg/100 g fw 2.48–5.54 mg/100 g fw	[12] [109]

fw: fresh weight; dw: dry weight; GAE: gallic acid equivalents; QCT: quercetin equivalents; RE: rutin equivalents; CE: catechin equivalents; CGA: chlorogenic acid equivalents; C3G: cyanidin-3-O-glucoside equivalents.

Peach pomace contains chlorogenic acid, rutin, cyanidin-3-glucoside, catechin, epicatechin, neochlorogenic acid, flavan-3-ol, procyanidins, quinic acid, fumaric acid, protocatechuic acid, nicotiflorin, isoquercitrin, quercetin, astragalin, hesperidin, and amentoflavone. The main phenolic ingredient in peach pomace extract is chlorogenic acid, a common hydroxycinnamic acid. According to Zuo et al. [116], chlorogenic acid has a variety of health advantages, including an antidiabetic effect, DNA protection effect, neuroprotective effect, and inhibitory activity against hepatitis B virus. Geduk and Atsız [113] noted that the ethanolic pomace extract had a substantially higher chlorogenic acid content than the methanolic extract. Neochlorogenic acid is another hydroxycinnamic acid present in peach pomace, with a concentration that is noticeably lower than that of chlorogenic acid [12,54,109,113].

The pulp contains a significant number of flavonoids in total. Flavonoids exhibit a wide range of biological activities, including leukocyte movement, antibacterial and anti-inflammatory effects, and glucose metabolism. According to Gutiérrez et al. [117], these substances are linked to favorable effects on coronary heart disease, hypertension, insulin resistance, glucose, and lipid metabolism. All of the peach cultivars studied by Liu et al. [110] were found to possess catechins in their pomace; however, the peels contain considerably more catechins than the pomace. According to Vizzotto et al. [112], cyanidin 3-rutinoside and cyanidin 3-glucoside are the two primary anthocyanins found in peach pomace. On the contrary, Andreotti et al. [54] reported low concentrations of these compounds in the peach cultivars tested, with the exception of a white cultivar with high levels of cyanidin-3-glucoside.

Carotenoids play a pivotal role in human nutrition and peach pomace stands out as a significant source of these natural pigments. The consumption of carotenoids has been linked to a reduced risk of various degenerative and chronic diseases, including certain types of cancer [118]. In peach pomace, several noteworthy carotenoids have been identified, including β-carotene, lutein, β-cryptoxanthin, phytoene, and zeaxanthin [110,112,115,119]. The composition of carotenoids in peach pulp undergoes significant changes during fruit maturation. According to Wu et al. [115], lutein and phytoene are the dominant carotenoids in most peach cultivars during the immature and mature stages, respectively. Loizzo et al. [110] reported that Tabacchiera peach pulp is rich in β-carotene, β-cryptoxanthin, and lutein. A similar trend was reported by Gil et al. [60] for peach pulp cultivars from California. The carotenoid content in peaches is also influenced by fruit flesh color, with yellow-flesh peaches generally containing higher levels of carotenoids compared to their light-colored counterparts. Genetic mutations have been identified as some of the key factors contributing to variations in carotenoid content among different peach cultivars [60,112,115,119]. Cao et al. [119] highlighted the role of carotenoid cleavage dioxygenase (PpCCD4) in regulating carotenoid degradation in white peaches. In contrast, yellow-pulp peach fruit exhibits a mutation in the CCD4 gene that impedes carotenoid breakdown, resulting in higher carotenoid accumulation in the pulp [120].

Peach pomace is a reservoir of sugars, surpassing the sugar content found in the peel fraction. The primary sugar detected in peach pulp is sucrose, while other sugars like fructose, glucose, and sorbitol are present in lower concentrations [12,114,121]. The precise sugar composition plays a pivotal role in determining the sweetness intensity of peach pomace. Notably, sorbitol (2.98 g/100 g pomace) plays a central role in shaping the peach aroma [122]. Moreover, according to Saidani et al. [12], sucrose (8.64–11.5 g/100 g pomace) is closely linked to the sweetness of the Big Top cultivar.

As far as carboxylic acids are concerned, malic, quinic, and citric acids dominate in peach pomace, with concentrations noticeably higher compared to those of the peel [12,114,121]. The composition of organic acids in peach pomace presents significant alterations during fruit maturation. Saidani et al. [12] observed that in mature peaches, the content of malic (0.40–1.03 g/100 g pomace) and quinic (0.11–0.27 g/100 g pomace) acids decreases in comparison to immature peaches. Conversely, regarding citric acid, the highest concentration (0.41 g/100 g pomace) was noted at intermediate maturities.

Regarding amino acids, the peach pomace contains asparagine, aspartic acid, glutamic acid, proline, and alanine. Additionally, glutamine, serine, and threonine have also been identified in peach pomace [121,123]. The deficiency of amino acids in the diet can lead to reduced protein production and consequent nutritional imbalances. Of particular note is asparagine, which has garnered attention for its potential health benefits, including the regulation of blood pressure, bronchitis management, antipeptic ulcer properties, gastric function enhancement, immune system regulation, infection prevention, and increased insulin secretion [123,124]. Aspartic acid plays an active role in reducing blood nitrogen and carbon dioxide levels, while enhancing liver function [125]. Glutamic acid and proline are associated with health benefits such as reducing blood ammonia levels and treating gastrointestinal diseases and scalds, respectively [126,127]. Alanine contributes to maintaining appropriate blood glucose levels and toxin removal, providing essential nutrients to the body [125,128]. Furthermore, as noted by Yu and Yang [125], aspartic acid and glutamic acid are responsible for the umami taste sensation, while proline, alanine, and serine contribute to the perception of sweetness.

Peach pomace stands out as a valuable source of ascorbic acid, which has been extensively investigated for its diverse biological and health-related benefits, primarily its potent antioxidant properties [109,129]. Notably, various studies, including those by Gil et al. [60], Saidani et al. [12], and Liu et al. [109], have identified significant variations in ascorbic acid content among different peach cultivars. Interestingly, while peach pomace is known for its ascorbic acid content, it is worth noting that peach peel contains approximately 1.5–2 times more ascorbic acid than the pomace [109].

Turning to minerals, potassium reigns as the most abundant mineral in peach pomace, with calcium, magnesium, manganese, and iron also present in measurable quantities. These minerals are integral components of essential nutrients in the human diet, with potassium playing a crucial role in maintaining cellular organization and permeability.

However, it is worth mentioning that these compounds are generally more concentrated in the peel compared to the pomace [2,12].

The impressive antioxidant potential of peach pomace is often linked to the cumulative levels of total phenolics, carotenoids, and ascorbic acid within it. These phytochemicals function as antioxidants by inhibiting oxidation processes, acting as free radical scavengers and metal chelators, and influencing cell signaling pathways and gene expression [110]. Loizzo et al. [110] specifically noted a significant correlation between carotenoids and the total antioxidant capacity of peach pomace. However, Gil et al. [60] reported that white-pulp peach cultivars exhibited stronger antioxidant activity than their yellow-pulp counterparts, even though white-pulp cultivars had a lower total carotenoid content. Additionally, both Gil et al. [60] and Vizzotto et al. [112] found that the correlation between total phenolics and antioxidant activity was stronger than that of ascorbic acid and carotenoids. Liu et al. [109] established a high correlation between total phenolics, total flavonoids, ascorbic acid content, and antioxidant activity in peach pomace. Saidani et al. [12] similarly discovered a significant positive relationship between total phenolics, total flavonoids, and antioxidant activity. Moreover, Ding et al. [130] emphasized that neochlorogenic and chlorogenic acids contributed notably to the antioxidant activity of peach pomace, surpassing the influence of other phenolic compounds. Finally, while Vizzotto et al. [112] and Manzoor et al. [2] observed a strong correlation between the total phenolic content of pulp and antioxidant activity, Saidani et al. [12] and Vizzotto et al. [112] suggested a slightly lesser contribution of anthocyanins to the antioxidant activity of pulp.

5.2. Valorization of Peach Pomace

While peach pomace is known to contain various bioactive components, research into the effective extraction of these phytochemicals has been somewhat limited. Mokrani and Madani [131] worked on the extraction of phenolic compounds from whole peach fruit, investigating the impact of various parameters, such as time, temperature, solvent type, acetone concentration, and solvent acidity, on the extraction yield. They determined that the optimal extraction conditions entailed using 60% acetone without acidification, with an extraction duration of 180 min at a temperature of 25 °C. In a different approach, Tsiaka et al. [132] focused on the extraction of phenolic components from peach skin and pomace using ultrasound- and microwave-assisted techniques. The ultrasound-assisted extraction was optimized at an extraction time of 15 min, a pulse duration/pulse interval ratio of 8/5, and a solvent/solid ratio of 35/1 mL/g. Conversely, the microwave-assisted extraction exhibited optimal performance with an extraction time of 20 min, an extraction temperature of 58 °C, and a solvent/solid ratio of 16/1 mL/g. Vargas et al. [133] worked on the extraction of peach pomace and peels through stirring at room temperature and concluded that the most effective recovery of carotenoids was achieved after four consecutive extractions, each lasting 10 min, employing 38.5 mL of ethanol.

6. Conclusions

In conclusion, this review has provided a comprehensive overview of the valorization potential of peach by-products, shedding light on their untapped value in various industries and sectors. The abundance of peach by-products generated by the global peach processing industry presents a unique opportunity for sustainable resource utilization, waste reduction, and economic growth.

Through the examination of diverse valorization strategies, it is evident that peach by-products can be transformed into high-value products such as bioactive compounds, functional foods, natural colorants, and biofuels. These applications not only contribute to reducing environmental burdens associated with waste disposal, but also have the potential to generate additional revenue streams for peach growers and processors. Furthermore, the nutritional and health-promoting properties of peach by-products underscore their potential in the development of functional foods and nutraceuticals, aligning with the growing consumer demand for natural and health-enhancing products.

While significant progress has been made in exploring the valorization pathways of peach by-products, there are still challenges that need to be addressed, including optimizing extraction and conversion processes, scaling up production, ensuring product safety and quality, and developing effective marketing strategies.

In the context of a circular and sustainable economy, the valorization of peach by-products represents a promising avenue for reducing waste, conserving resources, and promoting economic growth. Future research should focus on enhancing the efficiency and sustainability of valorization processes, while considering the broader environmental and socioeconomic impacts. Collaboration between researchers, industry stakeholders, and policymakers will be crucial in unlocking the full potential of peach by-products and advancing the concepts of circular agriculture and bioeconomy. Overall, the valorization of peach by-products holds great promise and should continue to be a subject of significant interest and investigation in the years to come.

Author Contributions: Conceptualization, N.S., I.M. and A.M.G.; writing—original draft preparation, N.S., A.M.D. and K.K.; writing—review and editing, I.M. and A.M.G.; supervision, A.M.G. All authors have read and agreed to the published version of the manuscript.

Funding: This research was funded by the Hellenic Foundation for Research and Innovation (HFRI) in the frame of the "2nd Call for H.F.R.I.'s Research Projects to Support Faculty Members & Re-searchers" under grant agreement No. 3308.

Institutional Review Board Statement: Not applicable.

Informed Consent Statement: Not applicable.

Data Availability Statement: Not applicable.

Conflicts of Interest: The authors declare no conflicts of interest.

References

1. Food and Agricultural Organization of the United Nations (FAOSTAT) Website. 2020. Available online: https://www.fao.org/faostat/en/#data/QCL (accessed on 25 July 2023).
2. Manzoor, M.; Anwar, F.; Mahmood, Z.; Rashid, U.; Ashraf, M. Variation in minerals, phenolics and antioxidant activity of peel and pulp of different varieties of peach (*Prunus persica* L.) fruit from Pakistan. *Molecules* **2012**, *17*, 6491–6506. [CrossRef] [PubMed]
3. Featherstone, S. *A Complete Course in Canning and Related Processes: Volume 1 Fundemental Information on Canning*; Woodhead Publishing: Cambridge, UK, 2015.
4. Nowicka, P.; Wojdyło, A. Content of bioactive compounds in the peach kernels and their antioxidant, anti-hyperglycemic, anti-aging properties. *Eur. Food Res. Technol.* **2019**, *245*, 1123–1136. [CrossRef]
5. Uysal, T.; Duman, G.; Onal, Y.; Yasa, I.; Yanik, J. Production of activated carbon and fungicidal oil from peach stone by two-stage process. *J. Anal. Appl. Pyrolysis* **2014**, *108*, 47–55. [CrossRef]
6. Kamenidou, I.; Tzimitra-Kalogianni, I.; Zotos, Y.; Mattas, K. Household purchasing and consumption behaviour towards processed peach products. *New Medit* **2002**, *1*, 45–49.
7. Plazzotta, S.; Ibarz, R.; Manzocco, L.; Martín-Belloso, O. Optimizing the antioxidant biocompound recovery from peach waste extraction assisted by ultrasounds or microwaves. *Ultrason. Sonochem.* **2020**, *63*, 104954. [CrossRef] [PubMed]
8. Vásquez-Villanueva, R.; Marina, M.L.; García, M.C. Revalorization of a peach (*Prunus persica* (L.) Batsch) byproduct: Extraction and characterization of ACE-inhibitory peptides from peach stones. *J. Funct. Foods* **2015**, *18*, 137–146. [CrossRef]
9. Wu, H.; Shi, J.; Xue, S.; Kakuda, Y.; Wang, D.; Jiang, Y.; Ye, X.; Li, Y.; Subramanian, J. Essential oil extracted from peach (*Prunus persica*) kernel and its physicochemical and antioxidant properties. *LWT-Food Sci. Technol.* **2011**, *44*, 2032–2039. [CrossRef]
10. Rudke, C.R.M.; Zielinski, A.A.F.; Ferreira, S.R.S. From biorefinery to food product design: Peach (*Prunus persica*) by-products deserve attention. *Food Bioproc. Technol.* **2023**, *16*, 1197–1215. [CrossRef]
11. Zhivkova, V. Evaluation of Nutritional and Mineral Composition of Apricot, Peach and Nectarine Wasted Peels. *Calitatea* **2020**, *21*, 144–146.
12. Saidani, F.; Giménez, R.; Aubert, C.; Chalot, G.; Betrán, J.A.; Gogorcena, Y. Phenolic, sugar and acid profiles and the antioxidant composition in the peel and pulp of peach fruits. *J. Food Compos. Anal.* **2017**, *62*, 126–133. [CrossRef]
13. Mihaylova, D.; Popova, A.; Desseva, I.; Dincheva, I.; Tumbarski, Y. Valorization of Peels of Eight Peach Varieties: GC–MS Profile, Free and Bound Phenolics and Corresponding Biological Activities. *Antioxidants* **2023**, *12*, 205. [CrossRef] [PubMed]
14. Michailidis, M.; Karagiannis, E.; Nasiopoulou, E.; Skodra, C.; Molassiotis, A.; Tanou, G. Peach, apple, and pear fruit quality: To peel or not to peel? *Horticulturae* **2021**, *7*, 85. [CrossRef]

15. Bassi, D.; Mignani, I.; Spinardi, A.; Tura, D. Peach (*Prunus persica* (L.) Batsch). In *Nutritional Composition of Fruit Cultivars*; Simmonds, M.S.J., Preedy, V.R., Eds.; Academic Press: Cambridge, MA, USA, 2016; pp. 535–571. [CrossRef]
16. Brown, A.F.; Yousef, G.G.; Guzman, I.; Chebrolu, K.K.; Werner, D.J.; Parker, M.; Gasic, K.; Perkins-Veazie, P. Variation of carotenoids and polyphenolics in peach and implications on breeding for modified phytochemical profiles. *J. Am. Soc. Hortic. Sci.* **2014**, *139*, 676–686. [CrossRef]
17. Mannino, G.; Ricciardi, M.; Gatti, N.; Serio, G.; Vigliante, I.; Contartese, V.; Gentile, C.; Bertea, C.M. Changes in the Phytochemical Profile and Antioxidant Properties of *Prunus persica* Fruits after the Application of a Commercial Biostimulant Based on Seaweed and Yeast Extract. *Int. J. Mol. Sci.* **2022**, *23*, 15911. [CrossRef] [PubMed]
18. Dabbou, S.; Maatallah, S.; Castagna, A.; Guizani, M.; Sghaeir, W.; Hajlaoui, H.; Ranieri, A. Carotenoids, phenolic profile, mineral content and antioxidant properties in flesh and peel of *Prunus persica* fruits during two maturation stages. *Plant Foods Hum. Nutr.* **2017**, *72*, 103–110. [CrossRef] [PubMed]
19. Bento, C.; Goncalves, A.C.; Silva, B.; Silva, L.R. Peach (*Prunus persica*): Phytochemicals and health benefits. *Food Rev. Int.* **2022**, *38*, 1703–1734. [CrossRef]
20. Redondo, D.; Gimeno, D.; Calvo, H.; Venturini, M.E.; Oria, R.; Arias, E. Antioxidant activity and phenol content in different tissues of stone fruits at thinning and at commercial maturity stages. *Waste Biomass Valorization* **2021**, *12*, 1861–1875. [CrossRef]
21. Patra, J.K.; Baek, K.H. Comparative study of proteasome inhibitory, synergistic antibacterial, synergistic anticandidal, and antioxidant activities of gold nanoparticles biosynthesized using fruit waste materials. *Int. J. Nanomed.* **2016**, *11*, 4691–4705. [CrossRef]
22. Petruccelli, R.; Bonetti, A.; Ciaccheri, L.; Ieri, F.; Ganino, T.; Faraloni, C. Evaluation of the fruit quality and phytochemical compounds in peach and nectarine cultivars. *Plants* **2023**, *12*, 1618. [CrossRef]
23. Lu, X.; Chen, Z.; Ma, Q.; Mu, J.; Li, X.; Liu, H. Preparation and characterization of yellow peach peel/sodium alginate/glycerol antioxidant film applicable for oil package. *Polymers* **2022**, *14*, 1693. [CrossRef]
24. de Escalada Pla, M.F.; González, P.; Sette, P.; Portillo, F.; Rojas, A.M.; Gerschenson, L.N. Effect of processing on physico-chemical characteristics of dietary fibre concentrates obtained from peach (*Prunus persica* L.) peel and pulp. *Food Res. Int.* **2012**, *49*, 184–192. [CrossRef]
25. Colaric, M.; Veberic, R.; Stampar, F.; Hudina, M. Evaluation of peach and nectarine fruit quality and correlations between sensory and chemical attributes. *J. Sci. Food Agric.* **2005**, *85*, 2611–2616. [CrossRef]
26. Zhao, X.; Zhang, W.; Yin, X.; Su, M.; Sun, C.; Li, X.; Chen, K. Phenolic composition and antioxidant properties of different peach [*Prunus persica* (L.) Batsch] cultivars in China. *Int. J. Mol. Sci.* **2015**, *16*, 5762–5778. [CrossRef]
27. Zhang, Y. Antioxidant effect of peach skin extracts from 13 varieties of South Carolina grown peaches. Ph.D. Dissertation, Clemson University, Clemson, SC, USA, 2014.
28. Liu, T.; Song, S.; Yuan, Y.; Wu, D.; Chen, M.; Sun, Q.; Zhang, B.; Xu, C.; Chen, K. Improved peach peel color development by fruit bagging. Enhanced expression of anthocyanin biosynthetic and regulatory genes using white non-woven polypropylene as replacement for yellow paper. *Sci. Hortic.* **2015**, *184*, 142–148. [CrossRef]
29. Chang, S.; Tan, C.; Frankel, E.N.; Barrett, D.M. Low-density lipoprotein antioxidant activity of phenolic compounds and polyphenol oxidase activity in selected clingstone peach cultivars. *J. Agric. Food Chem.* **2000**, *48*, 147–151. [CrossRef] [PubMed]
30. Zhou, H.; Yu, Z.; Ye, Z. Key proteins associated to coloured compounds of peach peel using iTRAQ proteomic techniques during development and postharvest. *Sci. Hortic.* **2018**, *239*, 123–132. [CrossRef]
31. Liu, H.; Jiang, W.; Cao, J.; Ma, L. Evaluation of antioxidant properties of extractable and nonextractable polyphenols in peel and flesh tissue of different peach varieties. *J. Food Proc. Preserv.* **2018**, *42*, e13624. [CrossRef]
32. Yusoff, I.M.; Taher, Z.M.; Rahmat, Z.; Chua, L.S. A review of ultrasound-assisted extraction for plant bioactive compounds: Phenolics, flavonoids, thymols, saponins and proteins. *Food Res. Int.* **2022**, *157*, 111268. [CrossRef]
33. Acosta-Estrada, B.A.; Gutiérrez-Uribe, J.A.; Serna-Saldívar, S.O. Bound phenolics in foods, a review. *Food Chem.* **2014**, *152*, 46–55. [CrossRef] [PubMed]
34. Verma, B.; Hucl, P.; Chibbar, R.N. Phenolic acid composition and antioxidant capacity of acid and alkali hydrolysed wheat bran fractions. *Food Chem.* **2009**, *116*, 947–954. [CrossRef]
35. Chen, G.L.; Zhang, X.; Chen, S.G.; Han, M.D.; Gao, Y.Q. Antioxidant activities and contents of free, esterified and insoluble-bound phenolics in 14 subtropical fruit leaves collected from the south of China. *J. Funct. Foods* **2017**, *30*, 290–302. [CrossRef]
36. Layne, D.; Bassi, D. *The Peach: Botany, Production and Uses*; CABI: Oxon, UK, 2008.
37. Pokorny, J.; Yanishlieva, N.; Gordon, M. Antioxidants in Food: Practical Application. In *The Development of Oxidative Rancidity in Foods*; Gordon, M., Ed.; Woodhead Publishing Limited: Orlando, FL, USA, 2001; p. 11.
38. Rossato, S.B.; Haas, C.; Raseira, M.D.C.B.; Moreira, J.C.F.; Zuanazzi, J.A.S. Antioxidant potential of peels and fleshes of peaches from different cultivars. *J. Med. Food* **2009**, *12*, 1119–1126. [CrossRef] [PubMed]
39. Kan, J.; Chen, C.; Huo, T.; Xie, W.; Hui, Y.; Liu, J.; Jin, C. Polyphenolic-enriched peach peels extract regulates lipid metabolism and improves the gut microbiota composition in high fat diet-fed mice. *J. Funct. Foods* **2020**, *72*, 104082. [CrossRef]
40. Kultys, E.; Kurek, M.A. Green extraction of carotenoids from fruit and vegetable byproducts: A review. *Molecules* **2022**, *27*, 518. [CrossRef]
41. Zhao, B.; Sun, M.; Li, J.; Su, Z.; Cai, Z.; Shen, Z.; Ma, R.; Yan, J.; Yu, M. Carotenoid profiling of yellow-flesh peach fruit. *Foods* **2022**, *11*, 1669. [CrossRef] [PubMed]

42. Wadhwa, M.; Bakshi, M.P.S.; Makkar, H.P.S. Wastes to worth: Value added products from fruit and vegetable wastes. *CABI Rev.* **2016**, *2015*, 1–25. [CrossRef]
43. Schilderman, P.A.E.L.; Ten Vaarwerk, F.J.; Lutgerink, J.T.; Van der Wurff, A.; Ten Hoor, F.; Kleinjans, J.C.S. Induction of oxidative DNA damage and early lesions in rat gastro-intestinal epithelium in relation to prostaglandin H synthase-mediated metabolism of butylated hydroxyanisole. *Food Chem. Toxicol.* **1995**, *33*, 99–109. [CrossRef]
44. Raturi, R.; Singh, H.; Bahuguna, P.; Sati, S.C.; Badoni, P.P. Antibacterial and antioxidant activity of methanolic extract of bark of *Prunus persica*. *J. Appl. Nat. Sci.* **2011**, *3*, 312–314. [CrossRef]
45. Yao, X.C.; Cao, Y.; Wu, S.J. Antioxidant activity and antibacterial activity of peach gum derived oligosaccharides. *Int. J. Biol. Macromol.* **2013**, *62*, 1–3. [CrossRef]
46. Nanni, V. Antimicrobial Activity of Peach and Grapevine Defensins. Ph.D. Dissertation, Università di Bologna, Bologna, Italy, 2012.
47. Kumar, D.R.; Kumar, M.A.; Naidu, P.B. Evaluation of Anthelmintic Activity of *Prunus persica* (L.). *Asian J. Pharm. Clin. Res.* **2015**, *4*, 716–721.
48. Angulo-López, J.E.; Flores-Gallegos, A.C.; Ascacio-Valdes, J.A.; Contreras Esquivel, J.C.; Torres-León, C.; Rúelas-Chácon, X.; Aguilar, C.N. Antioxidant dietary fiber sourced from agro-industrial byproducts and its applications. *Foods* **2022**, *12*, 159. [CrossRef]
49. Sousa, S.D.F.; da Silva, F.B.; de Araújo, A.C.; Gomes, J.P. Determinação das propriedades físicas e físico-químicas de pêssegos cultivar Rubimel. *Rev. Bras. Tecnol. Agroind.* **2018**, *12*, 2627–2644. [CrossRef]
50. Shahid, I.; Dildar, A. Nutritional and physicochemical studies on fruit pulp, seed and shell of indigenous *Prunus persica*. *J. Med. Plants Res.* **2011**, *5*, 3917–3921.
51. Zerva, E.; Abatis, D.; Skaltsounis, A.L.; Fokialakis, N. Development and application of a methodology for the recovery of high added value products from peach industry waste. *Planta Medica* **2012**, *78*, PJ98. [CrossRef]
52. Tomás-Barberán, F.A.; Gil, M.I.; Cremin, P.; Waterhouse, A.L.; Hess-Pierce, B.; Kader, A.A. HPLC−DAD−ESIMS analysis of phenolic compounds in nectarines, peaches, and plums. *J. Agric. Food Chem.* **2001**, *49*, 4748–4760. [CrossRef]
53. Mokrani, A.; Krisa, S.; Cluzet, S.; Da Costa, G.; Temsamani, H.; Renouf, E.; Mérillon, J.M.; Madani, K.; Mesnil, M.; Monvoisin, A.; et al. Phenolic contents and bioactive potential of peach fruit extracts. *Food Chem.* **2016**, *202*, 212–220. [CrossRef]
54. Andreotti, C.; Ravaglia, D.; Ragaini, A.; Costa, G. Phenolic compounds in peach (*Prunus persica*) cultivars at harvest and during fruit maturation. *Ann. Appl. Biol.* **2008**, *153*, 11–23. [CrossRef]
55. Hong, Y.; Wang, Z.; Barrow, C.J.; Dunshea, F.R.; Suleria, H.A. High-throughput screening and characterization of phenolic compounds in stone fruits waste by LC-ESI-QTOF-MS/MS and their potential antioxidant activities. *Antioxidants* **2021**, *10*, 234. [CrossRef]
56. de Ancos, B.; González, E.M.; Cano, M.P. Ellagic acid, vitamin C, and total phenolic contents and radical scavenging capacity affected by freezing and frozen storage in raspberry fruit. *J. Agric. Food Chem.* **2000**, *48*, 4565–4570. [CrossRef] [PubMed]
57. Nowicka, P.; Wojdyło, A.; Samoticha, J. Evaluation of phytochemicals, antioxidant capacity, and antidiabetic activity of novel smoothies from selected Prunus fruits. *J. Funct. Foods* **2016**, *25*, 397–407. [CrossRef]
58. Nüβlein, B.; Kreis, W. Purification and characterization of a cynaroside 7-O-β-D-glucosidase from *Cynarae scolymi* folium. In Proceedings of the IV International Congress on Artichoke, Bari, Italy, 17–21 October 2000; pp. 413–420. [CrossRef]
59. Campbell, O.E.; Padilla-Zakour, O.I. Phenolic and carotenoid composition of canned peaches (*Prunus persica*) and apricots (*Prunus armeniaca*) as affected by variety and peeling. *Food Res. Int.* **2013**, *54*, 448–455. [CrossRef]
60. Gil, M.I.; Tomás-Barberán, F.A.; Hess-Pierce, B.; Kader, A.A. Antioxidant capacities, phenolic compounds, carotenoids, and vitamin C contents of nectarine, peach, and plum cultivars from California. *J. Agric. Food Chem.* **2002**, *50*, 4976–4982. [CrossRef]
61. Giuffrida, D.; Torre, G.; Dugo, P.; Dugo, G. Determination of the carotenoid profile in peach fruits, juice and jam. *Fruits* **2013**, *68*, 39–44. [CrossRef]
62. Buratti, C.; Foschini, D.; Barbanera, M.; Fantozzi, F. Fermentable sugars production from peach tree prunings: Response surface model optimization of NaOH alkaline pretreatment. *Biomass Bioenergy* **2018**, *112*, 128–137. [CrossRef]
63. Conde-Mejia, C.; Jimenez-Gutierrez, A.; El-Halwagi, M. A comparison of pretreatment methods for bioethanol production from lignocellulosic materials. *Proc. Saf. Environ. Prot.* **2012**, *90*, 189–202. [CrossRef]
64. Shukla, R.K.; Kant, R. Assessment of phytochemical screening by Fourier Transform Infrared spectroscopic analysis of peach (*Prunus persica*) seed biomass from Uttarakhand region of India. *J. Appl. Nat. Sci.* **2020**, *12*, 519–524. [CrossRef]
65. Rahma, E.H.; Abd El-Aal, M.H. Chemical characterization of peach kernel oil and protein: Functional properties, in vitro digestibility and amino acids profile of the flour. *Food Chem.* **1988**, *28*, 31–43. [CrossRef]
66. Hao, E.; Pang, G.; Du, Z.; Lai, Y.H.; Chen, J.R.; Xie, J.; Zhou, K.; Hou, X.; Hsia, C.D.; Deng, J. Peach kernel oil downregulates expression of tissue factor and reduces atherosclerosis in ApoE knockout mice. *Int. J. Mol. Sci.* **2019**, *20*, 405. [CrossRef] [PubMed]
67. Sodeifian, G.; Sajadian, S.A. Antioxidant capacity, physicochemical properties, thermal behavior, and oxidative stability of nectarine (*Prunus persica* var. *nucipersica*) kernel oil. *J. Food Proc. Preserv.* **2021**, *45*, e15198. [CrossRef]
68. Gonzalez-Garcia, E.; Marina, M.L.; García, M.C.; Righetti, P.G.; Fasoli, E. Identification of plum and peach seed proteins by nLC-MS/MS via combinatorial peptide ligand libraries. *J. Proteom.* **2016**, *148*, 105–112. [CrossRef]
69. Gettens, C.S. Propriedades Funcionais, Nutricionais e Atividade Antimicrobiana de Subprodutos Agroindustriais de Pêssego e sua Aplicação em Cookies. Ph.D. Dissertation, Universidade Federal de Pelotas, Pelotas, Brazil, 2016.

70. Lazos, E.S. Composition and oil characteristics of apricot, peach and cherry kernel. *Grasas Aceites* **1991**, *42*, 127–131. [CrossRef]
71. Mezzomo, N.; Mileo, B.R.; Friedrich, M.T.; Martínez, J.; Ferreira, S.R. Supercritical fluid extraction of peach (*Prunus persica*) almond oil: Process yield and extract composition. *Biores. Technol.* **2010**, *101*, 5622–5632. [CrossRef]
72. Martínez, M.L.; Mattea, M.A.; Maestri, D.M. Pressing and supercritical carbon dioxide extraction of walnut oil. *J. Food Eng.* **2008**, *88*, 399–404. [CrossRef]
73. Oliveira, R.; Fátima Rodrigues, M.; Gabriela Bernardo-Gil, M. Characterization and supercritical carbon dioxide extraction of walnut oil. *J. Am. Oil Chem. Soc.* **2002**, *79*, 225–230. [CrossRef]
74. Özkal, S.G.; Salgın, U.; Yener, M.E. Supercritical carbon dioxide extraction of hazelnut oil. *J. Food Eng.* **2005**, *69*, 217–223. [CrossRef]
75. Reverchon, E.; De Marco, I. Supercritical fluid extraction and fractionation of natural matter. *J. Supercrit. Fluids* **2006**, *38*, 146–166. [CrossRef]
76. Kahla, N.E.; SafeKordi, A. Evaluation of temperature & solvent effect on peach kernel oil extraction & determination & quantification of its fatty. *Evaluation* **2012**, *2*, 1–7.
77. Mezzomo, N.; Martínez, J.; Ferreira, S.R. Supercritical fluid extraction of peach (*Prunus persica*) almond oil: Kinetics, mathematical modeling and scale-up. *J. Supercrit. Fluids* **2009**, *51*, 10–16. [CrossRef]
78. Barwick, V.J. Strategies for solvent selection—A literature review. *Trends Anal. Chem.* **1997**, *16*, 293–309. [CrossRef]
79. Markom, M.; Hasan, M.; Daud, W.R.W.; Singh, H.; Jahim, J.M. Extraction of hydrolysable tannins from *Phyllanthus niruri* Linn.: Effects of solvents and extraction methods. *Sep. Purif. Technol.* **2007**, *52*, 487–496. [CrossRef]
80. Jouyban, A.; Chan, H.K.; Foster, N.R. Mathematical representation of solute solubility in supercritical carbon dioxide using empirical expressions. *J. Supercrit. Fluids* **2002**, *24*, 19–35. [CrossRef]
81. Michielin, E.M.; Bresciani, L.F.; Danielski, L.; Yunes, R.A.; Ferreira, S.R. Composition profile of horsetail (*Equisetum giganteum* L.) oleoresin: Comparing SFE and organic solvents extraction. *J. Supercrit. Fluids* **2005**, *33*, 131–138. [CrossRef]
82. Vági, E.; Simándi, B.; Suhajda, A.; Hethelyi, E. Essential oil composition and antimicrobial activity of *Origanum majorana* L. extracts obtained with ethyl alcohol and supercritical carbon dioxide. *Food Res. Int.* **2005**, *38*, 51–57. [CrossRef]
83. Danielski, L.; Michielin, E.M.; Ferreira, S.R. Horsetail (*Equisetum giganteum* L.) oleoresin and supercritical CO2: Experimental solubility and empirical data correlation. *J. Food Eng.* **2007**, *78*, 1054–1059. [CrossRef]
84. Kitzberger, C.S.; Lomonaco, R.H.; Michielin, E.M.; Danielski, L.; Correia, J.; Ferreira, S.R. Supercritical fluid extraction of shiitake oil: Curve modeling and extract composition. *J. Food Eng.* **2009**, *90*, 35–43. [CrossRef]
85. Karadimou, C.C.; Koletti, A.E.; Moschona, A.; Gika, H.G.; Vlachos, D.; Assimopoulou, A.N. Peach Kernel: A Potential Source for Cosmeceuticals. In Proceedings of the Metrology Promoting Harmonization & Standardization in Food & Nutrition, Thessaloniki, Greece, 1–4 October 2017.
86. Li, W.; Xing, L.; Cai, Y.; Qu, H. Classification and quantification analysis of Radix scutellariae from different origins with near infrared diffuse reflection spectroscopy. *Vib. Spectrosc.* **2011**, *55*, 58–64. [CrossRef]
87. Ordoudi, S.A.; Bakirtzi, C.; Tsimidou, M.Z. The potential of tree fruit stone and seed wastes in Greece as sources of bioactive ingredients. *Recycling* **2018**, *3*, 9. [CrossRef]
88. Górnaś, P.; Mišina, I.; Olšteine, A.; Krasnova, I.; Pugajeva, I.; Lācis, G.; Siger, A.; Michalak, M.; Soliven, A.; Segliņa, D. Phenolic compounds in different fruit parts of crab apple: Dihydrochalcones as promising quality markers of industrial apple pomace by-products. *Ind. Crops Prod.* **2015**, *74*, 607–612. [CrossRef]
89. Rodrigues, F.; Pimentel, F.B.; Oliveira, M.B.P. Olive by-products: Challenge application in cosmetic industry. *Ind. Crops Prod.* **2015**, *70*, 116–124. [CrossRef]
90. de Jesus Raposo, M.F.; de Morais, R.M.S.C.; de Morais, A.M.M.B. Health applications of bioactive compounds from marine microalgae. *Life Sci.* **2013**, *93*, 479–486. [CrossRef] [PubMed]
91. Matthaeus, B.; Oezcan, M.M. Fatty acids and tocopherol contents of some *Prunus* spp. kernel oils. *J. Food Lipids* **2009**, *16*, 187–199. [CrossRef]
92. Qi, Z.X.; Chen, L. Effect of Chinese drugs for promoting blood circulation and eliminating blood stasis on vascular endothelial growth factor expression in rabbits with glucocorticoid-induced ischemic necrosis of femoral head. *J. Tradit. Chin. Med.* **2009**, *29*, 137–140. [CrossRef]
93. Kutlu, T.; Durmaz, G.; Ates, B.; Erdogan, A. Protective effect of dietary apricot kernel oil supplementation on cholesterol levels and antioxidant status of liver in hypercholesteremic rats. *J. Food Agric. Environ.* **2009**, *7*, 61–65.
94. Li, Y.; Wang, L. Genetic resources, breeding programs in China, and gene mining of peach: A review. *Hortic. Plant J.* **2020**, *6*, 205–215. [CrossRef]
95. Anagnostopoulou, M.A.; Kefalas, P.; Papageorgiou, V.P.; Assimopoulou, A.N.; Boskou, D. Radical scavenging activity of various extracts and fractions of sweet orange peel (*Citrus sinensis*). *Food Chem.* **2006**, *94*, 19–25. [CrossRef]
96. Ahn, J.H.; Kim, Y.P.; Seo, E.M.; Choi, Y.K.; Kim, H.S. Antioxidant effect of natural plant extracts on the microencapsulated high oleic sunflower oil. *J. Food Eng.* **2008**, *84*, 327–334. [CrossRef]
97. Redondo, D.; Venturini, M.E.; Oria, R.; Arias, E. Inhibitory effect of microwaved thinned nectarine extracts on polyphenol oxidase activity. *Food Chem.* **2016**, *197*, 603–610. [CrossRef] [PubMed]
98. Zheng, H.Z.; Kim, Y.I.; Chung, S.K. A profile of physicochemical and antioxidant changes during fruit growth for the utilization of unripe apples. *Food Chem.* **2012**, *131*, 106–110. [CrossRef]

99. Martins, C.M.; Ferro, D.M.; de Brito, E.S.; Ferreira, S.R.S. Industrial relevance of *Tamarindus indica* L. by-products as source of valuable active metabolites. *Innov. Food Sci. Emerg. Technol.* **2020**, *66*, 102518. [CrossRef]
100. Lara, M.V.; Bonghi, C.; Famiani, F.; Vizzotto, G.; Walker, R.P.; Drincovich, M.F. Stone fruit as biofactories of phytochemicals with potential roles in human nutrition and health. *Front. Plant Sci.* **2020**, *11*, 562252. [CrossRef] [PubMed]
101. Vásquez-Villanueva, R.; Orellana, J.M.; Marina, M.L.; García, M.C. Isolation and characterization of angiotensin converting enzyme inhibitory peptides from peach seed hydrolysates: In vivo assessment of antihypertensive activity. *J. Agric. Food Chem.* **2019**, *67*, 10313–10320. [CrossRef]
102. Kaltschmitt, M. *Renewable Energy from Biomass: Introduction*; Springer: New York, NY, USA, 2019; pp. 1–14.
103. Paraskevopoulou, C.; Vlachos, D.; Bechtsis, D.; Tsolakis, N. An assessment of circular economy interventions in the peach canning industry. *Int. J. Prod. Econ.* **2022**, *249*, 108533. [CrossRef]
104. Skoulou, V.; Zabaniotou, A. Investigation of agricultural and animal wastes in Greece and their allocation to potential application for energy production. *Renew. Sust. Energ. Rev.* **2007**, *11*, 1698–1719. [CrossRef]
105. Qiu, P.L.; Liu, S.Y.; Bradshaw, M.; Rooney-Latham, S.; Takamatsu, S.; Bulgakov, T.S.; Tang, S.R.; Feng, J.; Jin, D.N.; Aroge, T.; et al. Multi-locus phylogeny and taxonomy of an unresolved, heterogeneous species complex within the genus *Golovinomyces* (Ascomycota, Erysiphales), including *G. ambrosiae, G. circumfusus* and *G. spadiceus*. *BMC Microbiol.* **2020**, *20*, 51. [CrossRef]
106. Kul, A.R.; Aldemir, A.; Elik, H. Adsorption of Acid Blue 25 on peach seed powder: Isotherm, kinetic and thermodynamic studies. *Environ. Res. Technol.* **2019**, *2*, 233–242. [CrossRef]
107. Lima, B.N.B.; Lima, F.F.; Tavares, M.I.B.; Costa, A.M.M.; Pierucci, A.P.T.R. Determination of the centesimal composition and characterization of flours from fruit seeds. *Food Chem.* **2014**, *151*, 293–299. [CrossRef] [PubMed]
108. Pelentir, N.; Block, J.M.; Monteiro Fritz, A.R.; Reginatto, V.; Amante, E.R. Production and chemical characterization of peach (*Prunus persica*) kernel flour. *J. Food Proc. Eng.* **2011**, *34*, 1253–1265. [CrossRef]
109. Liu, H.; Cao, J.; Jiang, W. Evaluation and comparison of vitamin C, phenolic compounds, antioxidant properties and metal chelating activity of pulp and peel from selected peach cultivars. *LWT-Food Sci. Technol.* **2015**, *63*, 1042–1048. [CrossRef]
110. Loizzo, M.R.; Pacetti, D.; Lucci, P.; Núñez, O.; Menichini, F.; Frega, N.G.; Tundis, R. *Prunus persica* var. *platycarpa* (Tabacchiera Peach): Bioactive Compounds and Antioxidant Activity of Pulp, Peel and Seed Ethanolic Extracts. *Plant Foods Hum. Nutr.* **2015**, *70*, 331–337. [CrossRef] [PubMed]
111. Argun, H.; Dao, S. Bio-hydrogen production from waste peach pulp by dark fermentation: Effect of inoculum addition. *Int. J. Hydrogen Energy* **2017**, *42*, 2569–2574. [CrossRef]
112. Vizzotto, M.; Cisneros-Zevallos, L.; Byrne, D.H. Large Variation Found in the Phytochemical and Antioxidant Activity of Peach and Plum Germplasm. *J. Am. Soc. Hortic. Sci.* **2007**, *132*, 334–340. [CrossRef]
113. Geduk, A.S.; Atsız, S. LC-MS/MS phenolic composition of peach (*Prunus persica* (L.) Batsch) extracts and an evaluation of their antidiabetic, antioxidant, and antibacterial activities. *S. Afr. J. Bot.* **2022**, *147*, 636–645. [CrossRef]
114. Montevecchi, G.; Simone, G.V.; Masino, F.; Bignami, C.; Antonelli, A. Physical and chemical characterization of Pescabivona, a Sicilian white flesh peach cultivar [*Prunus persica* (L.) Batsch]. *Food Res. Int.* **2012**, *45*, 123–131. [CrossRef]
115. Wu, J.; Fan, J.; Li, Y.; Cao, K.; Chen, C.; Wang, X.; Fang, W.; Zhu, G.; Wang, L. Characterizing of carotenoid diversity in peach fruits affected by the maturation and varieties. *J. Food Compos. Anal.* **2022**, *113*, 104711. [CrossRef]
116. Zuo, J.; Tang, W.; Xu, Y. Chapter 68—Anti-hepatitis b virus activity of chlorogenic acid and its related compounds. In *Coffee in Health and Disease Prevention*; Preedy, V.R., Ed.; Academic Press: San Diego, CA, USA, 2015; pp. 607–613.
117. Gutiérrez-Grijalva, E.P.; Picos-Salas, M.A.; Leyva-López, N.; Criollo-Mendoza, M.S.; Vazquez-Olivo, G.; Heredia, J.B. Flavonoids and Phenolic Acids from Oregano: Occurrence, Biological Activity and Health Benefits. *Plants* **2017**, *7*, 2. [CrossRef]
118. Goula, A.M.; Ververi, M.; Adamopoulou, A.; Kaderides, K. Green ultrasound-assisted extraction of carotenoids from pomegranate wastes using vegetable oils. *Ultrason. Sonochem.* **2017**, *34*, 821–830. [CrossRef] [PubMed]
119. Cao, S.; Liang, M.; Shi, L.; Shao, J.; Song, C.; Bian, K.; Chen, W.; Yang, Z. Accumulation of carotenoids and expression of carotenogenic genes in peach fruit. *Food Chem.* **2017**, *214*, 137–146. [CrossRef] [PubMed]
120. Falchi, R.; Vendramin, E.; Zanon, L.; Scalabrin, S.; Cipriani, G.; Verde, I.; Vizzotto, G.; Morgante, M. Three distinct mutational mechanisms acting on a single gene underpin the origin of yellow flesh in peach. *Plant J.* **2013**, *76*, 175–187. [CrossRef]
121. Sun, M.; Zhao, B.; Cai, Z.; Yan, J.; Ma, R.; Yu, M. Amino Acid Profiles in Peach (*Prunus persica* L.) Fruit. *Foods* **2022**, *11*, 1718. [CrossRef]
122. Cirilli, M.; Bassi, D.; Ciacciulli, A. Sugars in peach fruit: A breeding perspective. *Hortic. Res.* **2016**, *3*, 12. [CrossRef]
123. Zhang, X.; Su, M.; Du, J.; Zhou, H.; Li, X.; Zhang, M.; Hu, Y.; Ye, Z. Analysis of the free amino acid content and profile of 129 peach (*Prunus persica* (L.) Batsch) germplasms using LC-MS/MS without derivatization. *J. Food Compos. Anal.* **2022**, *114*, 104811. [CrossRef]
124. Planchet, E.; Rannou, O.; Ricoult, C.; Boutet-Mercey, S.; Maia-Grondard, A.; Limami, A.M. Nitrogen metabolism responses to water deficit act through both abscisic acid (ABA)-dependent and independent pathways in Medicago truncatula during post-germination. *J. Exp. Bot.* **2011**, *2*, 605–615. [CrossRef]
125. Yu, Z.; Yang, Z. Understanding different regulatory mechanisms of proteinaceous and non-proteinaceous amino acid formation in tea (*Camellia sinensis*) provides new insights into the safe and effective alteration of tea flavor and function. *Crit. Rev. Food Sci. Nutr.* **2020**, *60*, 844–858. [CrossRef] [PubMed]

126. Urano, K.; Maruyama, K.; Ogata, Y.; Morishita, Y.; Takeda, M.; Sakurai, N.; Suzuki, H.; Saito, K.; Shibata, D.; Kobayashi, M.; et al. Characterization of the ABA-regulated global responses to dehydration in Arabidopsis by metabolomics. *Plant J.* **2009**, *57*, 1065–1078. [CrossRef]
127. Szabados, L.; Savoure, A. Proline: A multifunctional amino acid. *Trends Plant Sci.* **2010**, *15*, 89–97. [CrossRef]
128. Rocha, M.; Sodek, L.; Licausi, F.; Hameed, M.W.; Dornelas, M.C.; van Dongen, J.T. Analysis of alanine aminotransferase in various organs of soybean (Glycine max) and independence of different nitrogen fertilisers during hypoxic stress. *Amino Acids* **2010**, *39*, 1043–1053. [CrossRef] [PubMed]
129. Bae, J.; Shin, D.H.; Chun, B.Y.; Choi, B.Y.; Kim, M.K.; Shin, M.H.; Lee, Y.H.; Park, P.S.; Kim, S.K. The effect of vitamin C intake on the risk of hyperuricemia and serum uric acid level in Korean Multi-Rural Communities Cohort. *Jt. Bone Spine* **2014**, *81*, 513–519. [CrossRef]
130. Ding, T.; Cao, K.; Fang, W.; Zhu, G.; Chen, C.; Wang, X.; Wang, L. Evaluation of phenolic components (anthocyanins, flavanols, phenolic acids, and flavonols) and their antioxidant properties of peach fruits. *Sci. Hortic.* **2020**, *268*, 109365. [CrossRef]
131. Mokrani, A.; Madani, K. Effect of solvent, time and temperature on the extraction of phenolic compounds and antioxidant capacity of peach (*Prunus persica* L.) fruit. *Sep. Purif. Technol.* **2016**, *162*, 68–76. [CrossRef]
132. Tsiaka, T.; Lantzouraki, D.Z.; Polychronaki, G.; Sotiroudis, G.; Kritsi, E.; Sinanoglou, V.J.; Kalogianni, D.P.; Zoumpoulakis, P. Optimization of ultrasound-and microwave-assisted extraction for the determination of phenolic compounds in peach byproducts using experimental design and liquid chromatography–tandem mass spectrometry. *Molecules* **2023**, *28*, 518. [CrossRef]
133. de Vargas, E.F.; Jablonski, A.; Flôres, S.H.; Rios, A.O. Waste from peach (*Prunus persica*) processing used for optimisation of carotenoids ethanolic extraction. *Int. J. Food Sci. Technol.* **2017**, *52*, 757–762. [CrossRef]

sustainability

MDPI

Article

Consumer Perception of Food Product Packaging Materials Sustainability versus Life Cycle Assessment Results: The Case of Processed Tomatoes—A Quantitative Study in Germany

Marisa Bock and Stephan G. H. Meyerding *

Faculty of Life Sciences, Department of Nutrition and Home Economics, HAW Hamburg, 21033 Hamburg, Germany; marisa.bock@haw-hamburg.de

* Correspondence: stephan.meyerding@haw-hamburg.de

Abstract: Due to increasing environmental awareness, especially among the young German population, people are increasingly striving to buy food in the most environmentally friendly way. In this context, packaging is becoming the focus of sustainability assessment, not because of its protection against food waste but because of the increasing amount of packaging rubbish. The aim of this study is to investigate the influence of the packaging material on the environmentally friendly purchase decisions of consumers in Generations Y and Z and whether they can correctly assess the environmental impact of the different materials. For this purpose, an online choice experiment was conducted with a representative sample of 250 German consumers. The respondents could choose between products with different characteristics, such as price, packaging material, label, and origin. The results show that origin is the most important factor, followed by packaging material. With the help of a latent class analysis, the respondents were divided into three segments, which differ in whether origin or material is more important in the sustainability assessment of a product. Furthermore, a lack of knowledge about the environmental impact of specific product attributes among the respondents is evidenced, and a comparison with scientific data from product lifecycle assessments shows that they have difficulties correctly assessing the environmental impact of packaging material.

Keywords: sustainability knowledge; sustainable choices; choice-experiment; latent class analysis; consumer segments

1. Introduction

A growing awareness of the negative consequences of the current lifestyle is leading to efforts to manage and consume more sustainably, which is not an easy task [1].

According to a representative survey in Germany by the Federal Ministry for the Environment, Nature Conservation, Nuclear Safety, and Consumer Protection (BMUV) in 2020, two-thirds of those surveyed rated the topic of environmental and climate protection as very important [2]. The food production and consumption sectors are of great importance here, as around 30 percent of global greenhouse gas emissions are related to food [3]. In this sector, the participants see the greatest need for action in reducing packaging waste and ensuring that less food is thrown away [2]. This contains a contrast, as one way to prevent food waste is the use of optimized packaging [3]. In addition to maintaining quality and protection as the most elementary functions, packaging also plays an important role in storage and transport, handling of the product, and informing the consumer [4].

Packaging consumption by private consumers in Germany was 8.59 million tons in 2019, which corresponds to 103 kg per capita. Even though this consumption decreased by four percent compared to the previous year, the long-term trend shows a steady increase in packaging waste [5].

The research project "STOP waste—SAVE food" showed that only one-third of consumers perceive the shelf-life-extending function of food packaging and that options that are environmentally friendly from the consumer's point of view are preferred to the functionality of optimized packaging [3]. Especially younger age groups show a great willingness to change and exhibit a pronounced attitude toward climate protection, but show a comparatively low level of environmental behavior in contrast [2]. That is why the focus of the present study is on young consumers in Generation Y, also called Millennials, which includes people born from 1981 to 2000, and in Generation Z, including all those born after 1995 [6].

Consumers see actors in business, industry, and politics as being primarily responsible for securing a sustainable future [2]. Because of the purchasing power they have, their responsibility for the environment should not be underestimated. By making conscious decisions for more sustainable alternatives, individual consumers can significantly drive the change we seek [1]. Because the survey by the BMUV showed that the overall stated willingness to consume less is very low, reflective consumer choices are even more important [2]. In addition, this study reflects that 60 percent of the German participants feel well informed about the topic of climate protection [2]. Many studies from other countries show a contradiction between consumers' attitudes and purchase decisions [7–10].

Otto et al. [7] compared consumer perceptions of sustainable packaging with scientific assessments of environmental sustainability and showed that purchasing behavior is, in most cases, less ecological and sustainable than intended. Similar results are shown in studies by Tobler, Visschers, and Siegrist [8] in Switzerland, Steenis et al. [9] in the Netherlands, and Lindh et al. [10] in Sweden. In addition, studies by Klaiman et al. [11] in the United States of America, Steenis et al. [9] in the Netherlands, and Tobler et al. [8] from Switzerland show negative attitudes toward plastic as a packaging material and positive attitudes toward glass across countries. The literature review by Otto et al. [7] shows a similar picture and indicates that this does not correspond to the scientific results.

Despite their relevance to the consumer, there are only a few studies that investigate the role of different packaging materials in food purchases. Allegra et al. [12] conducted a survey in Italy in which consumers rated packaging materials without reference to a food product. Fernqvist et al. [13] included the packaging materials cardboard and plastic when examining the consumers' views on different packaging aspects for potatoes in Sweden. Tobler et al. [8] also had Swiss consumers rate the environmental friendliness of different packaged beans, tomatoes, and potatoes and compare the results to the ones of a life cycle assessment.

Results from Lindh et al. [10], Otto et al. [7], and the "STOP waste—SAVE food" project [3] show that packaging material plays a crucial role in consumers' assessment of sustainability. But there is a lack of research showing not only whether packaging material influences sustainable purchase decisions but also how much this influence is compared to other product attributes. Furthermore, a comparison between the different materials in previous studies was difficult because only a few different packages were available, as well as several unpackaged options [8,13].

The aim of the present study is to investigate the relevance of the common food packaging materials (glass, metal, plastic, and cardboard) for the food product choice decisions of German consumers in Generations Y and Z when they are asked to make the most sustainable choice-decision possible. There is no way to avoid packaging. For this purpose, an online choice experiment was conducted with a simulated shopping situation in which consumers had to decide between product alternatives. There are different types of environmental awareness among the German population, which differ in their willingness to act and interest in environmental protection [2]. Therefore, the importance of packaging material for an ecological decision is investigated for different consumer groups, which are identified in a latent class analysis. With the help of a factor analysis, attendees' attitudes will be examined, and, unlike in previous studies, knowledge of the environmental impact of packaging will be investigated in a quiz. Thus, choice behavior can be linked not only to the participants' frame of mind but additionally to the participants' level of information. Furthermore, the image of the materials among German consumers is investigated to discover if it is similar to the one in other countries. The results may have an effect on product management strategies and give an indication of whether it is necessary to inform consumers more and educate them about the environmental impacts of different packaging options. Passed tomatoes were chosen as the product to be selected, as these are available in all common packaging materials that are familiar to consumers.

This article is structured as follows: The next section describes the used materials and methods. After that, the results of the online survey are presented in the third part and critically discussed in the fourth section, before a short conclusion is drawn.

2. Materials and Methods

2.1. Data Collection

Sample data were collected in December 2021. Participants were recruited via email and social media by sending a link to the online survey. The sample of this study is a convenience sample, so it is likely that mostly consumers participated who have a general interest in sustainable consumption and might be more environmentally conscious than the general German population. Additionally, emails have been sent to consumers who are employees and students at HAW Hamburg. Participation was voluntary, and respondents were free to exit the survey at any time without negative consequences. To ensure that the sample resembled Generation Y and Z in Germany, sampling quotas were set for age and gender. In line with the generations, only people between the ages of 18 and 40 were allowed to participate. A total of 431 consumers participated in the survey, of which 250 answered the questionnaire in full. Five were discarded due to answering the questionnaire too quickly (under 5 min), so 245 complete datasets were included in the analysis.

2.2. Survey Design

The survey consisted of seven sections. In the first section, participants answered questions about their age and gender. These screener questions were used to ensure the representativeness of this study. Next was the choice experiment, in which the attendants were asked to make the most sustainable choice possible. They had to select their preferred product from three different options of strained tomatoes with different product characteristics and one non-option.

In the third section, attendees had to indicate their level of agreement with different statements on a five-point scale ranging from completely agree (1) to completely disagree (5). Respondents were then asked to rate the usefulness of six specific examples of packaging, on a scale from not at all useful (1) to very useful (5).

In the fifth section, a ranking of the materials should be made based on different attributes and environmental aspects. Before the participants had to answer some questions about employment, income, household size, and waste quantity, they were asked to answer five questions in a quiz about the impact of packaging and its disposal.

2.3. Design of the Choice-Experiment

A choice experiment was used in this study to investigate consumers' choice behavior and their preferences for different food product characteristics. For this purpose, different choice sets were created by randomization using Sawtooth Software (version 9.13.0). By selectively varying the product features, their influence on choice-decisions can be determined. The aim of the experiment was to evaluate the importance of individual product attributes on the respondent's choice behavior and whether these parameters vary between respondent groups to derive information on demand and acceptance. Even though it is not a real purchase situation, the results are assumed to show a high degree of correspondence between the hypothetical and the real decision, even for strongly socially desirable behaviors such as choosing sustainable products [14].

Choice experiments provide detailed information about decision-making and are therefore used in many areas of research [15]. For example, in the study by Muller, Lacroix, and Ruffieux [16], who used a choice experiment to investigate how environmental food labels influence consumer shopping behavior. In the study by de-Magistris and Gracia [17], a real-choice-experiment was used to investigate consumer preferences and willingness to pay for almonds with different sustainable labels.

When conducting the choice-experiment, the participant selects a product from various alternatives, each with different combinations of product attribute levels. A total of four product attributes were selected, which are assumed to influence consumers choice-behavior. These are price, origin, label, and packaging material, since this is decisive for the investigation. If we had asked the participants to indicate which product they would buy instead of choosing the one they perceive as being more sustainable, then it would be necessary to add more attributes that indicate food product quality, as it is expected for consumers to buy a product based on its quality rather than its environmental friendliness. To have a scientific basis for the later comparison between the consumer perception of product attribute sustainability and their real environmental impact, the packaging materials used in this study were inspired by the ones from a Europe-wide life cycle assessment (LCA) for packaging of durable foods, published by the Institute for Energy and Environmental Research (IFEU) on behalf of SIG Combibloc Services AG [18]. In the experiment, each attribute has five levels, and in the expression of the price, equal spacing was considered. All attributes and attribute levels used are listed in Table 1.

Table 1. Attributes and their levels were used in the choice-experiment.

Attributes	**Attribute Levels**				
Price	€0.39	€0.89	€1.39	€1.89	€2.39
Packaging material	Cardboard 	Plastic pouch 	Metal can 	Glass 	Plastic pot 
Label	Bio Germany	Blauer Engel	Pro Planet	WWF	No label
Origin	Regional	Germany	Netherlands	Italy	Morocco
Non-option	No purchase				

The product pictures were self-created and provided with a fictitious logo. The design was adapted so that all products look as similar as possible, except for the packaging material. This was to avoid visual influences, as Steenis et al. [9] showed that the packaging design also affects sustainability perceptions.

In the choice-experiment, participants were asked to select the product they considered to be the most sustainable choice from three alternatives. If none of the generated choices in the choice set were suitable for the respondent, he or she could select the non-option "No purchase". Since not all attribute levels can occur simultaneously in a choice set, several choice sets were generated by the software. A separate randomized choice-set design was created for each participant, containing a total of twelve choice-sets with images for illustration. Figure 1 shows an example of a selection set.

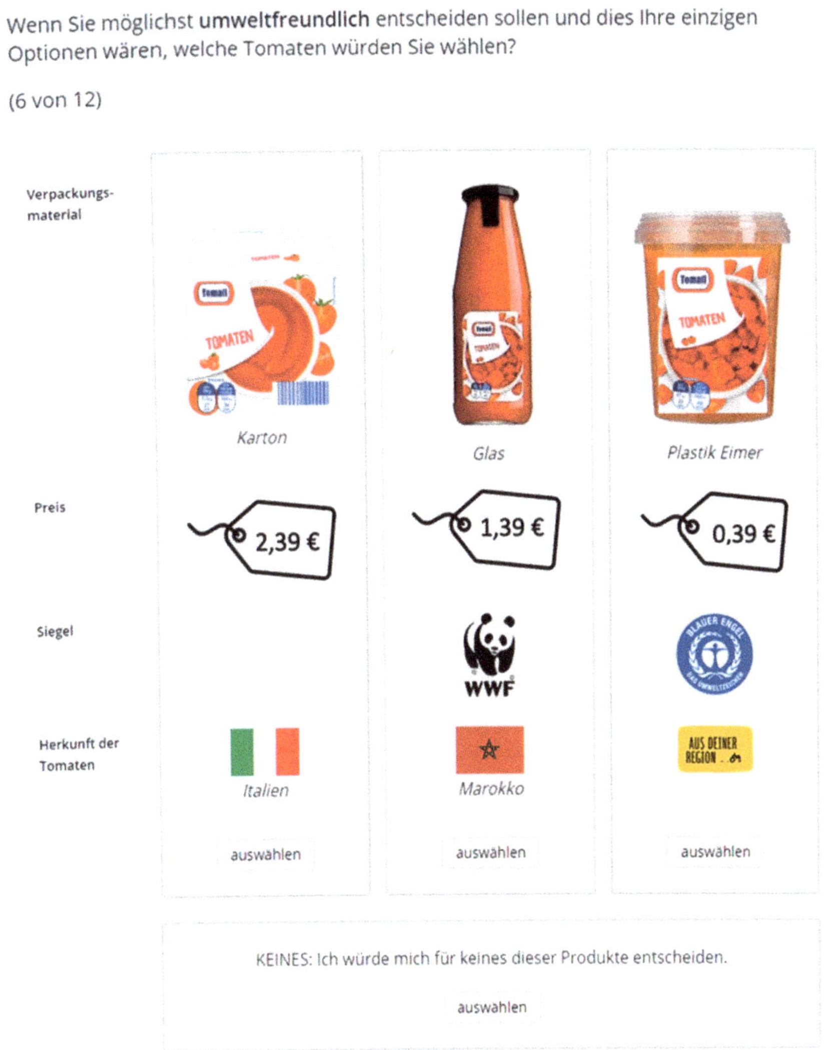

Note. Please note, that this figure is an original screenshot from the online questionnaire and is therefore in German. This means that commas are used as decimal signs instead of periods. The sentence on top of the choice-set says: "If you had to make the most environmentally friendly decision possible and these were your only options, which tomatoes would you choose?" On the left the vocables translate as follows: Verpackungsmaterial = packaging material, Preis = price; Siegel = label; Herkunft der Tomaten = origin of the tomatoes. "Auswählen" means select and "KEINES: Ich würde mich für keines dieser Produkte entscheiden" translates to: NONE: I would not choose any of these products.

Figure 1. Example choice-set in the choice-experiment.

2.4. *The Design of the Lifestyle Constructs*

The second part of this study included statements on various constructs used to capture the attitudes of the participants. The constructs are represented by statements that the attendees were asked to agree with on a Likert scale from 1 "Strongly disagree" to 5 "Strongly agree". Based on literature research, 18 statements were determined for six different constructs, from which a connection between the participants' attitude and their choice-decision can be assumed.

The first construct used deals with knowledge about the environment and comes from a study by Yadav and Pathak [19]. With the help of this construct, it was possible to investigate how the respondents themselves rated their knowledge about environmental aspects related to packaging waste. The second construct, taken from Biswas and Roy [20], describes the environmental behavior of the participants and reflects how environmentally aware they consider themselves to be. To capture the measure of concern for the environment, statements from Minton and Rose [21] were used that relate to both consumer and industry behaviors. The fourth construct, adopted from Suki [22], serves as an indicator of efforts to act environmentally conscious in relation to the choice of environmentally friendly products. Corresponding statements from a study by Chéron, Sudbury-Riley, and Kohlbacher [23] were used to analyze respondents' price consciousness. The last construct examines opinions about seals on food packaging and comes from a study by Van der Merwe, Bosman, and Ellis [24]. Table 6 in the results section provides an overview of all items.

2.5. Design of the Package Usefulness Evaluation

The agreement on statements was followed by an evaluation of six specific packaging examples. These contained the products cucumber, cress, and brioche braid, each in two different packaging options. The cucumber was available unpackaged or packaged in a plastic sleeve, the cress only on substrate in a tray or with additional foil packaging, and the brioche braid in a paper bag with a viewing strip or foil packaging.

The examples used were taken from a study by Denkstatt GmbH [25], in which the change in food waste after packaging changeover was investigated. The respondents were asked to rate the six examples on a scale from 1 (not at all useful) to 5 (very useful). The aim of this evaluation is to analyze whether the participants perceive the benefits of optimized packaging. In addition, the acceptance of packaging can be investigated when unpackaged options are available, and, in contrast to the products in the choice-experiment, fresh food is involved. Figure 2 shows an example from the package usefulness rating.

Note. Please note, that this figure is an original screenshot from the online questionnaire and is therefore in German. The sentence on top of the rating task says: "How useful do you find this packaging?". The rating is from Überhaupt nicht sinnvoll = not at all useful to Sehr sinnvoll = very useful".

Figure 2. Example from the usefulness rating.

2.6. Ranking Design

In the fourth part of the questionnaire, participants were asked to rank the packages of metal cans, glass bottles, cardboard containers, plastic pouches, and plastic pots in terms of various characteristics and environmental categories from 1 (best) to 5 (worst). Each rank could only be used once. The attributes in the first part of the ranking were sustainable, high quality, tasty, and convenient. These were taken from a study by Steenis et al. [9] that examined the role of packaging material in sustainability ratings in the Netherlands. The aim was to gain an impression of the image that the materials have among the participants and draw possible conclusions from this about the choice-behavior in the experiment.

The environmental categories, contribution to climate change, ozone layer depletion, and transport intensity in the second part of the ranking originate from the life cycle assessment (LCA) for the packaging of long-life foods [18]. Various resource-relevant categories (consumption of abiotic resources, fossil resources, and primary energy, renewable and non-renewable) were combined as 'consumption of energy and resources' and were integrated into the ranking [18].

The purpose of this task is to analyze whether the respondents can correctly assess the environmental impact of the materials in comparison with each other. The existing IFEU assessment, which also presents a ranking, makes it possible to compare the participants' assessment with LCA data. Figure 3 shows an example of a ranking task.

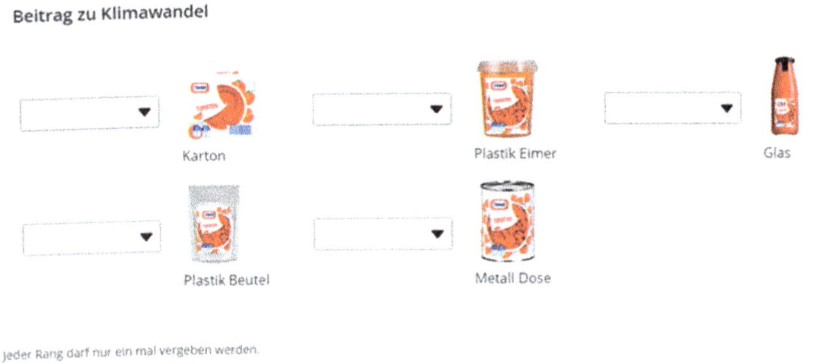

Note. Please note, that this figure is an original screenshot from the online questionnaire and is therefore in German. The sentence on top of the rating task says: "Contribution to climate change". On the bottom it says: "Each rank may only be awarded once".

Figure 3. An example from the ranking task.

2.7. Quiz Design

To analyze whether the respondents correctly assessed the benefits and environmental impacts of packaging, a quiz was included at the end of the questionnaire. It contained five questions on the climate impacts of food packaging.

The first three questions were taken from a study on ecological and economic aspects of packaging, which was conducted by the company for packaging market research (GVM) and the Denkstatt Institute in 2019 [4]. Question four and five were derived from the information in the guide to the research project "Stop Waste—Save Food" and self-authored [3].

When conducting the survey, participants were asked to answer five questions, each with four possible answers. All questions involved estimating numerical values related to the climate footprint, climate impact, or environmental benefit of packaging, and one point was awarded for each correct answer. The answer choices were coded from one to four. For each question, the first choice was right and reflected the lowest impact. As the number of options increased, the negative impacts of packaging in the answer option also increased numerically. Because of that, the evaluation could also measure whether the respondents overestimate the climatic impact of packaging by analyzing the average response number. Figure 4 shows an example of a question from the quiz.

2.8. Statistical Method

2.8.1. Analysis in Sawtooth Software

After downloading and cleaning the survey data, the choice-experiment was analyzed in Sawtooth software (version 9.13.0). To determine participants' preferences for the attribute levels of the attributes price, packaging material, label, and origin, the Hierarchical

Bayes (HB) Analysis was used first, which has gained significant and positive influence in the analysis of choice-based conjoint studies in recent years. With the help of HB Analysis, part-worth utilities can be calculated [26]. A high value reflects a large benefit for the consumer, which indicates a higher purchase probability for the selected product [27]. In addition, simple segmentation is made possible [26].

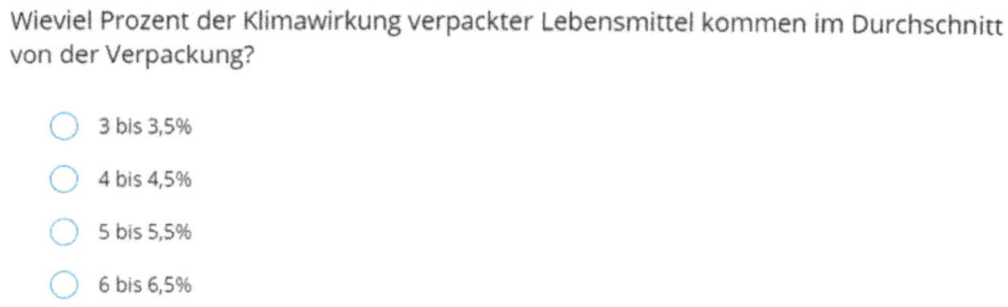
Wieviel Prozent der Klimawirkung verpackter Lebensmittel kommen im Durchschnitt von der Verpackung?

3 bis 3,5%

4 bis 4,5%

5 bis 5,5%

6 bis 6,5%

Note. Please note, that this figure is an original screenshot from the online questionnaire and is therefore in German. This means that commas are used as decimal signs instead of periods. The sentence on top of the answers says: "On average, what percentage of the climate impact of packaged food comes from the packaging?".

Figure 4. Example question from the quiz.

To identify relevant consumer segments, a latent class analysis (LCA) was performed in a second step. LCAs offer the possibility of identifying consumer segments that show a relative homogenous choice behavior in the choice experiment and can afterwards be characterized using several lifestyle constructs [28]. This method is used in many studies. For example, Leech et al. [29] used a Latent Class Analysis to divide Australian men and women into segments based on their eating behavior.

In this study, the Latent Classes were formed based on the choices in the choice experiment. Together with the individual part-worth utilities for attribute levels of price, material, label, and origin, three segments with similar preferences concerning these characteristics have been identified and are used for further investigation.

2.8.2. Analysis in SPSS

All further analyses based on HB and latent class analysis were conducted in the Statistical Package for Social Science (SPSS) program (version 27.0.1.0). First, participants' sociodemographic characteristics were examined in SPSS. Absolute frequencies and percentages in the sample were calculated for the variables gender, employment, and food waste. For gender and employment, the percentages of these variables for the population of generations Y and Z in Germany were reported in addition. For the variables age, available income, household size, waste generated per week, and efforts to avoid waste, the mean values and standard deviation were calculated. The mean of the German generations' population was added for age, income, and household size. The analysis was extended by including the three consumer segments by indicating the percentage of the variable in the groups as well as the entire sample. A summary of the sociodemographic analysis is later presented in Table 2 of the results section.

In the second step of the analysis, part-worth utilities for the attribute levels of price, packaging material, origin, label, and no purchase, as well as the relative importance of each attribute, were calculated. This was followed by the factor analysis, which included the items related to the constructs, knowledge about the environment, attitude toward environmental protection, measure of concern for the environment, environmental awareness, price awareness, and positive opinion about labels. Factor analysis was also used in Yadav and Pathak's study [19] on young consumers' green product purchasing behavior, as well as a study of green consumption behavior by Biswas and Roy [20], where it served as an appropriate means of analyzing statements (Analysis of Variance (ANOVA) and Post Hoc Test (Tukey) were used to identify significant differences between the mean factor scores for each construct and consumer segment). This made a comparison between the three groups possible and allowed a description of the segments in terms of the attitudes depicted. ANOVA and Post-Hoc Tests were also used in a study by Suki to investigate the effects of consumer values on the purchase behavior of environmentally friendly products [22].

To discover relationships between variables, regression analysis was used next. In this process, an attempt is made to explain a dependent variable by several independent variables, as an influence of these is suspected. In this study, it was examined to what extent the constructs as independent variables explain the dependent variable's waste prevention efforts.

The quiz was evaluated using the absolute frequencies for each answer choice as well as the percentage of frequencies for the entire sample and the three consumer groups. In addition, the mean values and standard deviations of the quiz scores were calculated.

For the evaluation of the assessed usefulness of the packaging examples, the mean values of the answer options coded from 1 (not at all useful) to 5 (very useful) were calculated for the entire sample and the groups. Similarly, the rankings were analyzed. Here, the mean rankings for each variable were again calculated for the entire sample and the consumer groups.

3. Results

In this results section, we first describe our sample and compare it to the population of Generations Y and Z in Germany to indicate the level of representation. After this, the results of the Hierarchical Bayes model for the whole sample based on the choice-experiment data are shown.

The results of the Hierarchical Bayes model include the relative importance of the attributes to the participants as well as the part-worth utilities of each attribute level within each attribute. This indicates which attribute is most important when choosing the most sustainable option and which attribute levels are perceived as more sustainable than others.

After the results of the Hierarchical Bayes model, we show the results of the latent class analysis to identify homogeneous consumer segments based on their choices in the experiment. The relative importance of the attributes as well as the part-worth utilities are then presented for each of the identified consumer groups. We additionally describe the socio-demographic variables for the estimated segments in the next subsection. This is conducted to identify significant differences between the segments regarding their socio-demographic variables. In our study, we also conducted items for different lifestyle constructs to describe the different consumer groups. First, the results of the factor analysis for these lifestyle constructs are presented. After that, we show the profiling of the latent consumer segments regarding these lifestyle constructs and indicate if there are significant differences between the segments. This helps us to better understand the values and motives of the different consumer segments. These lifestyle constructs are then used as predictors in a regression analysis to explain the efforts to avoid waste.

In the next subsection, the results of the quiz are shown for the different consumer segments. This helps to measure the knowledge of the consumer segments regarding environmental issues related to packaging. In another subsection, we present the results of the package usefulness evaluation for the whole sample and each consumer group. The results of the packaging image ranking are next and indicate how the different packaging materials are perceived by consumers. Here we are also comparing the results of this image ranking to the results of a life cycle assessment to finally see where these two differ.

3.1. Sample Description

The socio-demographic data were analyzed according to the pattern shown in Table 2. A total of 245 respondents between the ages of 18 and 40 were included in the sample. The average age of the participants is 26, which is slightly lower than the average of Generations

Y and Z in Germany. At 51.4%, more men than women (48.6%) participated in the survey, which corresponds to the distribution in the population. On average, respondents live in a two-person household that produces an average of 5.83 kg of estimated self-reported waste per week. Efforts to reduce this are high. Food is thrown away rarely or occasionally by 72.1% of respondents, and often or very often by only 7.8%. Compared to the population of 18- to 40-year-olds in Germany, the share of university students in this study is high at 69% and, in turn, a proportion of 25% employees are rather low. It is likely that most of the students in this study are studying at HAW Hamburg, as we used email lists from our university to distribute the link to the questionnaire. This is also reflected in the comparatively low average available income of €669.6 of this study participants.

Table 2. Summary of the socio-demographic analysis ($N = 245$).

Variable	Levels	Frequency Sample	Share (%) Sample	Share (%) Generation Y and Z
Gender [1]	Male	126	51.40	51.60
	Female	119	48.60	48.40
Employment	Student [2]	2	0.80	2.40
	Apprentice [2]	7	2.90	6.30
	University student [3]	169	69.00	12.00
	Employee [4]	60	24.50	69.50
	Without employment [4]	3	1.20	2.80
	Other [5]	4	1.60	7.00
Food waste	Rare	105	43.00	N.A.
	Occasionally	71	29.10	N.A.
	Now and then	49	20.10	N.A.
	Often	17	7.00	N.A.
	Very often	2	0.80	N.A.
Variable	**Unit of measurement**	**Mean**	**Standard deviation (SD)**	**Mean Generation Y and Z**
Age [1]	Years	25.70	5.17	29.60
Available income [6]	Euro	713.19	669.60	2372.00
Household size [7]	People	2.39	1.23	2.03
Amount of waste/week	Kilogram	5.83	6.80	N.A.
Efforts to avoid waste	1 very much, 5 absolutely not	1.84	0.82	N.A.

Note. [1] Source: Census Data in the version of 15 November 2021, Table 12411-0005 (Federal Statistical Office, 2021). [2] Source: Census Data in the version of 2 November 2021, Subject-matter series 11, series 1, 2, 4.1 and calculations by the German Center for Higher Education and Science Research (Federal Statistical Office, 2021) *for age group 18–30.* [3] Source: Census Data in the version of 5 August 2021, Subject-matter series 11, series 4.1, WS 2020/2021 (Federal Statistical Office, 2021) *for age group 18–37.* [4] Source: Census Data in the version of 26 January 2022, result 12211-9001 (Federal Statistical Office, 2022) *for age group 20–40.* [5] Percentages that could not be assigned to any of the above category. [6] Statistical Yearbook 2019 (Federal Statistical Office, 2019) *for age group 18–35.* [7] Census Data in the version of 7 September 2021, Subject-matter series 1, series 3 (Federal Statistical Office, 2021) *for all age groups.*

3.2. Results of the Hierarchical Bayes Model

The Hierarchical Bayes model was used to determine the average preferences (part-worth utilities) of the participants for the attributes, packaging material, price, label, and origin.

Figure 5 shows the estimated part-worth utilities for each attribute level, the non-option (no purchase), and the relative importance of the attributes. To enable a better comparison, one attribute level of each attribute was set to zero. This applies to the material metal, the highest price, the characteristic no-label, and Morocco as an origin.

The average participant prefers the packaging material glass to cardboard and shows a clear rejection of both plastic packagings. The part-worth utility values decrease slightly with increasing prices, with the exception that the price of €0.89 is preferred over the one of €0.39. In relation to the option no label, the labels show positive part-worth utilities, with the highest values for the German Bio (organic) label. Products from the region are clearly preferred by the respondents. The preferences for regional and German products are similarly close to those from the European countries of France and Italy. Morocco, on the other hand, is clearly lagging.

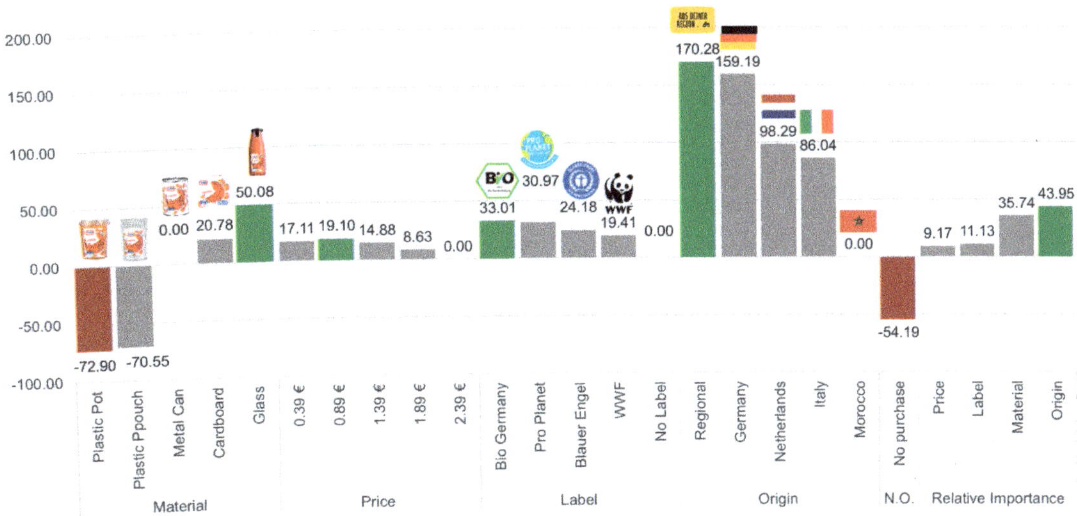

Figure 5. Resulting part-worth utilities for choosing perceived environmentally friendly strained tomatoes ($N = 245$).

For the average respondent, the origin is of the highest importance, followed by the packaging material, when choosing perceived environmentally friendly strained tomatoes. Price and label show lower values, whereas price is even less important when choosing the most sustainable alternative in the choice-experiment.

3.3. Results of the Latent Class Analysis

To detect heterogeneity in the sample and form meaningful groups that are similar in their choice of behavior, it is first necessary to decide how many classes to form. Nylund et al. [30] emphasize that the scientific community disagrees on what the best criteria are for determining the number of classes. Therefore, a combination of criteria, including statistical information criteria such as Akaike's information criterion (AIC) and Bayesian information criterion (BIC), was used. In addition, the consistent Akaike information criterion (CAIC), which is closely related to loglikelihood, was included in the decision because it was described as an appropriate criterion along with the BIC [30]. All information criteria have allowed for deeper investigation of a variety of content research areas in the past [30]. Table 3 shows the criteria calculated for a number of classes ranging from two to five.

Table 3. Model selection for latent class segmentation.

No. of Latent Classes	Log-Likelihood	AIC	CAIC	BIC	Average Max. Membership Probability
2	−2465.22	5000.44	5245.10	5210.10	0.98
3	−2341.92	4789.85	5160.33	5107.33	0.94
4	−2290.15	4722.30	5218.61	5147.61	0.94
5	−2226.77	4731.53	5253.66	5164.66	0.94

The values for CAIC and BIC decrease until the three-group solution and then increase again for the four-group solution. Since small values are preferred, three groups were formed for further analysis [27].

The results of the latent class analysis for the three group solution are presented in Table 4. The segment division is based on the decisions of the participants in the choice experiment. The results show the part-worth utilities for each attribute level and for each consumer group. The relative importance of each attribute is shown, as these give an indication of what has the greatest influence on the groups' decisions.

Table 4. Part-worth utilities for the three consumer groups (N = 245).

Attribute	Levels	Total Sample (100.00%)	Group 1: Plastic Hater (36.7%)		Group 2: Origin-Conscious Consumers (39.2%)		Group 3: Quality-Oriented Consumers (24.1%)	
Material	Plastic pot	−72.90	−112.79	a	−24.18	c	−91.34	b
	Plastic pouch	−70.55	−101.94	a	−29.58	b	−89.33	a
	Metal can	0.00	0.00	b	0.00	a	0.00	b
	Cardboard	20.78	22.29	c	24.58	a	12.32	b
	Glass	50.08	57.30	b	36.39	a	61.34	b
Price	€0.39	17.11	19.98	b	21.72	b	5.23	a
	€0.89	19.10	20.26	b	24.10	b	9.21	a
	€1.39	14.88	12.93	a	18.14	a	12.55	b
	€1.89	8.63	6.18	a	13.34	a	4.72	a
	€2.39	0.00	0.00	a,b	0.00	a	0.00	b
Label	Bio Germany	33.01	27.16	a,b	43.79	b	24.39	a
	Pro Planet	30.97	25.85	a	40.16	a	23.85	a
	Blauer Engel	24.18	10.79	a	38.08	b	21.99	b
	WWF	19.41	17.46	b	24.06	a	14.81	a,b
	No Label	0.00	0.00	b	0.00	a	0.00	b
Origin	Regional	170.28	122.58	a	216.50	c	167.82	b
	Germany	159.19	120.01	a	192.43	c	164.86	b
	Netherlands	98.29	68.11	a	127.17	b	97.36	a
	Italy	86.04	62.06	b	101.82	a	96.94	b
	Morocco	0.00	0.00	c	0.00	a	0.00	b
N.O.	No purchase	−54.19	−100.31	a	−79.42	a	57.20	b
Relative importance (%)	Price	9.17	10.11	b	9.24	a,b	7.63	a
	Label	11.13	10.19	a	13.36	b	8.94	a
	Material	35.74	46.38	c	23.12	a	40.06	b

Note. Superscripts stand for significant mean differences at the 0.05 level based on Tukey testing.

The consumer groups are very similar in the ranking of the individual attribute levels. For example, all groups have the highest part-worth utility for the material glass, followed by cardboard, and all favor the organic (Bio) label and products from the region and Germany.

Group 1 includes 36.7% of the respondents and is named 'Plastic Hater'. For the members of this group, the packaging material is most important in the choice of sustainable products. They prefer glass and show the greatest rejection of all groups for both packaging made of plastic, on which the name of this segment is based. For this group, the non-option no purchase has the lowest value, which suggests that the participants also decide on a product if it does not completely correspond to their conceptions.

The largest group, with 39.2% of the respondents, are Origin-Conscious consumers. The name was chosen because origin is of the utmost relative importance for its members. This segment favors, like the other groups, regional products, followed by those from Germany. But they also show the highest part-worth values for the Netherlands and Italy. Even though this group favors glass as a packaging material, the part-worth utility value is rather low in comparison with the other groups, and that for cardboard is the highest of the three segments. Group two is the only one that perceives the plastic pouch more negatively than the plastic pot, and the labels in general have the highest values in this consumer segment.

A mixture of the first two groups is described by the segment of Quality-Oriented consumers (24.1%). For these participants, origin and packaging material play a similar role in the choice of sustainable products. These group members have the highest utility value for glass and the lowest for cardboard. With the highest part-worth utility of 1.39€, this segment favors the highest price of all groups. It is also striking that its members are the only ones that show a positive part-worth utility value for the non-option (no purchase). It can be deduced that these respondents would rather choose no product than one that does not meet their requirements.

3.4. Results of the Socio-Demographic Variables for the Estimated Segments

The socio-demographic data of the participants were collected at the beginning and at the end of the questionnaire. Table 5 shows the results of these socio-demographic parameters for the three consumer groups formed in the LCA.

Table 5. Summary of socio-demographic attributes for latent class segments ($N = 245$).

Variable	**Levels**	**Share (%) Total Sample**	**Share (%) Group 1: Plastic Hater**	**Share (%) Group 2: Origin-Conscious Consumers**	**Share (%) Group 3: Quality-Oriented Consumers**
Gender	Male	51.40	56.70 [a]	50.00 [a]	45.80 [a]
	Female	48.60	43.30 [a]	50.00 [a]	54.20 [a]
Status	Student	0.80	0.00 [a]	2.10 [a]	0.00 [a]
	Apprentice	2.90	1.10 [a]	3.10 [a]	5.10 [a]
	University student	69.00	70.00 [a]	70.80 [a]	64.40 [a]
	Employee	24.50	27.80 [a]	19.80 [a]	27.10 [a]
	Without employment	1.20	1.10 [a]	2.10 [a]	0.00 [a]
	Other	1.60	0.00 [a]	2.10 [a]	3.40 [a]
Food waste produced per week	Rare	43.00	34.40 [a]	43.20 [a]	55.90 [a]
	Occasionally	29.10	33.30 [a]	29.50 [a]	22.00 [a]
	Now and then	20.10	23.30 [a]	20.00 [a]	15.30 [a]
	Often	7.00	8.90 [a]	6.30 [a]	5.10 [a]
	Very often	0.80	0.00 [a]	1.10 [a]	1.70 [a]
Variable	Unit of measurement	Mean Total Sample	Mean Group 1: Plastic Hater	Mean Group 2: Origin-Conscious Consumers	Mean Group 3: Quality-Oriented Consumers
Age	Years	25.70	26.02 [a]	25.43 [a]	25.64 [a]
Income per month	Euro	713.19	778.92 [a]	676.17 [a]	672.33 [a]
Household size	People	2.39	2.42 [a]	2.45 [a]	2.25 [a]
Amount of waste	Kilogram per week	5.83	6.21 [a]	5.86 [a]	5.22 [a]
Efforts to avoid waste	1 very much, 5 absolutly not	1.84	1.89 [a]	1.81 [a]	1.83 [a]

Note. Superscripts stand for significant mean differences at the 0.05 level based on Tukey testing.

It is noticeable that there are no significant, but just small, differences between the three consumer groups regarding their sociodemographic attributes.

Among the Plastic Haters, there are 13.4% more men than women, and the group of Quality-Oriented consumers consists of 8.4% more women than men. For Origin-Conscious consumers, the gender distribution is exactly half women and half men. Group 2 (Origin-Conscious consumers) is the only one with a small proportion of students, the most university students, and the fewest employees. This also results in the lowest average age. The group of Plastic Haters has the largest proportion of employees and the smallest proportion of trainees, which may also result in the highest average age and the highest average income. Even though the values of the segments differ only slightly from each other, small differences in food waste become apparent. A difference of one kilogram, concerning the stated amount of food waste per household and week, can be seen between the Plastic Haters (6.21 kg) and Quality-Oriented consumers (5.22 kg). The participants of Group 1 (Plastic Haters) also show slight deviations from the other groups in the indication of how often food is thrown away. According to their own statements, 8.9% of these group members often throw away spoiled food, which is the highest value of all groups. And with a share of 34.4%, this group shows the lowest value for rare food waste.

3.5. Results of the Factor Analysis for the Lifestyle Constructs

Two principal component factor analyses with varimax rotation were conducted using SPSS for the items measuring respondents' attitudes toward environmental issues and product attributes. The Kaiser–Meyer–Olkin (KMO) criterion and Bartlett's test for sphericity (BTS) were used to validate the results. The values of the KMO range from 0 to 1, whereas 0 is the worst and 1 is the best possible situation. Only values above 0.5 are considered acceptable [31].

In this study, the value of the KMO is 0.728 for the first factor analysis for the construct's 'Knowledge about the environment', 'Environmentally conscious actions', 'Price awareness', and 'Opinion on seals'. For the second factor analysis of the factors 'Attitude toward environmental protection' and 'Concern for the environment' the KMO value is 0.797. Both values are classified as 'Middling' according to Kaiser [31]. Using the BTS, the null hypothesis is tested to determine whether the sample belongs to a population with uncorrelated variables [27]. The BTS is significant for both analyses, which is why the null hypothesis can be rejected and a factor analysis is possible. Table 6 shows the results of the factor analysis and all the items used to measure the different psychographic constructs.

Table 6. Results of the factor analysis (N = 245).

Factors and the Corresponding Variables	Mean	SD	Factor Loading
Knowledge about the environment (Cronbach's alpha: 0.745)			
I know more about recycling than the average population.	3.12	0.967	0.847
I am very well informed about environmental issues.	3.26	0.917	0.787
I understand the different phrases and symbols about the environment on product packaging.	3.04	0.963	0.742
Attitude toward environmental protection (Cronbach's alpha: 0.723)			
I would describe myself as environmentally conscious.	3.76	0.766	0.850
It is important to me that the products I use do not harm the environment.	3.88	0.756	0.834
I am concerned about the waste of our planet's resources.	4.45	0.697	0.692
Concern for the environment (Cronbach's alpha: 0.757)			
I feel angry and frustrated when I think about the damage that pollution does to plant and animal life.	4.07	0.943	0.877
I feel angry and frustrated when I think about how industry pollutes the environment.	4.21	0.866	0.834
Consumers should care about the environmental impact of the products they buy.	4.28	0.688	0.577
Environmentally conscious actions (Cronbach's alpha: 0.702)			
When given a choice between two equivalent products, I buy the one that is less harmful to other people and the environment.	4.00	0.878	0.837
I have avoided buying a product because it had potentially harmful effects on the environment.	3.61	1.075	0.763
I make a special effort to buy paper and plastic products that are made from recycled materials.	3.60	0.965	0.711
Price awareness (Cronbach's alpha: 0.807)			
Price is the decisive factor when I buy a product.	2.96	1.049	0.850
Price plays an important role when I choose products.	3.82	0.996	0.847
I usually aim for the lowest possible price.	3.11	1.011	0.815
Positive opinion about labels (Cronbach's alpha: 0.808)			
A food label is a good source of information.	3.07	0.958	0.895
I think the information on food labels is useful.	3.43	0.901	0.834
Labels give me information about the food.	3.18	0.889	0.796

Note. Extraction method: Principal component analysis. Rotation method: Varimax with Kaiser normalization. Scale from 1 'totally disagree' to 5 'totally agree'. N = 245.

When items are used to form a scale, they must have internal consistency, which means that all items measure the same thing and correlate with each other [32]. To assess internal consistency, the Cronbach's alpha criterion was used. According to Bland and Altman [32], alpha values of 0.7 to 0.8 are considered satisfactory for comparing groups. In this study, the values for Cronbach's alpha range from 0.702 to 0.808, thus meeting the requirements.

Next, the three identified consumer segments are described using the extracted factors from Table 6. Table 7 presents the results with means and standard deviations for each group and factor (construct).

Table 7. Profiling the latent consumer segments (N = 245).

Factors	Sample Group Factor Means (SD)					
	Group 1: Plastic Hater (N = 90)		**Group 2: Origin-Conscious Consumers (N = 96)**		**Group 3: Quality-Oriented Consumers (N = 59)**	
Knowledge about the environment	−0.203 (1.05)	a	0.184 (0.90)	b	0.011 (1.03)	a,b
Attitude toward environmental protection	−0.164 (1.10)	a	0.087 (0.85)	a	0.110 (1.04)	a
Concern for the environment	−0.467 (1.11)	a	0.118 (0.84)	a	−0.119 (1.06)	a
Environmentally conscious actions	−0.101 (1.18)	a	0.121 (0.84)	a	−0.043 (0.94)	a
Price awareness	0.108 (0.96)	a	−0.052 (1.01)	a	−0.080 (1.05)	a
Opinion on seals	0.030 (1.12)	a	0.058 (0.90)	a	−0.141 (0.96)	a

Note. Items were assessed by means of Likert scales (1 = totally disagree; 5 = totally agree). Superscripts stand for significant mean differences at the 0.05 level based on Tukey testing.

The only factor in which the three groups differ significantly is perceived knowledge about the environment. The Plastic Hater segment has the lowest mean values, and the Origin-conscious consumers have the highest. Quality-Oriented consumers include members with both poor and good perceived environmental knowledge.

The Plastic Haters also have the lowest scores for the factors attitude toward environmental protection, concern for the environment, and environmentally conscious actions. Price consciousness is the highest among members of this group. In contrast, the mean values for concern for the environment and environmentally conscious actions are highest among Origin-Conscious consumers. This group has not only good environmental knowledge but also the greatest concerns and acts in the most environmentally conscious way, according to their own estimation. In addition, these participants also have the most positive opinions about labels. Quality-Oriented consumers have the highest value for their attitude toward environmental protection. As expected, price awareness is comparatively low in this segment because members rather focus on quality than on a low price. They also show little interest in labels.

3.6. Results of the Regression Analysis

With the help of regression analysis, relationships between different variables can be examined, and positive or negative correlations can be shown. The model consists of a dependent variable that is to be explained and several independent variables that are assumed to have an influence on the dependent variable [33].

This study examined the extent to which the six constructs 'knowledge about the environment', 'attitude toward environmental protection', 'concern for the environment', 'environmentally conscious actions', 'price awareness', and 'opinion on seals' as independent variables explain the dependent variable 'efforts to avoid waste'. Table 8 shows the standardized beta and significance level for each construct as well as the R^2 for this model.

The results show that 'knowledge about the environment', 'attitude toward environmental protection', 'concern for the environment', as well as 'environmentally conscious actions' have a significant positive influence on the participants' stated efforts to avoid waste. This means the higher the knowledge about the environment, attitude toward environmental protection, concern for the environment, or environmentally conscious actions of the respondent, the more he or she also tries to avoid waste. The strongest effect can be seen for environmentally conscious actions, with a standard beta of 0.224 at a significant level of 0.005. The R^2 represents the percentage of the variation in the outcome that can be explained by the model [33]. For this analysis, it means that the constructs explain the variance of the effort to avoid waste by 24.2%.

Table 8. Results of the regression analysis.

	Standardized Beta	Sig.
Knowledge about the environment	0.176	0.004
Environmentally conscious actions	0.224	0.005
Concern for the environment	0.163	0.014
Attitude toward environmental protection	0.177	0.020
Price awareness	−0.057	0.327
Opinion on seals	0.055	0.342

Dependent variable: Efforts to avoid waste, $N = 245$, $R^2 = 24.2\%$.

3.7. Results of the Quiz

One point was awarded for each correct answer in the quiz, so that a maximum score of five points could be achieved. Table 9 shows the frequencies of all point values and the percentage distribution for the entire sample and the three consumer groups.

Table 9. Quiz score for the entire sample and latent consumer segments ($N = 244$).

	Total Sample ($N = 244$)		Group 1: Plastic Hater ($N = 90$)		Group 2: Origin-Conscious Consumers ($N = 95$)		Group 3: Quality-Oriented Consumers ($N = 59$)	
Points	Frequency	Percent	Frequency	Percent	Frequency	Percent	Frequency	Percent
0	143	58.6	53	58.9 [a]	53	55.8 [a]	37	62.7 [a]
1	76	31.1	26	28.9 [a]	34	35.8 [a]	16	27.1 [a]
2	20	8.2	9	10.0 [a]	7	7.4 [a]	4	6.8 [a]
3	4	1.6	1	1.1 [a]	1	1.1 [a]	2	3.4 [a]
4	1	0.4	1	1.1 [a]	0	0.0 [a]	0	0.0 [a]
5	0	0.0	0	0.0 [a]	0	0.0 [a]	0	0.0 [a]
Mean	0.54		0.57		0.54		0.51	
SD	0.75		0.81		0.68		0.77	

Note. Superscripts stand for significant mean differences at the 0.05 level based on Tukey testing.

The average respondent scored 0.54 points, with a standard deviation of 0.75. More than half of the respondents (58.6%) could not answer any question correctly. The highest score of four points was achieved by only one respondent from the Plastic Hater group. Against expectations after the factor analysis, there are no significant differences between the latent classes when it comes to the results in the quiz regarding environmental knowledge. Table 10 shows the five questions and the frequencies of the chosen answer options in total numbers and percentages for the whole sample. As already described, the answer options were designed in such a way that the negative impact of packaging increases with the number of answers, and the least and first ones were always right. Therefore, a mean answer option was calculated for every question.

For all questions, participants chose the third answer option on average. The highest mean value is shown for the first question. Almost one-third of the attendees estimated the share of packaging in the climate footprint at 4.5–5.0%, which is more than double the correct answer of 1.5–2.0%. For this and the second question, the fewest respondents knew the correct answer (5.7%). For the question that relates packaging consumption to a flight distance, the mean answer of 2.5 is the best because more respondents (20.2%) knew the correct answer than for all other questions. The benefit of packaging through waste prevention was also significantly lower. Almost 30.0% of the sample estimates the environmental benefit to be as high as the environmental costs, although the benefits are 5 to 10 times higher. For the third question, just as many participants chose the correct and best answer as the worst answer option. For all other questions, a majority of the sample chose the worst scenario, followed by the real and best values.

Table 10. Distribution of answers in the quiz for each question (N = 244).

	Frequency	Percent	Mean
The climate footprint of the average European consumer in 2012 was around 15 tons of CO_2 equivalents per person and year. What is the share of packaging consumption in this, per person and year?			
(1) 1.5–2.0% *	14	5.7	
(2) 2.5–3.0%	50	20.5	
(3) 3.5–4.0%	100	41.0	
(4) 4.5–5.0%	80	32.8	3.0
Plastic bag consumption per person per year is equivalent to __ car kilometers.			
(1) 13 *	14	5.7	
(2) 23	60	24.6	
(3) 33	113	46.3	
(4) 43	57	23.4	2.9
The carbon footprint of a flight from Berlin to Paris and back (2 × 880 km) corresponds to a packaging consumption of _____.			
(1) 5 years *	49	20.2	
(2) 4 years	69	28.4	
(3) 3 years	76	31.3	
(4) 2 years	49	20.2	2.5
On average, what percentage of the climate impact of packaged food comes from packaging?			
(1) 3.0 to 3.5% *	35	14.3	
(2) 4.0 to 4.5%	75	30.7	
(3) 5.0 to 5.5%	93	38.1	
(4) 6.0 to 6.5%	41	16.8	2.6
The environmental benefit of packaging through avoided waste is usually _______ than/as the environmental cost of packaging.			
(1) 5 to 10 times higher *	20	8.2	
(2) 2 to 5 times higher	85	34.8	
(3) 1 to 2 times higher	66	27.0	
(4) just as high	73	29.9	2.8

Note. * right answer.

3.8. Results of the Package Usefulness Evaluation

The less packaging material there was, the more useful the packaging was rated by the participants. Especially the absence of plastic influenced the positive rating. Figure 6 shows the mean values of the evaluated usefulness and the images of the packaging options from the questionnaire.

The biggest difference can be seen in the cucumber. On average, the plastic tube packaging was rated 1.16 and the unpacked version 4.75 on a scale from 1 (not useful at all) to 5 (very useful). The result for cress is similar, although not as pronounced. Only in a plastic tray was it rated 3.61, and with additional foil, the value decreases to 1.76. The smallest difference can be seen in the packaging of the bakeware. Here, the plastic foil received the best rating of the pure plastic packaging with 1.84, and the paper bag with a window received the worst rating of the alternatives with 3.18. The mean scores for the latent classes show no significant differences and can be seen in Table 11.

3.9. Results of the Package Image Ranking

In the first part, the participants ranked the five types of packaging made of glass, plastic, cardboard, and metal in terms of their sustainability, quality, naturalness, consumer friendliness, and the extent to which they make the food appear tasty. Figure 7 shows the results in a spider web diagram. Here it is about the perception of the packaging material itself without relation to a particular product example. For example, metal can be perceived as natural.

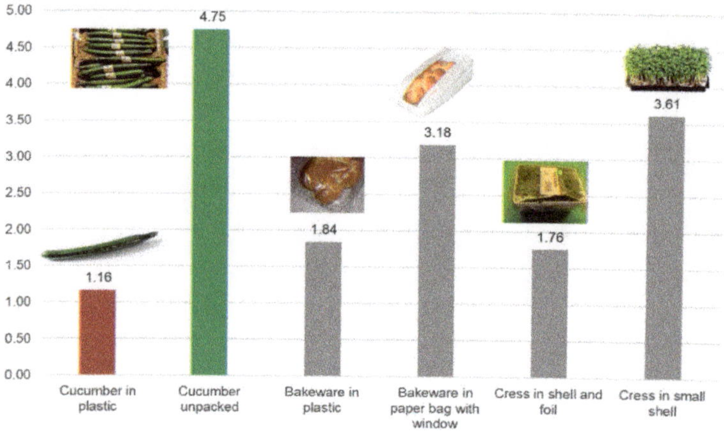

Figure 6. Means of the package usefulness evaluation (N = 245).

Table 11. Results of the package usefulness evaluation for the latent classes (N = 245).

Product and Packaging	Total Sample (N = 245)	Group 1: Plastic Hater (N = 90)	Group 2: Origin-Conscious Consumers (N = 96)	Group 3: Quality-Oriented Consumers (N = 59)
Cucumber in plastic	1.16	1.11 [a]	1.25 [a]	1.10 [a]
Cucumber unpacked	4.75	4.79 [a]	4.76 [a]	4.66 [a]
Bakeware in plastic	1.84	1.66 [a]	2.02 [a]	1.83 [a]
Bakeware in paper bag with window	3.18	3.21 [a]	3.08 [a]	3.27 [a]
Cress in shell and foil	1.76	3.76 [a]	3.60 [a]	3.41 [a]
Cress in small shell	3.61	1.70 [a]	1.85 [a]	1.69 [a]

Note. Items were assessed by means of scales (1 = not useful at all; 5 = very useful). Superscripts stand for significant mean differences at the 0.05 level based on Tukey testing.

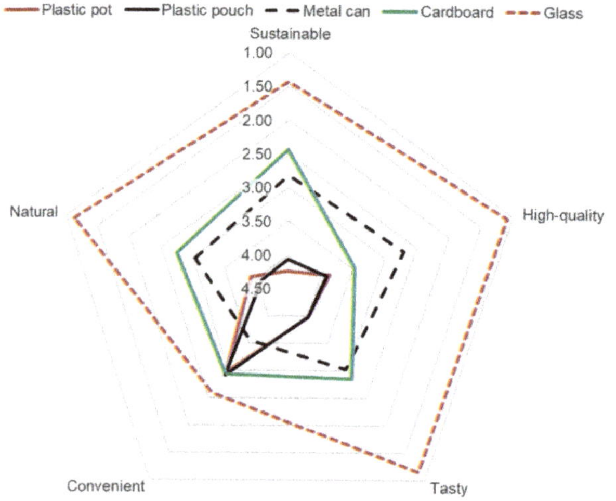

Note. (1) best, (5) worst.

Figure 7. Ranking results for the image of different packaging (N = 245).

Glass was ranked best in all categories. For the attributes natural, sustainable, high-quality, and tasty by a large margin over the alternatives, and for the attribute convenient only by a small margin over plastic and cardboard packaging. Cardboard packaging follows glass in most categories but shows declines in the perception of quality. When it comes to quality, the metal can is ranked after glass, which is otherwise behind cartons and even has the worst score for convenience. The plastic pouch and pot both show poor results and have a bad image except for consumer convenience, where they are about equal with the carton and ahead of the metal can. The mean ranks for the whole sample and the three latent consumer groups can be seen in Table 12.

Table 12. Mean ranks of the image of different packaging for the whole sample and latent classes (N = 245).

Attribute	Material	Mean Total Sample (N = 245)	Mean Group 1: Plastic Hater (N = 90)	Mean Group 2: Origin-Conscious Consumers (N = 96)	Mean Group 3: Quality-Oriented Consumers (N = 59)
Sustainable	Plastic pot	4.24	4.49 ^{b}	3.87 ^{a}	4.44 ^{b}
	Plastic pouch	4.06	4.09 ^{a}	3.94 ^{a}	4.22 ^{a}
	Metal can	2.82	2.72 ^{a}	2.94 ^{a}	2.78 ^{a}
	Cardboard	2.43	2.29 ^{a}	2.60 ^{a}	2.39 ^{a}
	Glass	1.43	1.41 ^{a,b}	1.61 ^{b}	1.17 ^{a}
High-quality	Plastic pouch	3.90	3.91 ^{a}	3.86 ^{a}	3.93 ^{a}
	Plastic pot	3.84	3.76 ^{a}	3.78 ^{a}	4.05 ^{a}
	Cardboard	3.46	3.47 ^{a,b}	3.61 ^{b}	3.19 ^{a}
	Metal can	2.70	2.79 ^{a}	2.62 ^{a}	2.71 ^{a}
	Glass	1.09	1.08 ^{a}	1.07 ^{a}	1.12 ^{a}
Tasty	Plastic pouch	3.98	3.97 ^{a}	3.98 ^{a}	3.98 ^{a}
	Plastic pot	3.97	4.07 ^{a}	3.81 ^{a}	4.05 ^{a}
	Metal can	3.03	3.01 ^{a}	3.01 ^{a}	3.08 ^{a}
	Cardboard	2.86	2.79 ^{a}	2.97 ^{a}	2.8 ^{a}
	Glass	1.15	1.17 ^{a}	1.17 ^{a}	1.12 ^{a}
Convenient	Metal can	3.55	3.51 ^{a}	3.63 ^{a}	3.49 ^{a}
	Plastic pot	2.97	2.97 ^{a}	2.88 ^{a}	3.10 ^{a}
	Cardboard	2.94	3.03 ^{a}	2.92 ^{a}	2.81 ^{a}
	Plastic pouch	2.92	2.98 ^{a}	2.77 ^{a}	3.07 ^{a}
	Glass	2.60	2.52 ^{a}	2.72 ^{a}	2.53 ^{a}
Natural	Plastic pouch	4.08	4.02 ^{a}	4.04 ^{a}	4.24 ^{a}
	Plastic pot	3.93	4.02 ^{a,b}	3.72 ^{a}	4.15 ^{b}
	Metal can	3.04	3.16 ^{a}	2.95 ^{a}	3.00 ^{a}
	Cardboard	2.77	2.68 ^{a,b}	3.04 ^{b}	2.47 ^{a}
	Glass	1.14	1.11 ^{a}	1.16 ^{a}	1.14 ^{a}

Note. Superscripts stand for significant mean differences at the 0.05 level based on Tukey testing. Scale from (1) best to (5) worst.

There are some significant differences in the image ranking between the consumer groups. The Origin-Conscious consumers rated the plastic pot more sustainable than the other two groups, and glass in the same category was worse than for the Quality-Oriented consumers. They also rated cardboard as less natural and high-quality than members of Group 3. On the other hand, Group 2 ranked the plastic pot as more natural than the Quality-Oriented consumers. The ranking continued for sustainability indicators also used in life cycle assessments. Table 13 provides an overview of the results for the entire sample and latent classes.

Table 13. Mean ranks for packaging options regarding environmental aspects for the whole sample and latent classes (N = 245).

Attribute	Material	Mean Total Sample (N = 245)	Mean Group 1: Plastic Hater (N = 90)	Mean Group 2: Origin-Conscious Consumers (N = 96)	Mean Group 3: Quality-Oriented Consumers (N = 59)
Climate change	Plastic pot	4.00	4.17 [a]	3.78 [a]	4.10 [a]
	Plastic pouch	3.91	3.89 [a]	3.86 [a]	4.02 [a]
	Metal can	2.85	2.73 [a]	2.97 [a]	2.84 [a]
	Cardboard	2.46	2.42 [a]	2.56 [a]	2.34 [a]
	Glass	1.75	1.80 [a]	1.75 [a]	1.69 [a]
Ozone layer depletion	Plastic pot	3.78	3.93 [a]	3.72 [a]	3.67 [a]
	Plastic pouch	3.71	3.69 [a]	3.71 [a]	3.74 [a]
	Metal can	2.85	2.76 [a,b]	2.72 [a]	3.19 [b]
	Cardboard	2.57	2.55 [a]	2.71 [a]	2.39 [a]
	Glass	2.05	2.06 [a]	2.07 [a]	2.02 [a]
Transport intensity	Glass	3.46	3.38 [a]	3.44 [a]	3.61 [a]
	Plastic pot	3.26	3.37 [a]	3.16 [a]	3.26 [a]
	Metal can	2.99	3.07 [a]	2.97 [a]	2.89 [a]
	Plastic pouch	2.97	2.81 [a]	3.11 [a]	2.98 [a]
	Cardboard	2.29	2.34 [a]	2.27 [a]	2.26 [a]
Consumption of energy and resources	Plastic pot	3.43	3.44 [a]	3.41 [a]	3.44 [a]
	Plastic pouch	3.21	3.30 [a]	3.05 [a]	3.35 [a]
	Metal can	3.05	2.97 [a]	3.09 [a]	3.12 [a]
	Glass	2.77	2.85 [a]	2.79 [a]	2.61 [a]
	Cardboard	2.52	2.45 [a]	2.60 [a]	2.47 [a]

Note. Superscripts stand for significant mean differences at the 0.05 level based on Tukey testing. Scale from (1) best to (5) worst.

The ranking of the packages according to the middle positions is the same for the contribution to climate change and ozone layer depletion, with glass at the top, followed by cardboard, metal cans, plastic pouches, and plastic pots at the end. Only the gaps between the ranks are clearer for climate change. For transport intensity, cardboard is in the lead with a middle rank of 2.29 and is followed by the plastic pouch (2.97). Glass is in last place, with a mean rank of 3.46. The middle ranks concerning consumption of resources and energy are closest together. Cardboard achieves the best average value with 2.52 and is followed by glass, metal, and plastic packaging. The only significant difference between the latent classes is that the Origin-Conscious consumers ranked the metal can better for the aspect of ozone layer depletion than the Quality-Oriented consumer group. Comparing the results of the survey, which can be seen in Figure 8a, to the results of the life cycle assessment, shown in Figure 8b, deviations become apparent.

It should be noted that the numerical values cannot be compared because they are based on different assessment methods. In the lifecycle assessment (LCA), the results for the packaging alternatives are given in relation to cardboard, which performs best in all categories [18]. Therefore, the comparison just refers to the ranking of the packaging materials.

The greatest deviations are shown for the packaging materials glass and plastic. For the categories 'consumption of energy and resources', 'ozone layer depletion', and 'contribution to climate change', the plastic pouch is in fourth place among the respondents and in second place in the lifecycle assessment. Glass, on the other hand, is rated too positively by the participants in the present study compared to the LCA results. It is in first place among consumers for both 'contribution to climate change' and 'ozone layer depletion', although it is in last place in the life cycle assessment. In general, the respondents' assessment in the category of which packaging materials contribute most to climate change differs the most from the actual circumstances.

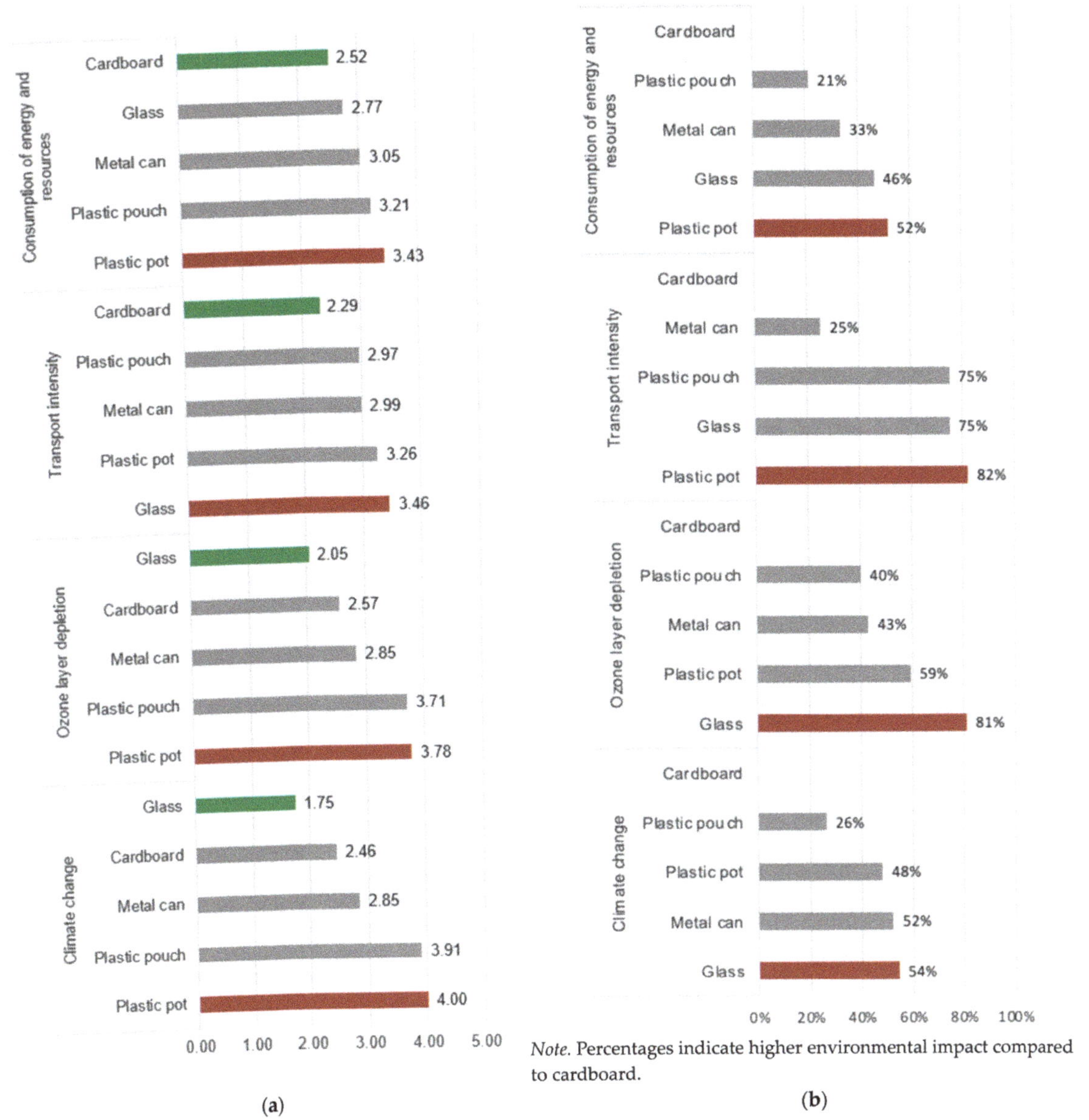

Figure 8. (**a**) Results of this study rank ($N = 245$). (**b**) Results of the life cycle assessment [18].

4. Discussion

In addition to the origin, packaging material plays a major role in the assessment of the sustainability of durable food products by consumers of Generations Y and Z in Germany. They prefer glass and rate it as the most positive in ecological terms. Plastic, on the other hand, is clearly rejected and is only convincing in terms of convenience. Environmental labels have a positive influence on the choice decision, and the respondents do not prefer the lowest price when it comes to a consciously sustainable decision. In general, the label is of greater importance than the price.

Based on the decision in the choice experiment, the respondents can be divided into three groups: (1) Plastic Haters (36.7%), (2) Origin-Conscious consumers (39.2%), and (3) Quality-Oriented consumers (24.1%). The biggest difference between the segments can be seen in the relevance of product attributes when evaluating the sustainability of the alternatives. For the Plastic Haters, the packaging material is of the utmost importance, as is the price. Whereas for the Origin-Conscious consumers, as the name already says, the origin is the most important aspect. Furthermore, they pay the most attention to the label. For the Quality-Oriented consumers, as the smallest group, material and origin are of the same importance, and they pay the least attention to price and label. Apart from the fact that the group of Plastic Haters assesses their own knowledge about the environment significantly worse than the members of group two, the consumer segments are very similar. They do not differ in terms of sociodemographic parameters and do not show any significant differences in their quiz answers or package usefulness ratings.

The image of packaging materials among Generations Y and Z in Germany largely corresponds to that of consumers in other countries [8,9,11,12]. Otto et al. [7] show that the sustainability of glass is overestimated and that of plastic packaging is underestimated. Similar results are shown by Tobler et al. [8], in whose study the environmental impact of packaging was generally overestimated and glass was given the most environmentally friendly rating, in contrast to the LCA results. When it comes to glass as a packaging material, it needs to be mentioned that the weight of glass packaging also affects the carbon footprint in transport [34]. These results are confirmed in the present study, as the ecological assessment of the participants deviates from the data of the LCA, especially for the materials glass and plastic. The overestimation of the environmental impact of packaging in general can also be verified by looking at the quiz results. Furthermore, the scores show that the knowledge of the respondents about the environmental impact of specific product attributes is worse than they themselves estimated by the evaluation of the factor 'knowledge about the environment'.

The research by Otto et al. [7] also shows that the function of packaging and the associated protection against food waste are often not considered, which leads to very sustainable ratings of unpacked food. This finding can also be confirmed by the results of the usefulness evaluation in this study. With the help of practical data on food waste in Austrian retail, the Denkstatt Institute carried out an ecological evaluation of packaged and unpackaged cucumbers. According to this, the carbon footprint of the packed one is better if the waste rate of the unpackaged variant in retail and among consumers is more than six percent higher than the waste rate of the cucumber packed in plastic [4]. It is important to emphasize that this calculation only refers to the carbon footprint and does not lead to a statement about which product is more sustainable. For an accurate sustainability assessment, not only one aspect but many can be used. But nevertheless, under certain circumstances, this packaging also has a benefit that was maybe not considered in the participants' assessment as 'not useful at all'.

4.1. Implications and Suggestions

The results of the present study show that consumers include the packaging material in a sustainable decision but have difficulties correctly assessing the environmental impact of the different materials. The sustainable choice corresponds mainly to what the respondents personally perceive as natural, appetizing, or ecological. In addition, consumers overestimate their own level of knowledge and do not act as environmentally conscious as their stated efforts would suggest.

The positive attitude towards environmental protection combined with concern for nature nevertheless indicates a willingness to act sustainably. One possibility why theoretical and actual actions differ could be that the information that is available to consumers is not suitable to support them in making sustainable purchasing decisions. Often, these are based on a variety of different evaluation methods that are hard to understand or cannot be comprehended at all. According to Tanner and Jungbluth [35], this misbehavior is not only a problem of lack of information but also the result of built-in cognitive mechanisms that lead people to translate environmental knowledge in such a way that it can lead to an incorrect assessment of the environmental friendliness of a product.

In addition, Tanner and Jungbluth [35], Otto et al. [7], and Lindh et al. [10] also see the need for guidance for consumers with the help of labels that enable them to make sustainable choices. As it is too complex to inform consumers about all environmentally friendly-relevant dimensions, Tobler et al. [8] recommend simple communication tools as suggested in the domain of nutrition labels. For example, a three-level ecolabel system adapts the design of a traffic light system [8]. The findings from this study, that all the labels investigated have a positive influence on choice behavior, hold potential for the development of new labels to assist consumers in making sustainable product choices.

4.2. Limitations

There are also limitations that need to be considered in this study. The sample is only representative of the German population for gender but not for age or employment. The average age of this study participants is lower than that of Generations Y and Z in Germany, and a large proportion of the respondents are students, which may also be the reason for the low available income of the sample. Future research should include more participants who are not university students. The gap between consumer perceptions of food product packaging material sustainability and life cycle assessment results might be even bigger if fewer students were surveyed. In addition, the participant's choice of behavior is based on the request to make it as sustainable as possible. That is why the data from the choice experiment does not reflect the usual purchasing behavior. Possible conclusions for action are therefore aimed at consumers who are willing to make sustainable decisions.

Assessing the sustainability of products proves to be complex, which leads to a limited selection of sustainability indicators that must be questioned critically. The life cycle assessment results used in this study were carried out by the IFEU on behalf of SIG Combibloc Services AG, a provider of aseptic cardboard packaging. The data published by SIG only gives an overview of the results, not of the methods and materials used for data collection. Furthermore, no precise information about the packages is given, although the type and quantity of material are decisive for the environmental impact [17]. Furthermore, a life cycle assessment is only one method of many to examine the environmental impact of products. Another possibility is the use of a utility value analysis, in which decision alternatives are compared with each other using weighted criteria [36]. Ms. Waldner, an employee of the Hamburg Packaging Institute (BFSV), presented several utility value analyses. However, since these are based on internal company evaluation benchmarks and precise information on packaging and material composition is necessary, an additional evaluation of the packaging alternatives with this method was not possible [37]. This once again makes clear how complex and individual the ecological assessment of different product characteristics is.

Another limitation of this study is that biodegradable packaging was not considered. This alternative to conventional plastic could evolve into a more environmentally friendly packaging alternative that has most of the useful characteristics of conventional plastic packaging without its negative impact on the environment [38].

5. Conclusions

Sustainable consumption that protects the environment and preserves resources is becoming increasingly important, especially among the younger generations. As a result, packaging and the waste it creates are getting the focus of consumers and, in the saturated food market, also the marketing strategies of companies. Against this background, the present study investigated the relevance of different product attributes in the evaluation of the sustainability of durable food in Generations Y and Z in Germany.

The results of the conjoint analysis show the highest relative importance for the origin of the product and the material of the packaging, which are thus decisive factors for a sustainable purchase decision. The participants favored regional or German products packaged in glass or cardboard. Packaging made of plastic shows by far the lowest part-worth utilities and has a poor image.

Comparing the respondents' evaluation to data from product lifecycle assessments, it becomes clear that the participants' choice behavior is not as sustainable as intended. The existing willingness to act environmentally friendly suggests that there is a lack of suitable opportunities for consumers to evaluate packaging in an ecologically correct way rather than according to personal impressions.

The development of transparent, universally valid evaluation schemes is therefore of great importance. As is the translation of these into communication tools that help consumers who want to make a sustainable choice. Further research needs to be conducted to investigate whether these tools will be accepted and deliver the desired results.

Author Contributions: Conceptualization, S.G.H.M.; Methodology, S.G.H.M.; Software, S.G.H.M.; Validation, S.G.H.M.; Formal analysis, S.G.H.M.; Investigation, M.B.; Data curation, M.B.; Writing—original draft, M.B.; Writing—review and editing, S.G.H.M.; Supervision, S.G.H.M.; Project administration, S.G.H.M. All authors have read and agreed to the published version of the manuscript.

Funding: This research received no external funding.

Institutional Review Board Statement: The study was conducted in accordance with the Declaration of Helsinki, and approved by the Institutional Review Board (or Ethics Committee) Ethikkommission des Competence Center Gesundheit CCG of the HAW Hamburg, Germany (protocol code 2020-08 and date of approval 28 September 2020).

Informed Consent Statement: Informed consent was obtained from all subjects involved in the study.

Data Availability Statement: Data are contained within the article.

Acknowledgments: We would like to thank Marcella Waldner and Bernd Sadlowsky from BFSV e.V., Verpackungsinstitut Hamburg and HAW Hamburg for their support and fruitful discussion about the life-cycle assessment of food packaging materials.

Conflicts of Interest: The authors declare no conflict of interest.

References

1. Hauk, P. Geleitwort. In *Nachhaltiger Konsum—Best Practices aus Wirtschaft, Unternehmenspraxis, Gesellschaft, Verwaltung und Politik*; Wellbrock, W., Ludin, D., Eds.; Springer Gabler: Wiesbaden, Germany, 2021; pp. VII–VIII.
2. BMUV. *Umweltbewusstsein in Deutschland 2020—Ergebnisse Einer Repräsentativen Bevölkerungsumfrage*; BONIFATIUS GmbH Druck-Buch-Verlag: Paderborn, Germany, 2022.
3. Ecoplus; BOKU; Denkstatt; OFI. Lebensmittel-Verpackungen-Nachhaltigkeit: Ein Leitfaden für Verpackungshersteller, Lebensmittelverarbeiter, Handel, Politik & NGOs. Entstanden aus den Ergebnissen des Forschungsprojektes "STOP Waste-SAVE Food". Wien, 2020. Available online: https://boku.ac.at/fileadmin/data/H03000/H81000/H81300/upload-files/Forschung/Lebensmittel/Leitfaden-Lebensmittel-Verpackungen-V1.pdf (accessed on 9 October 2023).
4. GVM; Denkstatt. Nutzen von Verpackungen—Ausgabe 2019. Mainz, 2019. Available online: https://www.agvu.de/wp-content/uploads/2019/12/Nutzen-von-Verpackungen-2019_Endbericht_18_07_2019.pdf (accessed on 9 October 2023).
5. Umweltbundesamt. *Aufkommen und Verwertung von Verpackungsabfällen in Deutschland im Jahr 2019—Abschlussbericht*; Umweltbundesamt: Dessau-Roßlau, Germany, 2021.
6. Eberhardt, D. *Generationen Zusammen Führen—Mit Millenials, Generationx und Babyboomern die Arbeitswelt Gestalten*; Haufe-Lexware: Freiburg, Germany, 2016.
7. Otto, S.; Strenger, M.; Maier-Nörth, A.; Schmid, M. Food packaging and sustainability—Consumer perception vs. correlated scientific facts: A review. *J. Clean. Prod.* **2021**, *298*, 126733. [CrossRef]
8. Tobler, C.; Visschers, V.H.; Siegrist, M. Organic tomatoes versus canned beans: How do consumers assess the environmental friendliness of vegetables? *Environ. Behav.* **2011**, *43*, 591–611. [CrossRef]
9. Steenis, N.D.; Van Herpen, E.; Van der Lans, I.A.; Ligthart, T.N.; Van Trijp, H.C. Consumer response to packaging design: The role of packaging materials and graphics in sustainability perceptions and product evaluations. *J. Clean. Prod.* **2017**, *162*, 286–298. [CrossRef]
10. Lindh, H.; Olsson, A.; Williams, H. Consumer perceptions of food packaging: Contributing to or counteracting environmentally sustainable development? *Packag. Technol. Sci.* **2015**, *29*, 3–23. [CrossRef]
11. Klaiman, K.; Ortega, D.L.; Garnache, C. Consumer preferences and demand for packaging material and recyclability. *Resour. Conserv. Recycl.* **2016**, *115*, 1–8. [CrossRef]

12. Allegra, V.; Zarbà, A.S.; Muratore, G. The post-purchase consumer behaviour, survey in the context of materials for food packaging. *Ital. J. Food Sci.* **2012**, *24*, 160–164. Available online: https://www.proquest.com/docview/1271624083?fromopenview=true&pq-origsite=gscholar (accessed on 9 October 2023).
13. Fernqvist, F.; Olsson, A.; Spendrup, S. What's in it for me? Food packaging and consumer responses, a focus group study. *Br. Food J.* **2015**, *117*, 1122–1135. [CrossRef]
14. Auspurg, K.; Liebe, U. Choice-Experimente und die Messung von Handlungsentscheidungen in der Soziologie. *Kölner Z. Soziologie Sozialpsychol.* **2011**, *63*, 301–314. [CrossRef]
15. Mariel, P.; Hoyos, D.; Meyerhoff, J.; Czajkowski, M.; Dekker, T.; Glenk, K.; Jacobsen, J.B.; Liebe, U.; Olsen, S.B.; Sagebiel, J.; et al. *Environmental Valuation with Discrete Choice Experiments—Guidance on Design, Implementation and Data Analysis*; Springer: Cham, Switzerland, 2021.
16. Muller, L.; Lacroix, A.; Ruffieux, B. Environmental labelling and consumption changes: A food choice experiment. *Environ. Resour. Econ.* **2019**, *73*, 871–897. [CrossRef]
17. de-Magistris, T.; Gracia, A. Consumers' willingness-to-pay for sustainable food products: The case of organically and locally grown almonds in Spain. *J. Clean. Prod.* **2016**, *118*, 97–104. [CrossRef]
18. SIG. Verantwortungsvolle Produkte. 2020. Available online: https://www.sig.biz/de/verantwortung/verantwortungsvolle-produkte (accessed on 20 April 2022).
19. Yadav, R.; Pathak, G.S. Young consumers' intention towards buying green products in a developing nation: Extending the theory of planned behavior. *J. Clean. Prod.* **2016**, *135*, 732–739. [CrossRef]
20. Biswas, A.; Roy, M. Leveraging factors for sustained green consumption behavior based on consumption value perceptions: Testing the structural model. *J. Clean. Prod.* **2015**, *95*, 332–340. [CrossRef]
21. Minton, A.P.; Rose, R.L. The effects of environmental concern on environmentally friendly consumer behavior: An exploratory study. *J. Bus. Res.* **1997**, *40*, 37–48. [CrossRef]
22. Suki, N.M. Consumer environmental concern and green product purchase in Malaysia: Structural effects of consumption values. *J. Clean. Prod.* **2016**, *132*, 204–214. [CrossRef]
23. Chéron, E.; Sudbury-Riley, L.; Kohlbacher, F. In pursuit of happiness: Disentangling sustainable consumption, consumer alienation, and social desirability. *J. Consum. Policy* **2021**, *45*, 149–173. [CrossRef]
24. Van der Merwe, D.; Bosman, M.; Ellis, S. Consumers' opinions and use of food labels: Results from an urban–rural hybrid area in South Africa. *Food Res. Int.* **2014**, *63*, 100–107. [CrossRef]
25. denkstatt. 2015, April. Nachhaltigkeit. Retrieved from Scheyer Verpackungstechnik. Available online: https://www.scheyer.at/de/verpackungstechnik/nachhaltigkeit (accessed on 20 April 2022).
26. Orme, B.; Howell, J. Application of covariates within Sawtooth Software's CBC/HB Program: Theory and Practical Example. *Sawtooth Softw. Res. Pap. Ser.* **2009**, *98382*, 1–20.
27. Meyerding, S.G.; Kürzdörfer, A.; Gassler, B. Consumer preferences for superfood ingredients—The case of bread in Germany. *Sustainability* **2018**, *10*, 4667. [CrossRef]
28. Lanza, S.T.; Rhoades, B.L. Latent class analysis: An alternative perspective on subgroup analysis in prevention and treatment. *Prev. Sci.* **2013**, *14*, 157–168. [CrossRef]
29. Leech, R.M.; Worsley, A.; Timperio, A.; McNaughton, S.A. Temporal eating patterns: A latent class analysis approach. *Int. J. Behav. Nutr. Phys. Act.* **2017**, *14*, 3. [CrossRef]
30. Nylund, K.L.; Asparouhov, T.; Muthén, B.O. Deciding on the number of classes in latent class analysis and growth mixture modeling: A Monte Carlo Simulation Study. *Struct. Equ. Model.* **2007**, *14*, 535–569. [CrossRef]
31. Kaiser, H.F. An Index of Factorial Simplicity. *Psychometrika* **1974**, *39*, 31–36. [CrossRef]
32. Bland, J.M.; Altman, D.G. Statistics notes—Cronbach's alpha. *BMJ* **1997**, *314*, 572. [CrossRef] [PubMed]
33. Field, A. *Discovering Statistics Using SPSS*, 3rd ed.; Sage: London, UK, 2009.
34. Golub, O.V.; Sanzharovskii, A.Y.; Mikhailidi, D.K.; Vartanyan, M.A. Carbon footprint of the life cycle of glass containers. *Glass Ceram.* **2022**, *79*, 306–311. [CrossRef]
35. Tanner, C.; Jungbluth, N. Evidence for the coincidence effect in environmental judgments: Why isn't it easy to correctly identify environmentally friendly food products? *J. Exp. Psychol.* **2003**, *9*, 3–11. [CrossRef]
36. Kühnapfel, J.B. *Scoring und Nutzwertanalysen—Ein Leitfaden für die Praxis*; Springer: Wiesbaden, Germany, 2021.
37. Waldner, M.; BFSV e.V., Verpackungsinstitut Hamburg, and HAW Hamburg, Hamburg, Germany; Bock, M.; Interviewer, HAW Hamburg, Hamburg, Germany. Information about Utility Value Analysis. Personal Communication, 3 May 2022.
38. Shaikh, S.; Yaqoob, M.; Aggarwal, P. An overview of biodegradable packaging in food industry. *Curr. Res. Food Sci.* **2021**, *4*, 503–520. [CrossRef]

sustainability

MDPI

Article

Integral Recovery of Almond Bagasse through Dehydration: Physico-Chemical and Technological Properties and Hot Air-Drying Modelling

Stevens Duarte, Ester Betoret, Cristina Barrera, Lucía Seguí and Noelia Betoret *

Instituto Universitario de Ingeniería de Alimentos para el Desarrollo, Universitat Politècnica de València, 46022 Valencia, Spain; steduase@doctor.upv.es (S.D.); mesbeval@upvnet.upv.es (E.B.); mcbarpu@tal.upv.es (C.B.); lusegil@upvnet.upv.es (L.S.)

* Correspondence: noebeval@tal.upv.es; Tel.: +34-963877000 (ext. 73658)

Abstract: Recovering waste from industrial food processes and developing new healthy foods as plant protein sources has been a major focus of scientific research and industrial innovation in food. Thus, the consumption of plant-based beverages from soy, oat, or almond has been promoted. In the case of almonds, the resulting solid bagasse has an interesting nutritional profile and its transformation into a powdered product could be a valuable option for the food industry. The main objective of this work was to determine the effect of hot air drying at 60 and 70 °C and freeze-drying on the physicochemical, water interaction, emulsifying and antioxidant properties of powdered almond bagasse. Furthermore, hot air-drying curves have been modelled and isotherms at 20 °C have been performed. The proximate composition of the powder revealed a protein content of 15% and a fat content of 25%, which makes it a remarkably different powder from those obtained from other vegetable residues such as fruits and vegetables. This composition was decisive in the effect of the drying method and drying temperature, and no significant differences were observed on the physico-chemical or antioxidant properties regardless of the drying method used. However, freeze-drying resulted in a powder with a more homogeneous particle size distribution and better oil-interaction properties, especially with higher emulsifying activity and stability.

Keywords: plant-based almond drink; almond; solid bagasse; air drying; freeze drying; sorption isotherms

1. Introduction

The food industry has become increasingly aware of the impact of food waste in economics and environment, and the need to reduce it. In fact, primarily motivated by the fulfilment of the Sustainable Development Goals (SDG), 71% of Spanish companies have a defined internal strategy to fight against food waste [1]. The production of new functional ingredients, biofuel production, or bioactive compounds extraction are some of the most considered strategies for food processing residues valorization [2–4]. The composition profile of the residues and their physico-chemical properties must be known in order to identify the opportunity for revalorization and to determine the possible uses [5].

The new trends in food development have been defined in the last years by the increased consumer awareness for health and sustainability and the growing incidence in allergies or food intolerances. Thus, the consumption of plant-based food has been promoted. Plant-based beverages or vegetable drinks are a clear example of this new orientation; more weight is being put behind them as an alternative to the consumption of dairy drinks. Among these, vegetable drinks such as soy, oats, rice, almond, and coconut stand out. In the manufacturing process, the raw material is soaked in water, milled, and filtered, resulting in a liquid phase that will constitute the vegetable beverage. The remaining solid material is usually referred to as press cake or bagasse, and it is usually discarded or used for animal feed or as fertilizer [6].

Regarding the almond, a relevant area is dedicated to its cultivation in Spain, only behind the olive and the grape [7]. Its consumption as a nut is growing due to the healthy properties associated with its unsaturated fatty acids (56%), proteins (23%), fiber (11%) and other carbohydrates (7%), minerals, and vitamins content [8]. Additionally, and motivated by new consumer trends, it is being increasingly used as the raw material for obtaining vegetable almond drink. The resulting solid bagasse has an interesting nutritional profile, which makes it very attractive for valorization. Its transformation into a powdered product with good nutritional properties for use as a functional food ingredient could be an option [9,10]. Determining its physico-chemical, technological, and functional properties is essential in determining its best use.

The main objective of this study was to determine the effect of hot air-drying at 60 and 70 °C and freeze-drying on the physico-chemical, water interaction, emulsifying, and antioxidant properties of powdered almond bagasse. Furthermore, hot air-drying curves have been modelled and isotherms at 20 °C have been performed.

2. Materials and Methods

2.1. *Process for Obtaining Almond Bagasse and Almond Bagasse Powder*

Natural peeled almonds were purchased from a local supermarket and ground with tap water in a ratio of 1/9 (*w/w*). A domestic food processor (Thermomix®, Vorwerk, Spain) at 10,000 rpm for 20 s was used. The grind was then filtered with a stainless steel 500 µm sieve and the almond bagasse was recovered for further characterization and processing. The recovered bagasse mass was about 82% of the rehydrated kernel mass.

For obtaining the dried almond bagasse, the moist almond bagasse was distributed homogeneously in plastic grids with a nominal opening of 2 mm and then introduced into the dryer until a water activity (a_w) below 0.3 was reached. A convective dryer (Pol-eko Aparatura, Katowice, Poland) with cross-flow air at a velocity of 10 m/s at 60 or 70 °C for 10 h and 7 h, respectively, was used to obtain air dried (HAD) bagasse, and a freeze-dryer (Telstar, Lioalta-g) was used to obtain the freeze-dried (LYO) one from almond bagasse previously frozen at −40 °C for 24 h. The inlet air to the convective dryer was ambient air at 25 °C and 25% of relative humidity. After that, the dried almond bagasse was ground using a food processor (Thermomix®, Vorwerk, Spain) at 4000 rpm for 20 s in intervals of 5 s and then at 10,000 rpm for 20 s in intervals of 5 s, thus obtaining almond bagasse powders with coarse granulometry. Finally, the powders were stored at 20 °C in light-opaque glass jars to prevent deterioration and oxidation reactions.

During the hot air-drying experiments, the samples weight change was registered. The evolution of the moisture content was determined from the initial moisture content and the mass of the samples at each time. Plotting the moisture on dry basis versus time made it possible to graph the drying curves and, from these, the drying rate curves. Data were modeled according to a lineal empirical and diffusional models. The goodness of fit was assessed by the coefficient of determination (R^2) (Equation (1)), the root mean square error (RMSE) (Equation (2)), and the mean relative error (MRE) (Equation (3)). For the best fit, the R^2 should be high and RMSE and MRE should be low.

$$R^2 = 1 - \frac{\sum_{i=1}^{n}\left(x_{exp,\,i} - x_{pred,\,i}\right)^2}{\sum_{i=1}^{n}\left(x_{exp,\,i} - \overline{x}\right)^2} \quad (1)$$

$$RMSE = \frac{\sqrt{\sum_{i=1}^{N}\left(x_{exp,\,i} - x_{pred,\,i}\right)^2}}{N} \quad (2)$$

$$MRE = \frac{1}{N}\sum_{i=1}^{N}\frac{\left|x_{exp,\,i} - x_{pred,\,i}\right|}{x_{exp,\,i}} \quad (3)$$

where x represents the variable under consideration, i.e., the velocity in the linear model and the reduced driving force in the diffusional model; $\overline{x}$: represents the mean value; N is the number of determinations; *exp.*; experimental. *pred*: predicted by the model.

2.2. Analytical Determinations

The water activity (a_w) of almond bagasse and almond bagasse powders (air dried at 60 °C and 70 °C and freeze-dried) was determined with a dew point hygrometer (DECAGÓN Aqualab 4TE) at 20 °C. The moisture content was determined following the official method in dried fruits established by the AOAC [11]. The total soluble solids (TSS) were determined by refractometry. For this, a dilution of the sample in distilled water was carried out in a ratio of 1:10 (m/v) and the Brix degrees were measured by means of a refractometer (ABBE ATAGO 3-T) thermostated at 20 °C. The fat content of almond bagasse was determined by Soxhlet extraction with petroleum ether according to the method established by the AOAC [12]. A relation of 5 g sample/90 mL solvent at 290 °C was used. The protein content was determined by the Kjeldahl method, considering 5.18 as the conversion factor from N to protein [13]. Different Van Soest fiber fractions, including neutral detergent fiber, which corresponds to the lignin, cellulose, and hemicellulose contents (NDF), acid detergent fiber, which corresponds to the lignin and cellulose contents (ADF), and lignin with acid detergent, which corresponds to the pure lignin content (LDF), were determined [14]. The values were used to estimate hemicellulose, cellulose, and lignin content. The ash determination was carried out by incineration of the material in a muffle at 550 °C [15].

Water Interaction and Emulsifying Properties

The solubility (SD) was determined following the procedure described by Mimouni et al. [16], in which the mass fraction of a dissolved solid (SS) in a rehydrated sample (TS) is determined. The hygroscopicity was evaluated according to the method described by Cai and Corke [17]; 0.5 g of each sample was weighed in glass crucibles and taken to an airtight chamber next to a saturated solution of sodium sulfate (Na_2SO_4) for 7 days at 25 °C. Wettability, defined by the time it takes for a sample to become wet in its entirety, was determined by weighing 2 g of each powder sample slowly poured into a beaker with 20 mL of distilled water, and measuring the time it took to become fully wet [18]. The swelling capacity (CS) was obtained from the ratio between volume occupied by 1 g of sample and that after hydration for 18 h at 25 °C [19,20]. Water holding capacity (WHC) is defined as the amount of water retained by the sample without applying any external force. It was determined by measuring the water content of the precipitate after mixing 0.2 g of sample and 10 mL of distilled water and left to stand for 18 h at 25 °C [19]. Water retention capacity (WRC) is defined as the ability of a sample to retain water after being subjected to an external force such as the centrifuge [19]. For its determination, 1 g of sample was weighed in a graduated conical tube and 10 mL of distilled water was added and left to stand for 18 h at 25 °C. After this time, it was centrifuged for 30 min at 2000 rpm, the supernatant was removed, and the sedimented residue was weighed. The oil retention capacity was evaluated following the methodology proposed by Garau et al. [21]. First, 0.2 g of sample and 1.5 g of sunflower oil were mixed and left to stand overnight at 20 °C. After that, the mix was centrifuged at 3416 rpm for 5 min, and with a Pasteur pipette the supernatant was removed and the weight of the residue was obtained. The oil retention capacity was evaluated based on the increase in the weight of the sample, and the results were expressed in g of oil absorbed by g of the initial sample. The emulsifying activity was determined following the methodology proposed by Yasumatsu et al. [22]. To carry out the procedure, a 2% (*w/v*) sample-water solution was prepared. Next, 7 mL of this solution was mixed with 7 mL of sunflower oil and homogenized for 5 min in a vortex at 2400 rpm. Finally, it was centrifuged at 10,000 rpm for 5 min and the volume of the emulsion formed was calculated by the ratio between the emulsion volume and the total fluid volume. Emulsifying stability was determined following the methodology proposed by Yasumatsu et al. [22]. For this, a 2% (*w/v*) sample-water solution was prepared. Then, 7 mL of this solution was mixed with 7 mL of sunflower oil and homogenized for 5 min in a vortex at 2400 rpm. Finally, it was heated to 80 °C for 30 min, allowed to cool, and centrifuged at 2000 rpm for 5 min. Emulsifying stability was calculated as the ratio between the emulsion volume and the total fluid volume.

2.3. Particle Size

The particle size of almond bagasse powders was determined by the wet method. Laser diffraction equipment (Masterizer, Malvern Instruments Limited, Worcester, UK) with a measurement range between 0.02 and 200 microns equipped with a blue light of 470 nm wavelength was used. A small amount of sample was diluted in deionized water until reaching an obscuration of 8–9%. Finally, the particle size distribution was obtained and was characterized by the mean diameter of equivalent volume (D [3,4]), equivalent diameter calculated from the area of the particles (D [2,3]), and, finally, d90, d50 and d10, representing the percentiles of the distribution, i.e., the volume of particles below 90%, 50%, and 10% of the particles analyzed, respectively.

2.4. Optical Properties

The CIE*L*a*b* coordinates were measured with a spectrocolourimeter (MINOLTA, CM-3600D, Japan), considering the standard light source D65, the 10° standard observer, and the surface reflectance spectra between 400 and 700 nm. The chroma (C_{ab}) and the color differences (ΔE) of the powders compared to almond bagasse were calculated using Equations (4) and (5), respectively.

$$C_{ab} = \sqrt{a^2 + b^2} \tag{4}$$

$$\Delta E = \sqrt{(\Delta L^*)^2 + (\Delta a^*)^2 + (\Delta b^*)^2} \tag{5}$$

2.5. Antiradical Capacity and Total Phenols Content

For the extraction of phenols and other components with antiradical capacity, a methanol-water mixture 80:20 (*v*/*v*) was prepared and used as a solvent in the relation 1 g sample/100 mL solvent. After 1 h of magnetic stirring, the mix was centrifuged (Selecta, "Medrifriger BL-S") at 10,000 rpm for 5 min at 20 °C. Determinations were made on the supernatant, hereinafter referred to as extract.

2.5.1. Total Phenol Content

The determination of total phenols was performed following the colorimetric method of Folin–Ciocalteu [23]. In a spectrophotometry bucket, 0.125 mL of extract, 0.125 mL of the Folin–Ciocalteu reagent (Sigma-Aldrich, Darmstadt, Germany) and 0.5 mL of bidistilled water were added in that order and allowed to react for 6 min. After this time, 1.25 mL of 7% (*m*/*v*) sodium carbonate solution and 1 mL of distilled water were added. As a reference, a target was used where the sample was replaced by bidistilled water and allowed to react for 90 min. Finally, the absorbance was measured at 765 nm in a spectrophotometer (Thermo Scientific, Helios Zeta U/Vis). The results obtained were compared to a standard curve of gallic acid (purity ≥ 98%, Sigma-Aldrich) and expressed as mg of gallic acid equivalents/g of dry matter (mg GAE/g dm).

2.5.2. Antiradical Capacity by DPPH and ABTS Methods

The antioxidant capacity was determined following the DPPH method described by Stratil et al. [24] with some modifications. First, 0.1 mL of the extract and 2.9 mL of the methanol-DPPH solution (0.394 of DPPH reagent/mL methanol) were mixed and absorbance was measured at 517 nm in a spectrophotometer (Thermo Scientific, Helios Zeta U/Vis). The results were expressed as mg of trolox equivalent/g of dry matter (mg TE/g dm) using the Trolox calibration line ($C_{14}H_{18}O_4$, purity ≥ 7%, Sigma-Aldrich) as the reference standard antioxidant, for the range of concentrations between 0 and 500 mg/L.

The antioxidant activity was also evaluated by the ABTS radical method (2,20-azobis-3-ethyl benzothiazolin-6-sulfonic acid) [25]. A solution including the radical ABTS 7 mM and potassium persulfate 2.45 mM in distilled water was prepared and incubated in darkness at room temperature for 16 h. Once this time had elapsed, a dilution with phosphate buffer was carried out to reach an absorbance of 0.7 ± 0.02 at 734 nm. Then, in a spectrophotometry bucket, 0.1 mL of extract with 2.9 mL of ABTS solution was reacted. As a reference, a white where the sample was replaced by bidistilled water was prepared. The absorbance was measured after 0, 3 and 7 min of reaction at a wavelength of 734 nm in a spectrophotometer (Thermo Scientific, Helios Zeta UV/Vis). The results were expressed as mg of trolox equivalent/g of dry matter (mg TE/g dm), using the Trolox calibration line ($C_{14}H_{18}O_4$, purity ≥ 7%, Sigma-Aldrich) as the reference standard antioxidant for the range of concentrations between 0 and 500 mg/L.

2.6. Sorption Isotherms

Sorption isotherms were determined following the gravimetric method described by Wolf et al. [26]. This method uses saturated salt solutions to keep a known and controlled humidity environment within a closed vessel at specific temperature conditions. First, 1 g of sample was placed in a closed jar at 20 °C together with one of the next saturated salt solutions: LiCl (a_w = 0.1), CH_3COOK (a_w = 0.23), $MgCl_2$ (a_w = 0.32), K_2CO_3 (a_w = 0.43), Mg $(NO_3)_2$ (a_w = 0.52), NaCl (a_w = 0.75), KCl (a_w = 0.85), and $BaCl_2$ (a_w = 0.90). The samples were weighed every eight days until a constant weight was reached.

2.7. Statistical Analysis

The results were statistically analyzed with Statgraphics software (Centurion XVI.I, Statpoint Technologies, Inc., Warrenton, VA, USA) at a 95 % confidence level (p-value ≤ 0.05). The normality of the data was tested with the Shapiro–Wilk test ($p > 0.05$). The data were processed by simple ANOVA after checking the normality of the data. For each processing treatment, three different experiments with three replicates each were carried out. Significant differences (p-value < 0.05) among groups were determined by Fisher's LSD test.

3. Results and Discussion

3.1. Properties of Almond Bagasse Powders

Table 1 shows the composition, physico-chemical, water interaction, emulsifying properties, and color of fresh, air dried, and freeze-dried almond bagasse powders. Dehydration, in all cases, reached a water activity lower than 0.3, which is the recommended limit to ensure the stability of powdered products [27,28]. Although the water activity limit for microbial growth is 0.90 for most bacteria and 0.87 and 0.75 for most yeasts and fungi, a water activity limit lower than 0.3 assures kinetic stability in powdered products since it is guaranteed that there is no free water that can participate in chemical and enzymatic reactions. The moisture content in the final samples was low, as isotherms showed a very low water binding capacity (see isotherms section). Thus, more than 98% of the water is easily removed during drying.

Considering the fat and protein content, the almond bagasse retained a high percentage of fat and protein from fresh almond and there were no significant differences between fresh and dehydrated samples. Fat content remained around 25% (0.25 g/g_{dm}) in the bagasse and protein reached 16–17% (0.16–0.17 g/g_{dm}); the initial values in fresh almond were around 54% and 25%, respectively [29]. In a study carried out with fresh baru almond, fat content around 39–43% was reported, and the protein content was around 23–28%, slightly higher than those obtained for almond bagasse [30]. Compared with other cereal by-products of interest to the food industry, protein content was similar to those obtained in rice bran (0.14 g protein/g) [31], oat bran (0.17 g protein/g) [32], by-product from tofu (0.15 g protein/g) [33], and soybean residue (0.15 g protein/g) [33], but lower than that for fresh okara (0.39 g protein/g) [34] and rice bran (0.22 g protein/g) [35]. Particular attention should be paid to the fat content. Stability in low-moisture, fat-containing foods is highly dependent on the characteristics of the matrix, its microstructure, and the presence of other macronutrients such as protein. Oxidation mechanisms are complex and need to be studied on a case-by-case basis to ensure proper packaging and storage [36].

Table 1. Composition, physico-chemical, water interaction, emulsifying properties and color of fresh almond bagasse and, air dried (HAD60: hot air dried at 60 °C; HAD70: hot air dried at 70 °C) and freeze-dried (LYO) almond bagasse powders. The values in brackets for fresh bagasse refer to the composition expressed in g/g of raw material. Mean ± standard deviation of three repetitions. Different superscripts letters in the same line indicate statistically significant differences with a confidence level of 95%. $_{dm}$, dry matter; $_w$, water; X_w, water content; X_{ss}, soluble solids content; WHC, water holding capacity; WRC, water retention capacity.

	FRESH	HAD60	HAD70	LYO	*p*-Value
		Physico-chemical properties			
a_w	0.99 ± 0.08^{a}	0.23 ± 0.04^{b}	0.20 ± 0.06^{bc}	0.13 ± 0.02^{c}	0.0000
Fat (g/g_{dm})	0.25 ± 0.002^{a} (0.11)	0.252 ± 0.002^{a}	0.253 ± 0.004^{a}	0.250 ± 0.006^{a}	0.7106
Protein (g/g_{dm})	0.15 ± 0.03^{a} (0.07)	0.16 ± 0.04^{b}	0.16 ± 0.03^{b}	0.165 ± 0.008^{b}	0.0030
X_w (g/g_{dm})	1.262 ± 0.011^{b} (0.558)	0.014 ± 0.002^{a}	0.015 ± 0.012^{a}	0.02 ± 0.08^{a}	0.0000
Ashes (g/g_{dm})	0.031 ± 0.011^{a} (0.014)	0.031 ± 0.007^{a}	0.03 ± 0.06^{a}	0.030 ± 0.012^{a}	0.0000
Fiber Van Soest (g/g_{dm})	0.47 ± 0.02^{a} (0.21)	0.45 ± 0.02^{a}	0.50 ± 0.03^{a}	0.50 ± 0.03^{a}	0.6605
Cellulose and lignine (g/g_{dm})	0.17 ± 0.02^{a} (0.08)	0.20 ± 0.05^{ab}	0.20 ± 0.15^{ab}	0.21 ± 0.02^{b}	0.0005
Hemicellulose (g/g_{dm})	0.23 ± 0.04^{a} (0.10)	0.260 ± 0.014^{a}	0.290 ± 0.012^{a}	0.295 ± 0.002^{a}	0.0008
X_{ss} (g_{ss}/g_{dm})	0.013 ± 0.003^{a} (0.006)	0.013 ± 0.004^{a}	0.013 ± 0.004^{a}	0.014 ± 0.004^{a}	0.6810
		Water interaction properties			
Solubility (%)	-	29 ± 1^{b}	26.2 ± 2.2^{a}	30.1 ± 1.1^{c}	0.0000
Hygroscopicity (g_w/g)	-	0.17 ± 0.06^{a}	0.17 ± 0.17^{a}	0.17 ± 0.03^{a}	0.9763
Wettability (s)	-	8.3 ± 1.1^{a}	8.9 ± 0.6^{a}	8.3 ± 1.1^{a}	0.7458
Swelling capacity (mL_w/g)	-	4.51 ± 0.08^{a}	4.51 ± 0.08^{a}	4.51 ± 0.08^{a}	1.0000
WHC (g_w/g_{dm})	-	2.9 ± 0.5^{a}	2.6 ± 0.2^{a}	8.4 ± 1.8^{b}	0.0009
WRC (g_w/g_{dm})	-	4.5 ± 0.2^{a}	4.6 ± 0.2^{a}	5.91 ± 0.08^{b}	0.0000
		Oil interaction properties			
Oil retention ability (g_o/g_s)	-	2.3 ± 0.5^{a}	2.6 ± 0.2^{a}	4.2 ± 0.06^{b}	0.0047
Emulsifying activity (%)	-	19 ± 2^{a}	20 ± 2^{a}	34 ± 2^{b}	0.0002
Emulsifying stability (%)	-	20 ± 2^{a}	24 ± 2^{a}	59 ± 2^{b}	0.0000
		Colour			
L	73.68 ± 0.07^{a}	62.358 ± 0.010^{c}	58.236 ± 0.002^{d}	66.561 ± 0.001^{b}	0.0010
a*	4.88 ± 0.02^{d}	4.999 ± 0.009^{c}	6.487 ± 0.009^{a}	6.039 ± 0.002^{b}	0.0039
b*	11.62 ± 0.04^{d}	14.279 ± 0.006^{c}	16.143 ± 0.017^{a}	15.026 ± 0.014^{b}	0.0030
C	12.61 ± 0.05^{d}	15.128 ± 0.08^{c}	17.398 ± 0.014^{a}	16.194 ± 0.012^{b}	0.0204
ΔE	-	11.625 ± 0.010^{b}	16.167 ± 0.003^{a}	7.971 ± 0.06^{c}	0.0001

According to the fiber content, no significant differences among treatments or fresh almond bagasse were detected. Van Soest fiber includes cellulose and lignin as insoluble fraction and the hemicellulose as a more soluble one. Values of total fiber were above those reported for okara fresh matter (13.84 g/100 g) [37], rice bran (28.6 g/100 g) [35], and oat bran (15.55 g/100 g of dry matter) [32], but they were below those of other by-products, such as by-product from tofu (58.6 g/100 g of dry matter) [33], bran fiber rice (53.25 g/100 g of dry matter) [35], or solid by-product tiger nut (59.71 g/100 g) [38], and similar to carrot skin (45.45–49.23 g/100 g of dry weight) [39]. Soluble dietary fibers such as pectin cannot be quantified by the fiber determination method used. However, this fraction could be estimated by the difference between the total mass and the total of the macronutrients considered. As the sum of fat, protein, fiber and water is 100%, the more soluble fiber including pectin can be considered negligible. Almond bagasse powders could be used as ingredients promoting intestinal transit more than an ingredient conferring viscosity since high content in soluble than insoluble was fiber observed in all cases. Soluble fiber is the one that confers viscosity properties, ability to form gels, and emulsifying capacity, while insoluble fiber with a greater porosity and lower density promotes intestinal transit [40].

Water solubility values ranged from 26.2% to 30.1%, without significant differences between hot air-dried and freeze-dried samples. The freeze-dried samples showed slightly high solubility levels, presumably caused by the more severe structural damage induced by the freezing and subsequent water sublimation during lyophilization. An increase in the air-drying temperature resulted in a decrease in solubility, probably due to physical changes affecting macromolecules during the drying process. These physical changes during hot air drying could promote the formation of a surface crust, which can hinder the interaction between molecules and water [41]. Solubility values were lower when compared to those from other fruit powders such as passion fruit (44.6% to 57.56%) [42] or pineapple juice powder (81.56%) [43]. Nevertheless, the results were closer to those obtained for oat bran (ranging from 11.70% to 26.32% depending on the drying process applied) [44]. Clearly, a higher percentage of macromolecules such as insoluble fiber and proteins in the composition of by-products such as bran or bagasse provides lower solubilities. Additionally, the presence of a high percentage of fat makes the interaction with water molecules even more difficult.

Hygroscopicity is the capacity of a material or powder to absorb moisture and come into equilibrium with the relative humidity of the environment. The low water content of food powders could contribute to their high hygroscopicity, which gives rise to sticky and caked powders with low porosity, therefore decreasing their ability to rehydrate and retain aromas [45]. Food powder is considered good if it has low hygroscopicity [45]. According to Callahan et al. [46], a material can be considered non-hygroscopic when an increase of less than 20% (w/w) in moisture content above 90% relative humidity is observed after one week. Almond bagasse powder gained 0.17 g of water/g (17%) when equilibrated at 97% relative humidity after one week and was therefore non-hygroscopic. Non-significant differences were observed among the samples.

Wettability, swelling capacity, the water holding capacity (WHC), and the water retention capacity (WRC) are largely conditioned by the particle size and composition, mainly the fiber type and fat content. Wettability and the swelling capacity of almond bagasse powders were not significantly affected by the drying method or air temperature. However, WHC and WRC are significantly higher in the lyophilized powders. A different size distribution (Figure 1) with a single peak indicative of a larger volume of larger particles could be the explanation for these differences. According to Bai et al. [44], the larger the particle size, the higher the wettability since water molecules can permeate through the larger voids left between the particles. Regarding composition, soluble fiber has a high capacity to retain water and expand to form a viscous solution, while insoluble fiber can also absorb and retain water in its fibrous matrix but in a lower quantity; fat, on the other hand, hinders any interaction with water. Lecumberry et al. [47] reported results for WRC in apple and orange pectin of 16.51 $\pm$ 3.77 and 28.07 $\pm$ 5.37 g water/g dry matter, respectively), these results being higher than those obtained for almond bagasse, which is consistent with the higher content of insoluble fiber and the presence of fat in the almond bagasse. Nevertheless, similar results were obtained in lulo bagasse (8.2 $\pm$ 0.7), a material also with a high content of insoluble fiber [48]. Bai et al. [44] provided data on WHC in oat bran (5.95 to 6.48 g of water/ g of dry matter), which were similar to those obtained in freeze-dried almond bagasse and slightly lower in hot air-dried powders.

	D [4,3]	D [3,2]	d (0.1)	d (0.5)	d (0.9)
HAD60	112 ± 3^{a}	25.8 ± 0.3^{a}	12.3 ± 0.1^{a}	81 ± 1^{a}	264 ± 9^{a}
HAD70	120 ± 4^{ab}	26.4 ± 0.3^{a}	12.6 ± 0.1^{a}	81 ± 2^{a}	291 ± 12^{b}
LYO	125 ± 11^{b}	26.0 ± 0.8^{a}	11.9 ± 0.4^{b}	94 ± 3^{b}	282 ± 25^{ab}
p-value	0.0334	0.1524	0.0032	0.0000	0.0451

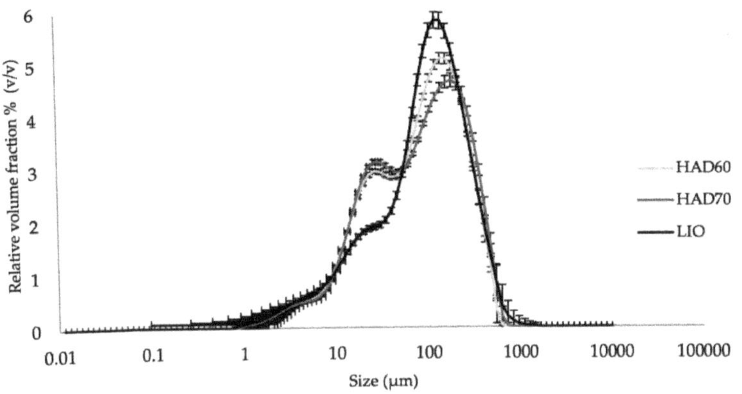

Figure 1. Particle size of hot air dried (HAD60: hot air dried at 60 °C; HAD70: hot air dried at 70 °C) and freeze-dried (LYO) almond bagasse powders. Mean ± standard deviation of five repetitions. Different superscripts letters in the same column indicate statistically significant differences with a confidence level of 95%.

Regarding oil interaction properties, the results obtained for almond bagasse powders showed good emulsifying properties, such as emulsifying activity and emulsifying stability. Significant differences were detected between freeze-dried and air-dried samples, regardless of air temperature. The values obtained for freeze-dried powders were higher (Table 1). The emulsifying capacity is associated with the presence of hydrophilic and hydrophobic groups. The high protein content present in the almond bagasse justifies its good oil-interaction properties. In freeze-dried samples, the increased structural damage caused by freezing and sublimation contributes to the breakdown of complex molecules, leaving more hydrophilic and hydrophobic groups available for interaction and consequently improving the oil-interaction properties [49]. Regarding oil retention ability, similar values were reported for commercial fibers from lemon, orange, peach, apple, and persimmon (2.5 to 2.9 g oil/ g sample) [50]. Similar emulsion stability to that of freeze-dried almond bagasse powder was obtained for peas (59.4% ± 1.0) and lentils (55.0% ± 2.5). The emulsifying activity of almond bagasse could be compared with the results obtained for peas (40.9% ± 0.7) and lentils (39.9% ± 1.0) [51].

Associated with browning and oxidation reactions, all samples experienced color differences when compared to the fresh almond bagasse (Table 1). These changes gave the samples more yellowish-red tones, denoted by higher values of the a^* and b^* coordinates. The saturation (C) in all cases shows a low value, being lower in the fresh bagasse. According to Bodart et al. [52], color differences are imperceptible to the human eye when they are $\Delta E < 1$. Small differences can be seen when $1 < \Delta E < 3$ and will be visibly evident when the value of $\Delta E > 3$. Since in all samples the values were higher than 3, the changes were clearly perceptible. However, the color difference in the freeze-dried powder was smaller since freeze-drying occurs under vacuum and at low temperatures, minimizing oxidation processes.

Figure 1 shows the particle size distribution and characteristic parameters of hot air-dried and freeze-dried almond bagasse powders. Practically, a monomodal distribution for the freeze-dried powder and a bimodal distribution for air-dried powders were observed. In the freeze dried, the structural breakdown induced by/100 g water freezing and sublimation resulted in a more homogeneous particle size distribution and a slight shift in the maximum towards a smaller particle size. Probably, in hot air-dried samples, phase transitions in macromolecules such as carbohydrates and proteins, and their different mechanical resistance to crushing, resulted in a more heterogeneous distribution, and specifically a bimodal one. This distribution is quite common in carbohydrate and fiber-rich powders produced by hot-air drying, such as blueberry powder and the tangerine skin powder dried at 70 °C [53,54].

Regarding antioxidant properties (Table 2), the highest values of antiradical activities were obtained for fresh samples. No significant differences were observed between the different drying methods and temperatures used. The total phenols of the freeze-dried samples were very similar to those of the fresh samples. Freeze-drying occurs at low temperature and in vacuum conditions, which contributes to maintaining bioactive compounds with anti-radical activity such as phenols [40]. In hot air-drying treatments, structural damage and the presence of oxygen at high temperature resulted in higher degradation. However, for the inactivation of enzymes involved in some of the degradation reactions, the difference between 60 and 70 °C could be decisive. In terms of interaction with other molecules, it has been shown that dehydration can increase polyphenolic compounds, despite some degradation, because it can improve extraction and lead to a greater release of these compounds [55]. In almond bagasse, the macronutrient composition, consisting mainly of fiber and fat, could interact with the polyphenols and prevent them from getting released after processing. Comparing results from the DPPH and ABTS methods, the ABTS radical reacted with more antioxidant compounds. The lower reaction time of ABTS radical and its more hydrophilic nature enabled it to react in both organic and aqueous media.

Table 2. Total phenols content and antiradical capacity by DPPH and ABTS methods of fresh almond bagasse and, hot air-dried (HAD60: hot air-dried at 60 °C; HAD70: hot air-dried at 70 °C) and freeze-dried (LYO) almond bagasse powders. Mean ± standard deviation of three repetitions. Different superscripts letters for the same determination indicate statistically significant differences with a confidence level of 95%. $_{dm}$, dry matter; GAE, acid gallic equivalents; TE, Trolox equivalent.

	FRESH	HAD60	HAD70	LYO	*p*-Value
Total phenols (mg GAE/g_{dm})	0.59 ± 0.03^{a}	0.291 ± 0.012^{b}	0.33 ± 0.02^{b}	0.5 ± 0.2^{ab}	0.0000
DPPH (mg TE/g_{dm})	0.67 ± 0.06^{a}	0.296 ± 0.007^{b}	0.31 ± 0.03^{b}	0.32 ± 0.05^{b}	0.0154
ABTS (mg TE/g_{dm})	2.9 ± 0.2^{a}	0.96 ± 0.03^{b}	1.03 ± 0.07^{b}	1.121 ± 0.012^{b}	0.0000

The values for total phenols were quite similar to those reported for almond shell, ranging from 0.86 to 1.16 mg GAE/gdm [56], but higher values were found in fresh almonds (2.87 mg GAE/gdm), brazil nuts (2.44 mg GAE/gdm), hazelnuts (6.87 mg GAE/gdm), and pecans (1,81 mg GAE/gdm) [57]. Similar results were obtained in peach (0.51 mg GAE/gdm), fig (0.59 mg GAE/gdm), macadamias (0.46 mg GAE/gdm), and pines (0.32 mg GAE/gdm) [58,59].

3.2. Air Drying Kinetics

Figure 2 shows the drying and drying rate curves of thin-layer air-drying of almond bagasse at 60 and 70 °C. The almond bagasse was dried from the initial moisture of 55% to a final value of around 5.5%. The time needed to reduce the water content was 4.5 and 3.7 h at 60 and 70 °C, respectively. As expected, the statistical analysis revealed the significant effect (p-value < 0.05) of air temperature on water content removal during the process. When the air temperature increased, it had a greater capacity to retain water, promoting the drying process. At the same time, the temperature of the bagasse increased significantly, increasing the water diffusivity from the inner layers to the surface [60]. Furthermore, Ling et al. [61] suggested that in pasty products, such as sludge, this temperature increase was linked to a porosity reduction. Fresh almond bagasse was slightly pasty, so the reduction in porosity could have also contributed to the increase in the drying rate.

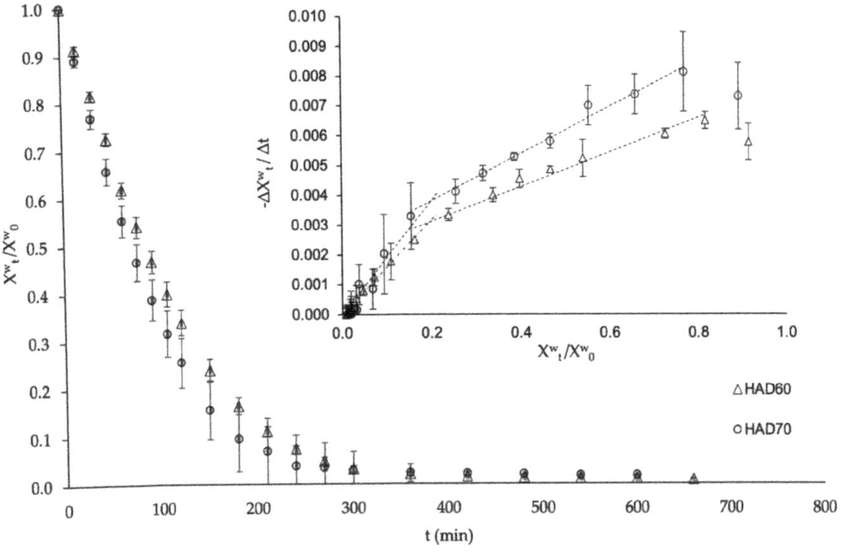

Figure 2. Air drying curves of almond bagasse at 60 and 70 °C. In the nested figure, the drying rate curves and the linear fits of the experimental data at 60 and 70 °C have been plotted. HAD: Hot air drying; X^w: water content (g water/g dry matter); 0, t are referring to initial and any other time. Mean values and standard deviation of three repetitions have been represented.

During the first few minutes of the air-drying process, the drying rate increased until it reached the highest value (Figure 2). This increase was associated with the progressive heating of the product when it comes into contact with the hot air. The experimental data revealed that this stage, which corresponds to the induction stage, had a duration of 20–30 min, depending on the drying temperature. After the induction period, in high moisture foods, a water-free layer over the entire surface of the food usually results in a constant drying rate [62]. However, the initial moisture content of the fresh bagasse was around 55%, which was low enough that there was no longer a free layer of water. Thus, drying rate curves revealed that the process at the temperature values took place in the falling rate period entirely. Two periods of declining drying rate were observed; in both cases, the decrease in drying rate was linear with the reduction in moisture ratio $\frac{Xw}{Xw_0}$. Therefore, it could be said that the drying process was controlled by internal water diffusion. In the first stage, when the bagasse had the higher water content, the reduction in velocity was lower than in the second stage when the bagasse was almost dry. This behavior was largely influenced by the composition and structural characteristics of the bagasse. Considering that the main components of the bagasse do not have a high water-holding capacity (this will be discussed later in the sorption isotherms section), it can be stated that its structural characteristics, in particular its porosity and particle size, determined the facility with which water molecules were removed. Additionally, the extent of compartmentalization associated with the crushing level influences physical interactions that also affect the rate of the process.

Modelling the drying curves and obtaining the kinetic parameters provides information on the mechanisms involved. Furthermore, it makes it possible to control the process by improving energy consumption and subsequently optimize the drying process for greater efficiency and a better quality final product. Numerous models have been used by researchers [63]. These are theoretical, semi-theoretical, and empirical models that usually correlate the moisture ratio with the drying time. Theoretical models provide insights to the mechanisms involved in water loss but offer complex mathematical solutions that are difficult to fit and manage. On the other hand, empirical models provide simple and fast solutions that are effective for practical operation management when the experimental conditions under which they are obtained correspond to the real operating conditions. Semi-theoretical models are the most applied and are generally derived from a direct solution of Fick's second law by assuming some simplifications.

In this study, the experimental data were fitted to an empirical model that establishes a linear correlation between the drying rate and the moisture ratio and to the simplified diffusional model, considering a single term of the serial progression from the integration of Fick's second law (Table 3). The simplified diffusional model usually fits well when drying occurs in the falling rate, as this is when the predominant mechanism is the diffusion of water from the innermost layers of the food samples to the surface. In the application of the equation, it was assumed that water diffusion occurred in a single direction and remained constant, the material was isotropic, and the moisture distribution uniform. The external resistance to water transport was negligible compared to the internal resistance and there was no shrinkage or swelling of the food material. The adjustment allowed the calculation of the effective moisture diffusivity (De) as a kinetic parameter to compare the facility with which water diffuses from the inner part of the bagasse to the outer part. The values obtained were 1.97×10^{-9} and 2.18×10^{-9} m^2/s for the temperatures of 60 and 70 °C, respectively. These values are within the range generally given for the moisture diffusion of food materials (10^{-11} to 10^{-6} m^2/s) [64,65].

Table 3. Air drying kinetics of almond bagasse at 60 and 70 °C. $\frac{X^w - X^w_\infty}{X^w_c - X^w_\infty}$: Dimensionless moisture ratio, $\frac{Xw}{Xw_0}$: Moisture ratio, $\frac{\Delta Xw}{\Delta t}$: Drying rate, De: Effective water diffusivity, L: Half-thickness of bagasse thin layer, t: time, R^2: Correlation coefficient, RMSE: Root mean square error, MRE: Mean relative error.

	Model equation	60 °C	70 °C
Linear empirical model	Stage 1 $\frac{\Delta Xw}{\Delta t} = k_1 \frac{Xw}{Xw_0} + k_2$	$\frac{Xw}{Xw_0} \in \lfloor 0.816,\ 0.2 \rfloor$	$\frac{Xw}{Xw_0} \in \lfloor 0.769,\ 0.18 \rfloor$
	k_1	0.006	0.008
	k_2	0.002	0.002
	R^2	0.971	0.983
	RMSE	6.40×10^{-4}	9.36×10^{-5}
	MRE	0.049	0.031
	Stage 2 $\frac{\Delta Xw}{\Delta t} = k'_1 \frac{Xw}{Xw_0}$	$\frac{Xw}{Xw_0} \in \lfloor 0.2,\ 0.02 \rfloor$	$\frac{Xw}{Xw_0} \in \lfloor 0.18,\ 0.022 \rfloor$
	k'_1	0.016	0.019
	R^2	0.995	0.921
	RMSE	1.26×10^{-5}	1.05×10^{-4}
	MRE	0.194	0.207
Difusional model	$\frac{X^w - X^w_\infty}{X^w_c - X^w_\infty} = \frac{8}{\pi^2} e^{\left(-\frac{D \cdot \pi^2 \cdot t}{4 \cdot L^2}\right)}$	$\frac{Xw}{Xw_0} \in \lfloor 0.816,\ 0.02 \rfloor$	$\frac{Xw}{Xw_0} \in \lfloor 0.769,\ 0.022 \rfloor$
	De (m^2/h)	7.11×10^{-6}	7.88×10^{-6}
	R^2	0.993	0.983
	RMSE	0.039	0.033
	MRE	0.331	0.310

Saravacos and Maroulis [66] investigated the effect of food properties on the drying kinetics of non-cellular structured food. They established the important effect of food structure and hygroscopicity and reported typical values of effective water diffusivity, varying from 50 to 0.01×10^{-10} m^2/s depending on hygroscopicity. Xiong et al. [67] showed that the effective diffusivity (De) was higher in pregelatinized samples and was found to be much higher through porous puffed pasta than regular pasta. Ruimin et al. [68] found that the total drying time of sludge particles with a diameter of 10 mm is not much different from that of particles with a diameter of 6 mm, while the total drying time of particles with a diameter of 18 mm increases significantly.

The goodness of the fit was determined by the correlation coefficient (R^2), the root mean square error (RMSE), and the mean relative error (MRE). It is generally accepted that an R^2 value higher than 0.93 and an MRE lower than 0.1 are good fits. Although, the MRE of the fit to the simplified diffusional Fick's model is too high, the correlation coefficient is good and could be accepted as an acceptable approach. In the case of the linear empirical model, the fit was more accurate.

3.3. Sorption Isotherms

Figure 3 shows the moisture sorption isotherms at 20 °C of hot air-dried almond bagasse powder at 60 °C (HAD60), at 70 °C (HAD70), and the freeze-dried one (LYO).

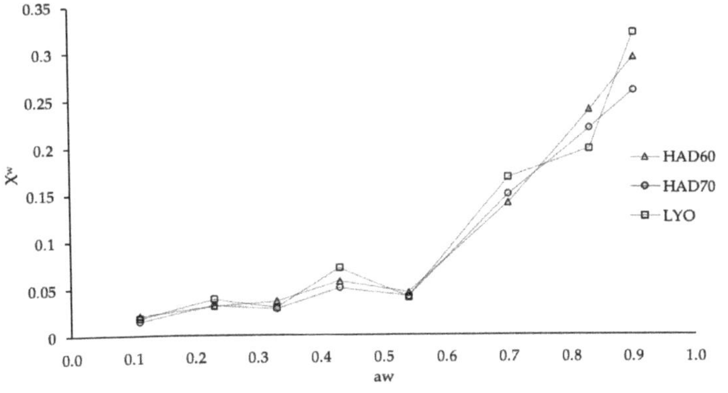

Figure 3. Sorption isotherms of hot air-dried (HAD60: hot air-dried at 60 °C; HAD70: hot air-dried at 70 °C) and freeze-dried (LYO) almond bagasse powders at 20 °C. X^w: water content (g water/gdm).

It can be observed that at rather low moisture values (~0.3 g water/g dry matter), water activity values of 0.9 are reached. The isotherm is very close to the x-axis, which indicates that the product has a very low water binding capacity, possibly influenced to a large extent by its fat content. Two practically linear sections can be identified; a rather flat first section for water activities equal to or less than 0.54, and a second section with a positive slope for water activities equal to or greater than 0.54 This results in a type III isotherm, which is quite common in non-porous foods. This shape appears when the net heat of sorption is small (specifically with a BET C value of less than 2). A small net heat of sorption indicates that the interactions between the water and the other components are weak and more linked to physical than chemical phenomena [69].

When comparing the isotherm with that obtained for raw almond powder [70], the typical plateau at very low a_w has disappeared. This plateau is associated with high water adsorption by complex molecules with many active points, such as carbohydrates or soluble proteins. These have been extracted during the production process of vegetable almond drink and are no longer present in the bagasse.

In powdered products, physical and chemical sorption phenomena are largely conditioned by the macromolecular structure of the product as well as by its chemical composition and the physical state of its components [71]. Regarding the macromolecular structure, in all cases, a powder with large and slightly caked particles was obtained, which greatly limits the adsorption phenomena. Considering the composition, the fat content, which is hydrophobic in nature, together with insoluble long-chain carbohydrates (insoluble fibre constituted mainly of cellulose and lignin) is high, and the water adsorption capacity is low.

Furthermore, the drying processes applied, such as hot air drying and freeze-drying, may have induced phase transitions aimed at the crystallization of some molecules, resulting in very small or zero stoichiometric hydration contents.

4. Conclusions

Hot air drying and freeze-drying were found to be suitable processes for obtaining a plant-based powder from the bagasse resulting from the production of vegetable almond drink. In all cases, a nutritious powder was obtained with low water binding capacity and therefore good properties for packaging and storage. However, due to its high fat content, it is worth studying its stability when stored. No clear trend was observed for the effect of the drying method (hot air or freeze-drying) on total phenolic content, antiradical capacity, physico-chemical properties, or interaction with water or oil. However, faster kinetics at 70 °C resulted in higher industrial productivity. Freeze-drying resulted in a powder with a more homogeneous particle size distribution and better oil-interaction properties, especially with higher emulsifying activity and stability. It would be the most recommended process to obtain a powder with emulsifying properties.

Author Contributions: Conceptualization, N.B.; methodology, N.B., S.D. and E.B.; data curation, S.D.; writing—original draft preparation, N.B. and S.D.; writing—review and editing, N.B., E.B., C.B. and L.S.; funding acquisition, N.B. All authors have read and agreed to the published version of the manuscript.

Funding: This research was funded by Universitat Politècnica de València, Ayuda PAID-11-21.

Conflicts of Interest: The authors declare no conflict of interest. The funders had no role in the design of the study; in the collection, analyses, or interpretation of data; in the writing of the manuscript; or in the decision to publish the results.

References

1. MAPA Ministerio de Agricultura, Pesca y Alimentación. Available online: https://www.mapa.gob.es/en/alimentacion/temas/desperdicio/ (accessed on 20 January 2023).
2. Feng, J.Y.; Wang, R.; Thakur, K.; Ni, Z.J.; Zhu, Y.Y.; Hu, F.; Zhang, J.G.; Wei, Z.J. Evolution of Okara from Waste to Value Added Food Ingredient: An Account of Its Bio-Valorization for Improved Nutritional and Functional Effects. *Trends Food Sci. Technol.* **2021**, *116*, 669–680. [CrossRef]
3. Trabold, T.A.; Rodríguez Alberto, D. Valorization of Food Processing By-Products via Biofuel Production. In *Sustainability of the Food System*; Academic Press: Cambridge, MA, USA, 2020; pp. 53–69. [CrossRef]
4. de Souza, T.S.P.; Dias, F.F.G.; Koblitz, M.G.B.; de Moura Bell, J.M.L.N. Effects of Enzymatic Extraction of Oil and Protein from Almond Cake on the Physicochemical and Functional Properties of Protein Extracts. *Food Bioprod. Process.* **2020**, *122*, 280–290. [CrossRef]
5. Saba, B.; Bharathidasan, A.K.; Ezeji, T.C.; Cornish, K. Characterization and Potential Valorization of Industrial Food Processing Wastes. *Sci. Total Environ.* **2023**, *868*, 161550. [CrossRef]
6. Lorente, D.; Duarte Serna, S.; Betoret, E.; Betoret, N. Opportunities for the Valorization of Waste Generated by the Plant-Based Milk Substitutes Industry. In *Advanced Technologies in Wastewater Treatment*; Elsevier: Amsterdam, The Netherlands, 2023; pp. 25–66, ISBN 9780323885102.
7. FAO. El Director General de la Industria Alimentaria Valora la Estabilidad de los Datos de Desperdicio Alimentario en 2019. Available online: https://www.mapa.gob.es/es/prensa/ultimas-noticias/el-director-general-de-la-industria-alimentaria-valora-la-estabilidad-de-los-datos-de-desperdicio-alimentario-en-2019/tcm:30-543935 (accessed on 4 June 2023).
8. Espinosa-Puerta, J.M. Aprovechamiento Integral del Residuo de la Industria de la Almendra. Available online: http://efaidnbmnnnibpcajpcglclefindmkaj/https://crea.ujaen.es/bitstream/10953.1/12079/1/Memoria.JM.Espinosa.pdf (accessed on 20 January 2023).
9. Castellanos Fuentes, A.P.; Bengoa, A.; Gagliarini, N.; Abraham, A.; de Escalada Pla, M.F.; Flores, S.K. Physicochemical and Functional Characterisation of a Food Ingredient Based on Okara Containing Probiotics. *Food Bioprod. Process.* **2022**, *135*, 74–86. [CrossRef]
10. Martins, Z.E.; Pinho, O.; Ferreira, I.M.P.L.V.O. Food Industry By-Products Used as Functional Ingredients of Bakery Products. *Trends Food Sci. Technol.* **2017**, *67*, 106–128. [CrossRef]
11. AOAC 934.06, 1934 AOAC 934.06-1934(1996), Loss on Drying (Moisture) in Dried Fruit: AOAC Official Method. Available online: http://www.aoacofficialmethod.org/index.php?main_page=product_info&products_id=695 (accessed on 11 February 2023).

12. AOAC 991.36, 1996 AOAC 991.36-1996, Fat(Crude) in Meat and Meat Products–Solvent: AOAC Official Method. Available online: http://www.aoacofficialmethod.org/index.php?main_page=product_info&cPath=1&products_id=2528 (accessed on 11 February 2023).
13. AOAC 950.48, 1950 AOAC 950.48-1950, Protein (Crude) in Nuts and Nut Products. AOAC Official Method. Available online: http://www.aoacofficialmethod.org/index.php?main_page=product_info&products_id=836 (accessed on 11 February 2023).
14. Mertens, D.R.; Collaborators. Gravimetric Determination of Amylase-Treated Neutral Detergent Fiber in Feeds with Refluxing in Beakers or Crucibles: Collaborative Study. *J. AOAC Int.* **2002**, *85*, 1217–1240. [CrossRef]
15. AOAC 940.26, 1940 AOAC 940.26-1940, Ash of Fruits and Fruit Products: AOAC Official Method. Available online: http://www.aoacofficialmethod.org/index.php?main_page=product_info&cPath=1&products_id=1447 (accessed on 11 February 2023).
16. Mimouni, A.; Deeth, H.C.; Whittaker, A.K.; Gidley, M.J.; Bhandari, B.R. Rehydration Process of Milk Protein Concentrate Powder Monitored by Static Light Scattering. *Food Hydrocoll.* **2009**, *23*, 1958–1965. [CrossRef]
17. Cai, Y.Z.; Corke, H. Production and Properties of Spray-Dried Amaranthus Betacyanin Pigments. *J. Food Sci.* **2000**, *65*, 1248–1252. [CrossRef]
18. Freudig, B.; Hogekamp, S.; Schubert, H. Dispersion of Powders in Liquids in a Stirred Vessel. *Chem. Eng. Process. Process Intensif.* **1999**, *38*, 525–532. [CrossRef]
19. Raghavendra, S.N.; Rastogi, N.K.; Raghavarao, K.S.M.S.; Tharanathan, R.N. Dietary Fiber from Coconut Residue: Effects of Different Treatments and Particle Size on the Hydration Properties. *Eur. Food Res. Technol.* **2004**, *218*, 563–567. [CrossRef]
20. Robertson, J.A.; De Monredon, F.D.; Dysseler, P.; Guillon, F.; Amadò, R.; Thibault, J.F. Hydration Properties of Dietary Fibre and Resistant Starch: A European Collaborative Study. *LWT-Food Sci. Technol.* **2000**, *33*, 72–79. [CrossRef]
21. Garau, M.C.; Simal, S.; Rosselló, C.; Femenia, A. Effect of Air-Drying Temperature on Physico-Chemical Properties of Dietary Fibre and Antioxidant Capacity of Orange (*Citrus aurantium* v. Canoneta) by-Products. *Food Chem.* **2007**, *104*, 1014–1024. [CrossRef]
22. Yasumatsu, K.; Sawada, K.; Moritaka, S.; Misaki, M.; Toda, J.; Wada, T.; Ishii, K. Whipping and Emulsifying Properties of Soybean Products. *Agric. Biol. Chem.* **2014**, *36*, 719–727. [CrossRef]
23. Wolfe, K.; Wu, X.; Liu, R.H. Antioxidant Activity of Apple Peels. *J. Agric. Food Chem.* **2003**, *51*, 609–614. [CrossRef] [PubMed]
24. Stratil, P.; Klejdus, B.; Kubáň, V. Determination of Total Content of Phenolic Compounds and Their Antioxidant Activity in Vegetables–Evaluation of Spectrophotometric Methods. *J. Agric. Food Chem.* **2006**, *54*, 607–616. [CrossRef]
25. Re, R.; Pellegrini, N.; Proteggente, A.; Pannala, A.; Yang, M.; Rice-Evans, C. Antioxidant Activity Applying an Improved ABTS Radical Cation Decolorization Assay. *Free Radic. Biol. Med.* **1999**, *26*, 1231–1237. [CrossRef] [PubMed]
26. Wolf, W.; Spiess, W.E.L.; Jung, G. Standardization of Isotherm Measurements (Cost-Project 90 and 90 BIS). In *Properties of Water in Foods: In Relation to Quality and Stability*; Springer: Dordrecht, The Netherlands, 1985; pp. 661–679. [CrossRef]
27. Inglett, G.E.; Chen, D.; Liu, S.X. Physical Properties of Gluten-Free Sugar Cookies Made from Amaranth–Oat Composites. *LWT-Food Sci. Technol.* **2015**, *63*, 214–220. [CrossRef]
28. Barbosa-Cánovas, G.V.; Fontana, A.J.; Schmidt, S.J.; Labuza, T.P. *Water Activity in Foods: Fundamentals and Applications*; John Wiley & Sons: Hoboken, NJ, USA, 2008; pp. 1–435. [CrossRef]
29. BEDCA Base de Datos BEDCA. Available online: https://www.bedca.net/bdpub/ (accessed on 4 June 2023).
30. Fernandes, D.C.; Freitas, J.B.; Czeder, L.P.; Naves, M.M.V. Nutritional Composition and Protein Value of the Baru (*Dipteryx alata* Vog.) Almond from the Brazilian Savanna. *J. Sci. Food Agric.* **2010**, *90*, 1650–1655. [CrossRef]
31. Amagliani, L.; O'Regan, J.; Kelly, A.L.; O'Mahony, J.A. Composition and Protein Profile Analysis of Rice Protein Ingredients. *J. Food Compos. Anal.* **2017**, *59*, 18–26. [CrossRef]
32. Nedeljković, N.; Hadnađev, M.; Dapčević Hadnađev, T.; Šarić, B.; Pezo, L.; Sakač, M.; Pajin, B. Partial Replacement of Fat with Oat and Wheat Bran Gels: Optimization Study Based on Rheological and Textural Properties. *LWT* **2017**, *86*, 377–384. [CrossRef]
33. Lu, F.; Liu, Y.; Li, B. Okara Dietary Fiber and Hypoglycemic Effect of Okara Foods. *Bioact. Carbohydr. Diet. Fibre* **2013**, *2*, 126–132. [CrossRef]
34. Castellanos Fuentes, A.P.; Genevois, C.E.; Flores, S.K.; De Escalada Pla, M.F. Valorisation of Soy By-Products as Substrate for Food Ingredients Containing *L. casei* through Solid State Fermentation. *LWT-Food Sci. Technol.* **2020**, *132*, 109779. [CrossRef]
35. Gul, K.; Yousuf, B.; Singh, A.K.; Singh, P.; Wani, A.A. Rice Bran: Nutritional Values and Its Emerging Potential for Development of Functional Food—A Review. *Bioact. Carbohydr. Diet. Fibre* **2015**, *6*, 24–30. [CrossRef]
36. Hu, M. Chapter 9—Oxidative Stability and Shelf Life of Low-Moisture Foods. In *Oxidative Stability and Shelf Life of Foods Containing Oils and Fats*; Hu, M., Jacobsen, C., Eds.; AOCS Press: Urbana, IL, USA, 2016; pp. 313–371, ISBN 9781630670566. [CrossRef]
37. Guimarães, R.M.; Silva, T.E.; Lemes, A.C.; Boldrin, M.C.F.; da Silva, M.A.P.; Silva, F.G.; Egea, M.B. Okara: A Soybean by-Product as an Alternative to Enrich Vegetable Paste. *LWT* **2018**, *92*, 593–599. [CrossRef]
38. Sánchez-Zapata, E.; Díaz-Vela, J.; Pérez-Chabela, M.L.; Pérez-Alvarez, J.A.; Fernández-López, J. Evaluation of the Effect of Tiger Nut Fibre as a Carrier of Unsaturated Fatty Acids Rich Oil on the Quality of Dry-Cured Sausages. *Food Bioprocess Technol.* **2013**, *6*, 1181–1190. [CrossRef]
39. Chantaro, P.; Devahastin, S.; Chiewchan, N. Production of Antioxidant High Dietary Fiber Powder from Carrot Peels. *LWT-Food Sci. Technol.* **2008**, *41*, 1987–1994. [CrossRef]

40. Elleuch, M.; Bedigian, D.; Roiseux, O.; Besbes, S.; Blecker, C.; Attia, H. Dietary Fibre and Fibre-Rich by-Products of Food Processing: Characterisation, Technological Functionality and Commercial Applications: A Review. *Food Chem.* **2011**, *124*, 411–421. [CrossRef]
41. Jafari, S.M.; Ghalegi Ghalenoei, M.; Dehnad, D. Influence of Spray Drying on Water Solubility Index, Apparent Density, and Anthocyanin Content of Pomegranate Juice Powder. *Powder Technol.* **2017**, *311*, 59–65. [CrossRef]
42. Oliveira, V.M.; Jorge, E.C.; Borges, S.V. Empleo de un Secador por Atomización a Escala Piloto en la Producción de Maracuyá en Polvo y su Aceptabilidad para Elaborar Jugo Reconstituido. *Aliment. Rev. Tecnol. Hig. Aliment.* **2003**, *342*, 83–88.
43. Gay, J.; Campos, F.R.; Oliveira, V.M.; Borges, S.V.; Francisoni, A.D.; Pereira, D.B. Propiedades Físicas del Jugo de Maracuyá en polvo 1: Efecto de la Velocidad de Atomización y Concentración de Maltodextrina. *Aliment. Rev. Tecnol. Hig. Aliment.* **2003**, *346*, 97–100.
44. Bai, X.; Zhang, M.L.; Zhang, Y.; Zhang, J.; Zhang, Y.; Wang, C.; Liu, R. Effects of Steaming, Microwaving, and Hot-Air Drying on the Physicochemical Properties and Storage Stability of Oat Bran. *J. Food Qual.* **2021**, *2021*, 4058645. [CrossRef]
45. Bhusari, S.N.; Muzaffar, K.; Kumar, P. Effect of Carrier Agents on Physical and Microstructural Properties of Spray Dried Tamarind Pulp Powder. *Powder Technol.* **2014**, *266*, 354–364. [CrossRef]
46. Callahan, B.J.; McMurdie, P.J.; Rosen, M.J.; Han, A.W.; Johnson, A.J.A.; Holmes, S.P. DADA2: High-Resolution Sample Inference from Illumina Amplicon Data. *Nat. Methods* **2016**, *13*, 581–583. [CrossRef] [PubMed]
47. Lecumberri, E.; Mateos, R.; Izquierdo-Pulido, M.; Rupérez, P.; Goya, L.; Bravo, L. Dietary Fibre Composition, Antioxidant Capacity and Physico-Chemical Properties of a Fibre-Rich Product from Cocoa (*Theobroma cacao* L.). *Food Chem.* **2007**, *104*, 948–954. [CrossRef]
48. Hinestroza-Córdoba, L.I.; Serna, S.D.; Seguí, L.; Barrera, C.; Betoret, N. Characterization of Powdered Lulo (*Solanum quitoense*) Bagasse as a Functional Food Ingredient. *Foods* **2020**, *9*, 723. [CrossRef]
49. Özdemir, E.E.; Görgüç, A.; Gençdağ, E.; Yılmaz, F.M. Physicochemical, Functional and Emulsifying Properties of Plant Protein Powder from Industrial Sesame Processing Waste as Affected by Spray and Freeze Drying. *LWT* **2022**, *154*, 112646. [CrossRef]
50. Martínez-Las Heras, R.; Landines, E.F.; Heredia, A.; Castelló, M.L.; Andrés, A. Influence of Drying Process and Particle Size of Persimmon Fibre on Its Physicochemical, Antioxidant, Hydration and Emulsifying Properties. *J. Food Sci. Technol.* **2017**, *54*, 2902–2912. [CrossRef]
51. Cheng, F.; Ding, K.; Yin, H.; Tulbek, M.; Chigwedere, C.M.; Ai, Y. Milling and Differential Sieving to Diversify Flour Functionality: A Comparison between Pulses and Cereals. *Food Res. Int.* **2023**, *163*, 112223. [CrossRef]
52. Bodart, M.; de Peñaranda, R.; Deneyer, A.; Flamant, G. Photometry and Colorimetry Characterisation of Materials in Daylighting Evaluation Tools. *Build. Environ.* **2008**, *43*, 2046–2058. [CrossRef]
53. Calabuig-Jiménez, L.; Hinestroza-Córdoba, L.I.; Barrera, C.; Seguí, L.; Betoret, N. Effects of Processing and Storage Conditions on Functional Properties of Powdered Blueberry Pomace. *Sustainability* **2022**, *14*, 1839. [CrossRef]
54. Saez, R. Caracterización de Polvos de Piel de Mandarina para su Uso como Ingrediente Funcional en Alimentos. Available online: https://www.google.com.hk/url?sa=t&rct=j&q=&esrc=s&source=web&cd=&cad=rja&uact=8&ved=2ahUKEwiKzbzB0Pv_AhXimFYBHaoRDhMQFnoECAsQAQ&url=https%3A%2F%2Friunet.upv.es%2Fbitstream%2Fhandle%2F10251%2F93422%2FS%25C3%2581EZ%2520-%2520CARACTERIZACI%25C3%2593N%2520DE%2520POLVOS%2520DE%2520PIEL%2520DE%2520MANDARINA%2520PARA%2520SU%2520USO%2520COMO%2520INGREDIENTE%2520FUNCIONAL%2520EN%2520A....pdf%3Fsequence%3D1&usg=AOvVaw02o4F1-7vzzC40ZjullLzV&opi=89978449 (accessed on 20 January 2023).
55. Bas-Bellver, C.; Andrés, C.; Seguí, L.; Barrera, C.; Jiménez-Hernández, N.; Artacho, A.; Betoret, N.; Gosalbes, M.J. Valorization of Persimmon and Blueberry Byproducts to Obtain Functional Powders: In Vitro Digestion and Fermentation by Gut Microbiota. *J. Agric. Food Chem.* **2020**, *68*, 8080–8090. [CrossRef] [PubMed]
56. Garrido, I.; Monagas, M.; Gómez-Cordovés, C.; Bartolomé, B. Extracción de Antioxidantes a Partir de Subproductos del Procesado de la Almendra. *Grasas Aceites* **2007**, *58*, 130–135. [CrossRef]
57. Bolling, B.W.; Chen, C.Y.O.; McKay, D.L.; Blumberg, J.B. Tree Nut Phytochemicals: Composition, Antioxidant Capacity, Bioactivity, Impact Factors. A Systematic Review of Almonds, Brazils, Cashews, Hazelnuts, Macadamias, Pecans, Pine Nuts, Pistachios and Walnuts. *Nutr. Res. Rev.* **2011**, *24*, 244–275. [CrossRef]
58. Ribarova, F.; Marinova, D.; Ribarova, F.; Atanassova, M. Total Phenolics and Flavonoids in Bulgarian Fruits and Vegetables. *J. Univ. Chem. Technol. Metall.* **2005**, *40*, 255–260.
59. Kornsteiner, M.; Wagner, K.H.; Elmadfa, I. Tocopherols and Total Phenolics in 10 Different Nut Types. *Food Chem.* **2006**, *98*, 381–387. [CrossRef]
60. Inyang, U.E.; Oboh, I.O.; Etuk, B.R.; Inyang, U.E.; Oboh, I.O.; Etuk, B.R. Kinetic Models for Drying Techniques—Food Materials. *Adv. Chem. Eng. Sci.* **2018**, *8*, 27–48. [CrossRef]
61. Ling, W.; Xing, Y.; Hong, C.; Zhang, B.; Hu, J.; Zhao, C.; Wang, Y.; Feng, L. Methods, Mechanisms, Models and Tail Gas Emissions of Convective Drying in Sludge: A Review. *Sci. Total Environ.* **2022**, *845*, 157376. [CrossRef]
62. Srikiatden, J.; Roberts, J.S. Moisture Transfer in Solid Food Materials: A Review of Mechanisms, Models, and Measurements. *Int. J. Food Prop.* **2007**, *10*, 739–777. [CrossRef]
63. Onwude, D.I.; Hashim, N.; Janius, R.B.; Nawi, N.M.; Abdan, K. Modeling the Thin-Layer Drying of Fruits and Vegetables: A Review. *Compr. Rev. Food Sci. Food Saf.* **2016**, *15*, 599–618. [CrossRef] [PubMed]
64. Maskan, A.; Kaya, S.; Maskan, M. Hot Air and Sun Drying of Grape Leather (Pestil). *J. Food Eng.* **2002**, *54*, 81–88. [CrossRef]

65. Aghbashlo, M.; Kianmehr, M.H.; Samimi-Akhijahani, H. Influence of Drying Conditions on the Effective Moisture Diffusivity, Energy of Activation and Energy Consumption during the Thin-Layer Drying of Berberis Fruit (Berberidaceae). *Energy Convers. Manag.* **2008**, *49*, 2865–2871. [CrossRef]
66. Saravacos, G.D.; Maroulis, Z.B. *Transport Properties of Foods*; CRC Press: Boca Raton, FL, USA, 2001. [CrossRef]
67. Xiong, X.; Narsimhan, G.; Okos, M.R. Effect of Composition and Pore Structure on Binding Energy and Effective Diffusivity of Moisture in Porous Food. *J. Food Eng.* **1992**, *15*, 187–208. [CrossRef]
68. Zhao, G.; Yin, F.; Liang, X.; Yuan, D.; Geng, W.; Wang, L.; Sun, R. Drying Experiment and Drying Model Analysis of Dehydrated Sludge Particles. *IOP Conf. Ser. Mater. Sci. Eng.* **2020**, *768*, 022031. [CrossRef]
69. Labuza, T.P.; Kaanane, A.; Chen, J.Y. Effect of Temperature on the Moisture Sorption Isotherms and Water Activity Shift of Two Dehydrated Foods. *J. Food Sci.* **1985**, *50*, 385–392. [CrossRef]
70. Pahlevanzadeh, H.; Yazdani, M. Moisture adsorption isotherms and isosteric energy for almond. *J. Food Process Eng.* **2005**, *28*, 331–345. [CrossRef]
71. See, X.Y.; Dupas-Langlet, M.; Forny, L.; Meunier, V.; Zhou, W. Physical Stability of Co-Freeze-Dried Powders Made from NaCl and Maltodextrins—Impact of NaCl on Glass Transition Temperature, Water Vapour Sorption Isotherm and Water Vapour Sorption Kinetics. *Food Hydrocoll.* **2023**, *136*, 108238. [CrossRef]

sustainability

MDPI

Article

Characterization of a New Powdered, Milk-Based Medicinal Plant (*Alcea rosea*) Drink Product

Mustafa Mortas

Food Engineering Department, Faculty of Engineering, Ondokuz Mayıs University, Samsun 55139, Turkey; mustafa.mortas@omu.edu.tr; Tel.: +90-362-312-19-19-1502

Abstract: *Alcea rosea*, known as hollyhock, is an ornamental dicot flower in the Malvaceae family, and it has been used for different purposes, ranging from traditional medicine to food applications, through the use of its leaves, roots, and seeds. The hollyhock flowers possess several properties, including a diuretic, cooling, demulcent, emollient, febrifuge, and astringent effects. Hollyhock flowers were commonly included in a traditional medicine formulation for hypoglycemic or hypolipidemic treatments. Along with its use in traditional medicine, it has also been considered a valuable ingredient in some traditional food preparations; however, the processing of hollyhock into a new food product has not been studied. Accordingly, this study aimed to evaluate the production of a new product, a milk-based Hollyhock (*Alcea rosea*) powder, and its powder product characterization via particle size, water activity, density, flowability, etc., in addition to the determination of its chemical composition (with 5.73% ash and 29.12% protein). In this paper, we report the application of spray-dried milk-based hollyhock flower extract to produce a new ready-to-drink product of this medicinal plant for food sustainability.

Keywords: hollyhock; *Alcea rosea*; powdered product; spray-dried extract

1. Introduction

Hollyhock (*Alcea rosea*) belongs to the Althaea genus and Malvaceae (Mallow) family. It is commonly known as marshmallow plant but can be called different names depending on the location. It is called hollyhock in English; malva rosa and rosoni in Italian; shi kui in Chinese; passe rose and rose paple in French; khatmae in Arabic; jeop-si-kkot in Russian; and rishak hatmi, khatmi, and khaira in Japanese [1]. The reason why it is so well known is that it has been used in traditional treatment processes from the past to the present. Traditional medicine is the oldest method of curing diseases and infections, a practice that uses various plants, and hollyhock is one of them [2]. Although hollyhock was reported to have originated from China or tropical areas [3], similarly to other common medicinal plants, its different parts, including the leaves, roots, and seeds, have be used in various applications all over the globe [4].

The health-promoting attributes of different parts of the hollyhock plant have been investigated; its roots are used against a wide range of health problems, such as diarrhea, constipation, inflammation, bronchitis, severe cough, and angina [5]. The whole plant has favorable effects on asthma, coughing, throat pain, jaundice, swelling, stomach irritation, kidney pain, and urinary irritation [6]. Furthermore, its flowers, ranging from white to dark red [7], present some significant properties, such as diuretic, cooling, demulcent, emollient, febrifuge, and astringent effects [8]. Lastly, the anti-influenza properties of hollyhock were investigated in mice, and the data suggested promising results for its use as an anti-influenza drug [9].

In addition to its health benefits, the hollyhock flower can also be consumed in different ways [10]. It can be one of the agents in herbal tea mixtures for brightening [2] and can be used as a cooking material for different purposes [10]. Moreover, a novel product that has been developed [11] is an edible film made of hollyhock flower gum. The flower petals, flower buds, and hollyhock leaves are also used in salads [10] and can be used with milk to obtain a milky extract for health benefits. Although hollyhock flowers were usually prepared with milk for use in traditional medicinal formulas for hypoglycemic or hypolipidemic treatment, there are currently no similar industrial products, such as a milky drink or powdered formulas. For this purpose, a new possible product was produced from hollyhock flower (Figure 1) mixed with milk by using a spray dryer, and the quality parameters of the product, in terms of chemical composition and physical properties, were investigated.

Figure 1. Hollyhock flower (*Alcea rosea*).

2. Materials and Methods

2.1. Materials

For producing the traditional milky hollyhock extract, the milk (M) with 1.5% fat (Birsah, Selçuklu/Konya, Turkey) and dried hollyhock flowers (*Alcea rosea*) (Toroslar Naturel Aktar Organik, Toroslar/Mersin, Turkey) used in the study were obtained from local markets in Samsun, Turkey. The chemicals used for all the listed methods were Sigma-Aldrich and Merck brands.

2.2. Preparation Methods

Traditional milky hollyhock extract was produced with milk and hollyhock flower; a spray dryer (Bushi, B-290) was used to create the powdered product after the extract was obtained. In addition, reconstitution was applied to the powdered product to compare it to the traditional extract. The traditional milky hollyhock drink was produced from milk and dried hollyhock flowers by adding about 25 g of the flowers into 500 mL of milk according to the traditional preparation ratio. The mixture was stirred and heated (50 °C) on a hot plate for 30 min; then, the mixture was filtered using filter paper. The dried milky hollyhock drink was produced from the milky hollyhock drink using a spray dryer with a 150 °C inlet temperature. The milk powder was produced from milk under the same spray dryer conditions in order to compare the effect of hollyhock in the milk. After spray drying was complete, the reconstituted hollyhock drink was produced from the dried milky hollyhock drink by dissolving the powder in water at the same Brix as the milky hollyhock drink. Through this process, the differences between the milk powder and dried new products, and between traditional milky drinks and reconstituted samples from the new product with 1–9% ratio were able to be compared.

2.2.1. Process Yield

The production process yield from milky hollyhock drink to dried milky hollyhock drink was calculated as the ratio of the total powder weight after spray drying to the initial amount of solid-liquid feed

2.2.2. Proximate Analysis

Proximate analysis as moisture contents, ash contents, pH levels, and protein contents were measured in both the milk powder and dried milky hollyhock drinks according to the Association of Office Analytical Chemists [12]. For water activity measurements, samples were filled in the special containers at 2/3 and measured at 25 °C (Aqualab Dewpoint Water Activity Meter 4TE, Pullman, WA, USA).

2.2.3. Total Phenolic Content

Total phenolic content (TPC) was determined according to Singleton and Rossi (1965). The extract (0.5 mL) was mixed with 2.5 mL Folin Ciocalteau's phenol reagent (0.2 N) and 2 mL Na_2CO_3 (7.5%) and incubated at room temperature. After thirty-minute incubation, absorbance was measured at 760 nm using a UV/VIS spectrophotometer (Shimadzu UV-1800, Kyoto, Japan). TPC was expressed as gallic acid equivalent [13] and calculated according to Equation (1) as below;

$$\text{TPC (mg/L)} = [\text{Absorbance} - 0.0166)/0.0102] \times \text{Dilution Factor} \quad (1)$$

2.2.4. Antioxidants Capacity

The antioxidant capacity analysis was conducted to determine the free DPPH radical scavenging capacity [14]. A total of 0.1 mL extract was mixed with 4.9 mL of DPPH solution (0.1 mM) in ethanol. After being incubated at room temperature for half an hour, the absorbance was measured at 517 nm. The antiradical activity (ARA, %) can be calculated using the following equation:

$$\text{ARA}(\%) = ((\text{Ac} - \text{As})/\text{Ac}) \times 100 \quad (2)$$

Ac represents the absorbance of the control (ethanol and DPPH), while As represents the absorbance of the sample. The results were expressed as Trolox equivalent and calculated using the following equation:

$$\text{Trolox equivalent (mM)} = [(\text{ARA}\% + 0.5998)/56.608] \times \text{Dilution Factor} \quad (3)$$

2.2.5. Measurement of Color Properties

The color properties were assessed using a Minolta colorimeter, which utilizes the CIELAB scale (L*, a*, and b*). The color parameters range from L* = 0 (representing dark) to L* = 100 (representing light), −a* (representing greenness) to + a* (representing redness), and −b* (representing blueness) to +b* (representing yellowness).

2.2.6. Viscosity Measurement

The rheological measurements of the samples were conducted using a rheometer (HAAKE Mars III; Thermo Scientific, Karlsruhe, Germany) equipped with a cone and plate system (diameter: 25 mm; cone angle: 2°; gap between cone; and plate: 0.106 mm). The samples were allowed to equilibrate for 5 min at the desired temperature (25 °C), and the measurements were performed at this temperature. The shear rate was increased linearly from 0.1 to 100 s^{-1} over a period of 3 min.

2.2.7. FTIR

The molecular differentiation of the milk powder and dried milky hollyhock drink samples was determined using an MN 115 Bruker Tensor 27 FTIR (Rheinstetten, Germany) with a wavelength range of 4000–400 cm^{-1}. Prior to measuring the samples, background spectra of the medium were collected and recorded using OPUS software 6.5 (Bruker Corporation, Ettlingen, Germany).

2.2.8. Sensory Evaluation

The samples were evaluated by a panel of 15 individuals consisting of semi-trained staff and graduated students. The evaluation criteria included color, taste aroma, viscosity and mouthfeel, which were assessed using a numerical scale ranging from 1 to 9. The samples were enumerated with different three-digit numbers.

2.2.9. Mineral Contents

Approximately 1 g of powder sample was weighed. It was turned into ashes at 500 °C in the furnace (Nuve MF 120). Then, the ash was dissolved in 3N HCl by stirring in a heater for 10 min. Subsequently, the ash was filtered through filter paper (Whatman no.1). Sodium (Na), potassium (K), and calcium (Ca) contents were determined using the BWB-1 Flame Photometer. Phosphorus (P) was determined using the vanadomolybdophosphoric acid colorimetric method [15]. The measurement of phosphorus was performed at 420 nm using a UV/Vis spectrophotometer (Agilent Technologies, Cary 60, Victoria, Australia). The phosphorus contents of the samples were calculated based on a calibration curve (y = 0.0397x + 0.014, R2: 0.9997) constructed using KH_2PO_4.

2.2.10. Powder Characterization

The loose density (ρL) of the milk-based Alcea powder was determined by pouring it into a 25 mL graduated cylinder and measuring the corresponding weight. Tapped density (ρT) was determined after completing the tapping process 125 times. Apparent density (ρP) was measured using a gas steropycnometer (Quantachrome Instruments, Boynton Beach, FL). The samples were placed into sample cells and degassed by purging with helium gas. The porosity (ε) was calculated based on the relationship between the tapped bulk density (ρT) and apparent density (ρP) as follows:

$$\varepsilon = ((\rho P - \rho T)/\rho P) \times 100 \quad (4)$$

Cohesiveness (Hausner ratio, HR) properties of powders were characterized by a ratio of the two density types.

$$HR = \rho T/\rho L \quad (5)$$

where ρT is tapped density, and ρL is loose density.

The flowability properties of the powders were evaluated using Carr's index (Cl) and angle of repose (AOR) approach.

$$Cl = (\rho T - \rho L)/\rho T \times 100 \quad (6)$$

where ρT is tapped density, and ρL is loose density.

The AOR value was measured using a powder AOR device (Torontech, ON, Canada). The angle of repose (AOR) (θ) was calculated using the following formula:

$$AOR(\theta) = \arctan\ h/r \quad (7)$$

where h is the height of powder after dropping;, and r is the average radius of powder after dropping.

The physical properties of the powder are presented in Table 1.

Table 1. Specification for powder physical properties.

Flowability	Carr's Index (%)	Hausner Ratio	Angle of Repose (°)
Excellent	0–10	1.00–1.11	25–30
Good	11–15	1.12–1.18	31–35
Fair	16–20	1.19–1.25	36–40
Passable	21–25	1.26–1.34	41–45
Poor	26–31	1.35–1.45	46–55
Very poor	32–37	1.46–1.59	56–65
Very, very poor	>38	>1.60	>66

2.2.11. Solubility

The total solubility of the powder was assessed by determining the total solids remaining after dissolution and centrifugation [16]. A total of 0.1 g of the powder was dispersed in 24.9 g of distilled water and stirred for 30 min to ensure proper dispersion. The dispersions were then transferred to 50 mL conical tubes and centrifuged (Hettich 320R, Germany) at 5000 rpm for 20 min. The supernatant was carefully transferred to a preweighed moisture dish and dried overnight at 105 °C. The solubility was calculated using the following the equation:

$$\text{Solubility (\%)} = (\text{Weight of the dry supernatant})/(\text{Weight of the supernatant} \times 0.4\%) \times 100 \quad (8)$$

2.2.12. Particle Size Distribution (PSD)

The particle size distribution (PSD) levels of the samples were determined using a static laser light (Malvern Mastersizer 2000 with Hydro 2000S (A), Malvern Instruments Ltd., Worcestershire, UK). The powder samples were dispersed into ultrapure water at a ratio of 1:100 for measurement. The refractive index used for the powders was 1.57, while that for water was 1.33 [17]. The mean diameter of the powdered sample was evaluated using Equation (9) for volume-weighted mean diameter (d4,3) and Equation (10) for particle surface area (d3,2). This approach is useful when the particles are not ideal spheres, as the (d4,3) value is more influenced by the larger particles, while the d3,2 value is more influenced by the smaller particles [18]. The values of d0.1, d0.5 and d0.9 represent the cumulative percentiles and indicate that 10%, 50% and 90% of the particles, respectively, fell below the specified diameter [19],

$$\text{d4,3} = (\textstyle\sum \text{inidi4})/(\sum \text{inidi3}) \quad (9)$$

$$\text{d3,2} = (\textstyle\sum \text{inidi3})/(\sum \text{inidi2}) \quad (10)$$

2.2.13. Scanning Electron Microscopy (SEM) Analysis

For SEM analysis, the morphological properties of the powder samples were examined using a scanning electron microscope (JEOL JSM-7001FTTLS LV, Peabody, MA, USA). The images of the samples were captured at 5 kV with magnifications of ×100 and ×2000.

2.2.14. Statistical Analysis

All the measurements occurred triplicated and statistical analysis, comparison of Tukey's test results, was analyzed with Statistical Package for Social Sciences (SPSS), v23.0 (IBM SPSS Statistics, Armonk, NY, USA).

3. Results and Discussion

3.1. Process Yield

Process yield is an important factor for production cost as it is closely related to the morphologies of the particles and is critical for powder flowability, redispersibility, and density. All the mentioned morphological properties are affected by the operating conditions of the spray dryer spray, such as drying air temperature, feed rate, and viscosity [20].

Additionally, the presence of sugars in the extract is closely related to process yield [19], and phenolic content is another factor that affects the stickiness of powders [21]. In this situation, the stickiness of herbal extract on the dryer walls was similar to the spray drying process of herbal medicinal powders (*P. boldus* and *C.asiatica*), and the process yield of the mentioned process was found to be 46% by Gallo et al. [19]. Our process yield percentage for dried hollyhock drink was found to be 53%, which is within the mentioned range, and the main reason for the lower process yield is closely related to the presence of high-chain carbohydrates in the sample.

3.2. Proximate Analysis

In this study, proximate analysis of milk powder and dried milky hollyhock drink samples were conducted to compare differences at the same process conditions in terms of adding hollyhock into the milk. The dry matter of the dried milky hollyhock drink sample was found to be 98.40 ± 0.29, while the milk powder had a dry matter of 99.42 ± 0.43. The dry matter value of the dried milky hollyhock drink was higher than that of the MP because the soluble dry matter of the dry hollyhock flowers in the milk contributed to an increase in the dry weight of the dried milky hollyhock drink. The situation was similar for the ash content as the value of the dried milky hollyhock drink (5.73 ± 0.28) was higher than the milk powder value (4.67 ± 0.21) because the mineral content of the dried milky hollyhock drink was more than the milk powder sample. The pH value of the milk powder (6.693 ± 0.006) was found to be a higher value than the dried milky hollyhock drink one (6.340 ± 0.053). In addition to this, the pH value of the reconstituted hollyhock drink was determined as 6.347 ± 0.006. It was statistically different but closer to each other, and changes in pH and acidity can affect the stability of the structural network of the drink [22]. There was no statistical importance between the protein content of the milk powder (29.92 ± 2.40) and dried milky hollyhock drink (29.12 ± 1.26) samples. Water activity (aw) results of the samples milk powder (0.1463 aw) and dried milky hollyhock drink (0.0931 aw) are given in Table 2, and the result is similar to other research (aw 0.20) [23].

Table 2. Analysis results of milk powder and dried hollyhock extract.

Analysis		Milk Powder	Dried Hollyhock Extract
Dry Matter (%)		98.40 ± 0.29 ^{b}	99.42 ± 0.43 ^{a}
Ash Content (%)		4.67 ± 0.21 ^{b}	5.73 ±0.28 ^{a}
Water Activity (aw)		0.1463 ± 0.0033 ^{a}	0.0931 ± 0.0009 ^{b}
Protein content		29.92 ± 2.40 ^{a}	29.12 ± 1.26 ^{a}
pH		6.693 ± 0.006 ^{a}	6.340 ± 0.053 ^{b}
Mineral Contents	Na (mg/100 g)	429.52 ± 22.14 ^{b}	502.71 ± 17.37 ^{a}
	K (mg/100 g)	1652.18 ± 41.51 ^{a}	1346.91 ± 41.44 ^{b}
	Ca (mg/100 g)	724.02 ± 13.84 ^{a}	704.31 ± 15.92 ^{a}
	P (mg/100 g)	418.96 + 3.48 ^{b}	747.45 ± 1.78 ^{a}

Values are means ± standard deviation. (a,b) Different letters on the same line show the significant differences ($p < 0.05$) between samples.

3.3. Total Phenolic Content (TPC)

The spray drying process can decrease the total phenolic content. It has been reported that using an inlet temperature of 150 °C (air outlet temperature: 80 °C) for the spray drying process can recover approximately 94% of bioactive antioxidant components, such as phenolic content and total anthocyanins in bayberry juice [24]. Prior to the spray drying process, the milky hollyhock drink had a TPC value of 681.44 ± 9.88 mg GAE/kg, whereas the dried milky hollyhock drink had a TPC value of 603.56 ± 10.52 mg GAE/kg after the process. The spray-dried process yield of the TPC recovery was calculated as 88%, which was acceptable for this kind of heat-based process. In a study involving 56 medicinal plants, the total phenolic content ranged from 0.12 ± 0.01 to 59.43 ± 1.03 mg GAE/g, and our finding for phenolic content of the hollyhock were consistent with the literature [25].

3.4. Antioxidant Capacity

One of the factors contributing to diseases, such as atherosclerosis, cancer, aging, and coronary heart diseases, is oxidative stress [26–28]. Minimizing oxidative stress is crucial for promoting our physical condition and preventing degenerative diseases. The total antioxidant capacity of the milky hollyhock drink was calculated to be 1.49 ± 0.07 mmol Trolox/g, while the dried milky hollyhock drink had a total antioxidant capacity of 1.32 ± 0.04 mmol Trolox/g. The recovery ratio of antioxidant capacity value for the spray drying process was calculated as 89%. Similarly, the antioxidant capacity values of some medical plants ranged from 0.61 ± 0.05 to 326.87 ± 7.17 μmol Trolox/g [25]. Furthermore, the results for the other medicinal plants were consistent with our results [29,30].

3.5. Color Measurements

Color is a crucial quality attribute for food products because the appearance can significantly impact consumer acceptability as it is the first thing consumers judge. Dried hollyhock flower can impart color to the milk. The L*, a*, and b* values for milk powder were found to be 91.80 ± 0.01, 0.41 ± 0.00, and 10.98 ± 0.00, respectively. For the milky hollyhock drink, the values were 66.79 ± 0.03, 67.17 ± 0.08, and 0.02 ± 0.00, respectively. Kalusevic et al. [31] observed the highest L* and a* values in their spray-dried black soybean coat extract, indicating that this sample had the darkest color with the highest proportion of red. Hollyhock flowers contain carbohydrates, cyanides, tannins and alcea mucilage, and kaempferol, which is present in all flower varieties. Additionally, different flower colors such as pink and orange, mauve and red, white and yellow contain herbacetin, quercetin, and undefined pigments, respectively [8]. After extracting hollyhock flowers into milk, the spray drying process affects color values of the milky hollyhock drink and dried hollyhock drink samples, resulting in L*, a*, and b* values of 77.75 ± 1.37, 1.22 ± 0.03, and 3.03 ± 0.06, respectively, for the dried hollyhock drink. The observed color differences between the samples were primarily due to variations in the b* values, which is typical for the spray-dried product in terms of increased yellowness. This could be attributed to the presence of sugars [32], mucilage and pigment degradation. After preparing the dried hollyhock drink sample, it was reconstituted in the same ratio to prepare the reconstituted hollyhock drink, and the L*, a*, and b* values were measured as 67.17 ± 0.08, 0.26 ± 0.01, and 2.37 ± 0.02, respectively. Furthermore, there were statistical differences in the L* value between dried milky hollyhock and reconstituted hollyhock drink samples, but there were none in the a* and b* values.

3.6. Viscosity

The viscosity range for milky hollyhock drink was measured as 1.916–9.597 mPas, while for the reconstituted hollyhock drink samples, it was 0.848–2.269 mPas (Table 3). Generally, the viscosity values of the milky hollyhock drink samples were higher than those of the reconstituted hollyhock drink samples. The viscosity values of all reconstituted hollyhock drink samples, with different concentrations (1–9%), were higher than the viscosity value of the milk used as a control. This can be attributed to the direct effect of the polysaccharide structure of alcea mucilage on viscosity; however, after applying the reconstitution step using different concentration scale (1–9%) for the reconstituted hollyhock drink, the viscosity values of some samples (1, 3, 5%) were found to be lower than that of the milk sample (1.556 mPas). The main reason for the lower viscosity in these cases is the heat treatment during the spray drying process, which leads to the breakdown of the polysaccharide chains in the alcea mucilage.

Table 3. Sensory and appearance of the milky hollyhock extract and reconstituted milky hollyhock drink.

Samples	Hollyhock Amount %	Viscosity	Color Values			Sensory Evaluation				
			L*	a*	b*	Color	Consistency	Aroma	Sandiness	Overall Accep.
Milk	0	1.556 ± 0.073 ^{efg}	81.84 ± 0.03 ^{a}	−2.80 ± 0.01 ^{h}	5.41 ± 0.00 ^{a}	9.22 ± 1.71 ^{a}	9.00 ± 1.58 ^{a}	4.66 ± 1.53 ^{a}	9.55 ± 0.72 ^{a}	8.88 ± 1.27 ^{a}
Milky hollyhock extract	1	1.916 ± 0.195 ^{ef}	75.25 ± 0.03 ^{b}	−0.33 ± 0.00 ^{g}	4.62 ± 0.01 ^{b}	7.11 ± 2.31 ^{ab}	6.00 ± 2.44 ^{a}	6.22 ± 1.43 ^{a}	9.33 ± 1.11 ^{a}	5.77 ± 2.05 ^{a}
	3	3.417 ± 0.282 ^{d}	68.71 ± 0.01 ^{c}	1.64 ± 0.03 ^{e}	3.55 ± 0.00 ^{c}	5.44 ± 1.08 ^{ab}	6.44 ± 1.94 ^{a}	6.55 ± 1.81 ^{a}	9.11 ± 1.53 ^{a}	5.44 ± 1.87 ^{a}
	5	4.899 ± 0.359 ^{c}	64.98 ± 0.01 ^{d}	2.55 ± 0.00 ^{c}	2.67 ± 0.01 ^{d}	4.33 ± 1.06 ^{b}	5.22 ± 2.68 ^{a}	6.55 ± 1.81 ^{a}	8.88 ± 1.45 ^{a}	5.11 ± 1.52 ^{a}
	7	6.641 ± 0.541 ^{b}	62.18 ± 0.24 ^{e}	3.09 ± 0.02 ^{b}	2.03 ± 0.01 ^{g}	4.45 ± 1.12 ^{b}	6.33 ± 1.23 ^{a}	7.11 ± 2.14 ^{a}	9.22 ± 1.45 ^{a}	4.33 ± 2.34 ^{a}
	9	9.597 ± 0.741 ^{a}	59.43 ± 0.48 ^{f}	3.55 ± 0.03 ^{a}	1.37 ± 0.04 ^{h}	4.66 ± 1.80 ^{b}	7.33 ± 1.65 ^{a}	7.55 ± 1.74 ^{a}	9.22 ± 1.39 ^{a}	3.88 ± 2.47 ^{a}
Reconstituted hollyhock drink	1	0.848 ± 0.086 ^{fg}	45.87 ± 0.02 ^{h}	−1.28 ± 0.01 ^{h}	−2.29 ± 0.01 ^{j}	5.66 ± 1.57 ^{ab}	4.55 ± 1.69 ^{a}	5.11 ± 1.16 ^{a}	8.88 ± 2.31 ^{a}	3.88 ± 1.96 ^{a}
	3	1.025 ± 0.069 ^{fg}	58.89 ± 0.02 ^{g}	−0.37 ± 0.00 ^{g}	1.02 ± 0.01 ^{i}	6.11 ± 1.31 ^{ab}	5.44 ± 1.96 ^{a}	6.22 ± 2.98 ^{a}	9.00 ± 1.80 ^{a}	5.00 ± 1.39 ^{a}
	5	1.431 ± 0.036 ^{efg}	64.50 ± 0.01 ^{d}	1.15 ± 0.00 ^{f}	2.21 ± 0.01 ^{f}	4.78 ± 1.27 ^{b}	5.33 ± 1.95 ^{a}	6.00 ± 1.41 ^{a}	9.22 ± 1.30 ^{a}	5.55 ± 2.06 ^{a}
	7	1.861 ± 0.066 ^{ef}	64.65 ± 0.01 ^{d}	2.21 ± 0.01 ^{d}	2.23 ± 0.00 ^{f}	5.11 ± 1.31 ^{ab}	5.66 ± 2.69 ^{a}	6.33 ± 1.93 ^{a}	9.11 ± 1.69 ^{a}	4.66 ± 1.54 ^{a}
	9	2.269 ± 0.071 ^{e}	64.91 ± 0.01 ^{d}	2.23 ± 0.02 ^{d}	2.55 ± 0.00 ^{e}	4.77 ± 1.22 ^{b}	6.22 ± 1.11 ^{a}	6.33 ± 1.65 ^{a}	8.66 ± 2.00 ^{a}	5.00 ± 1.87 ^{a}

Values are means ± standard deviation. Different letters on the same column show the significant differences ($p < 0.05$) between samples.

3.7. FTIR

The FTIR spectra of the milk powder and powdered hollyhock in the 4000–400 cm^{-1} spectroscopic region are shown in Figure 2. The figure indicates that the O-H groups, which belong to bound water, and N-H streches of proteins were observed in the 3000–3700 cm^{-1} and 2800–3000 cm^{-1} regions, respectively. The vibrational modes of CH groups were represented by the centered peaks at 2917.61–2988.22 cm^{-1}. The protein amide groups (-CONH-) appeared as a peak at 1640 cm^{-1} [33], and the C=O stretch assigned to acetyl groups was centered at 1646 cm^{-1} [33]. Furthermore, the peaks centered between 1300 and 1450 cm^{-1} correspond to CH_2 or C=O-H groups, and OH in plane bending [34]. The peaks between 800 and 1200 cm^{-1} were attributed to the stretching of CO, CC, COC and to the skeletal modes of vibration of sugar residues [35,36]. In a study related to *Alcea rosea* flower extract encapsulated with nanoparticles, the main peaks in the FTIR spectrum were attributed to oxygen-bearing functional groups [37].

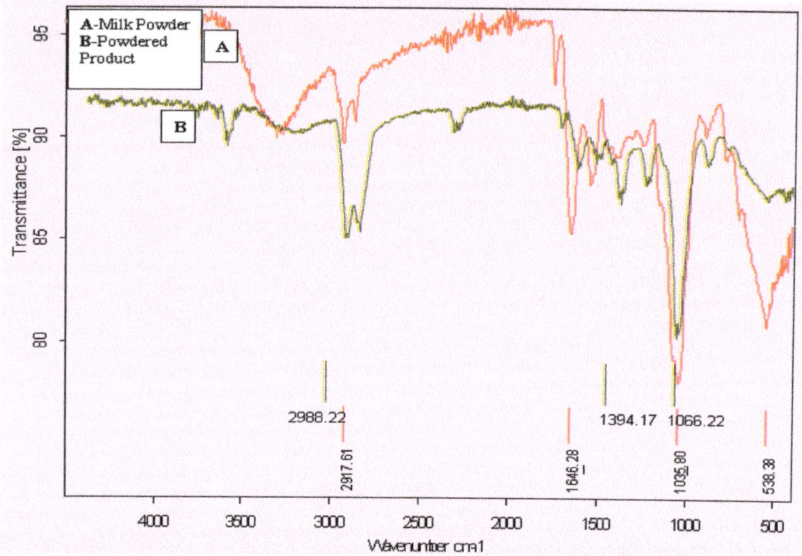

Figure 2. FTIR spectra of the milk powder and hollyhock extract powder.

3.8. Sensory Evaluation

Adding hollyhock extract up to 3–5% to milky hollyhock drink and hollyhock powder to reconstituted hollyhock drink samples has a negative effect on sensory scores. The color values of low concentration samples, both milky hollyhock drink and reconstituted hollyhock drink, were acceptable to sensory analysts. As the concentration increased, the sensory scores decreased, which was consistent with the changes in L* and b*, while the a* values. The overall acceptability scores were higher for samples with concentration between 1 and 5% compared to the other samples. It was observed that as the concentration increased, the acceptability of the samples improved due to the presence of polysaccharides and mouthfeel, but the sensory scores decreased as the color deviated from the color of milk (Table 3). The situation is similar to the browning observed in watermelon powder due to presence sugar [32]. *Alcea rosea* contains mucilage, which is a high molecular weight acidic polysaccharides ranging from 1.3 to 1.6 million Dalton and is abundant in flowers. The mucilage is composed of glucoronic acid, galacturonic acid, rhamnose, and galactose. Some of the acidic polysaccharides also contain carboxyl groups and/or sulfuric ester groups; therefore, the sulfuric ester groups can be a major factor contributing to the aroma acceptability of the product [38].

3.9. Mineral Contents

Mineral contents (Na, K, Ca, and P) of the milk powder and dried milky hollyhock drink samples are given in Table 2. The mineral contents of the milk powder were found to be consistent with the values reported in the literature [39]; however, for the dried milky hollyhock drink, the addition of hollyhock resulted in increased mineral contents in the product. The dried milky hollyhock drink exhibited higher levels of P, Ca, and Na, with values of 747 mg/100 g, 724 mg/100 g and 502 mg/100 g, respectively, compared to regular milk powder. Hollyhock can be utilized to enhance the mineral contents in milk due to its natural mineral composition, which includes calcium, sodium, potassium, and phosphorus [1].

3.10. Powder Characterization and Particle Size Distribution

While powder characterization is important for assessing a new spray-dried product, the shape of the particles generally tends to be spherical with a size range of 10–250 microns, which is primarily influenced by the properties of the spray dryer nozzle [40]; however, the overall properties of the food samples are equally significant. The D [4,3] (the volume-weighted mean) and D [3,2] (the surface weighted mean) diameters values of the milk powder were determined as 55.35 and 25.00 microns, respectively. For the dried milky hollyhock drink samples, the corresponding values were calculated as 29.1 and 21.3 microns, respectively (Table 4). Fitzpatrick et al. [41] also reported values of 53 and 99 microns for skim milk and whole milk, respectively. Additionally, the D [4,3] value for infant milk formula was found to be 155.4 microns by Murphy et al. [42]. Powder flowability is directly influenced by the drying process and is affected by both the size distribution and interparticle relationships [43]. Narrower size distributions tend to result in better flow properties [44]. Carr's index, also known as the compressibility index, measures a powder's ability to reduce in volume when tapped [43]. According to the classification in Table 1 based on Carr's index (CI) and the Hausner ratio (HR), the CI value for milk powder was determined to be 40.60, indicating very poor flowability. Similarly, the dried milky hollyhock drink exhibited a CI value of 36.93, also indicating very poor flowability. The poor flowability can be attributed to the increased contact surface area between powder particles, which enhances frictional and cohesive forces impeding powder flow [45]. While, both samples demonstrated similar flow tendencies, there were statistical differences between them.

Table 4. Powder properties of the milk powder and dried hollyhock extract.

		Milk Powder	Dried Milky Hollyhock Extract
Particle Properties	Bulk Density (*ρT*) (kg/m³)	0.222 ± 0.003 [b]	0.315 ± 0.003 [a]
	Tapped Density (*ρT*) (kg/m³)	0.378 ± 0.003 [b]	0.490 ± 0.001 [a]
	Carr's Index (Cl) (%)	40.606 ± 0.525 [a]	36.932 ± 0.964 [b]
	Powder Cohesiveness Hausner Ratio (HR)	1.684 ± 0.015 [a]	1.569 ± 0.023 [b]
	Angle of Repose (AOR) (°)	35.725 ± 2.043 [b]	42.325 ± 2.489 [a]
	Apparent density	1.481 ± 0.050 [b]	3.073 ± 0.039 [a]
	Porosity (epsilon)	64.123 ± 0.141 [b]	84.360 ± 0.631 [a]
Particle Size	D_{10} µm	0.091 ± 0.001 [a]	0.068 ± 0.001 [b]
	D_{50} µm	0.532 ± 0.005 [b]	3.315 ± 0.049 [a]
	D_{90} µm	249.5 ± 3.535 [a]	27.35 ± 0.212 [b]
	D [4,3] µMm	55.35 ± 1.484 [a]	29.1 ± 0.84 [b]
	D [3,2] µm	0.25 ± 0.001 [a]	0.213 ± 0.008 [b]

[a,b] Means within a row with different superscripts differ ($p < 0.05$).

Another parameter that can affect the powder flowability is the Hausner ratio (HR), for which the values were calculated as 1.68 and 1.56 for milk powder and dried milky hollyhock drink, respectively. Similar results were observed for HR values as for Carr's index (CI). Ilari and Mekkoui [46] calculated HR values of 1.59 for skim milk and 1.26 for whole milk, which align with the findings in the present study. The presence of large agglomerates and minimal fines can contribute to improved flow properties of powders [46]. Lower cohesion due to weaker Van der Waals forces and reduced friction is one of the main reasons for larger particle size [46]. The density of the powder is also associated with the economic factors, such as packaging, transportation and storage cost in the dairy industry [47]; therefore, bulk density, also known as apparent or packing density, is used as a measure of powder mass. It depends on particle density, internal porosity of particles, their porosity of particles, and their arrangement within the container. In addition to that, another factor can be listed as the volume of solids/liquids and open/closed pores [48]. In this study, the bulk density (ρT) was determined as 0.22 kg/m^3 for milk powder and 0.31 kg/m^3 for dried milky hollyhock drink. Literature reports various bulk density values for milk powders ranging from 0.30 to 0.62 kg/m^3 [44]. Bulk density can be categorized in four ways [48], and the tapped density, one of the density groups, is particularly useful in describing the powder behavior during compaction [44].

Powder density is also linked to economic challenges in the dairy industry, such as packaging, transportation and storage costs. In this study, the bulk density values were 0.37 for MP and 0.49 kg/m^3 for DMHFEP samples. While the presence of milk fat decreased the bulk density and the flowability of cow milk powder [9], the inclusion of starch in powder form can increase the tapped density [15]. Another important physical property, known as the angle of repose, is used to characterize the bulk behavior of particulate foods characterization and design processing, storage, and conveying systems. A high angle of repose is indicative of very fine and sticky food, while a low angle of repose suggest highly flowable food [49]. In this study, the angle of repose values was calculated as 35° for MP and 42° for DMFEP, with the hollyhock-based milk powder having a higher value than the milk powder. The addition of an extra component, such as *Alcea rosea*, can increase the apparent density, which is consistent with findings from previous research [50].

For dairy powders, the solubility is based on the remaining amount of total solids in the supernatant after the stirring and centrifugation process [16]. This technique can be applied to different dairy powder products, such as cheese powder [51]. The solubility values of the milk powder and dried milky hollyhock drink samples were calculated as 90 and 84%, respectively. The lower solubility of the dried milky hollyhock drink sample compared to the milk powder can be attributed to the fact that small hydrophilic molecules promote dissolution [52]. Achieving high solubility in milk powder production is important for its future applications, as solubility is a key factor on the solubility of milk powder [53]; however, milk fat content can be as a significant influencing factor on the solubility of milk powder [47]. Additionally, the presence of mucilage has a negative effect on solubility due to its long-chain molecular structure.

3.11. Particle Appearance

The morphological characteristics of the particles in the samples are described in Figure 3. As observed in the micrographs, milk particles with hollyhock extract exhibit larger particles and have smaller materials surrounding them. The presence of agglomerates in samples, as confirmed by scanning electron microscopy (SEM) and light microscopy, aligns with the specifications for powder physical properties. The morphological changes observed in the samples produced by spray dryer are consistent with previous findings [54].

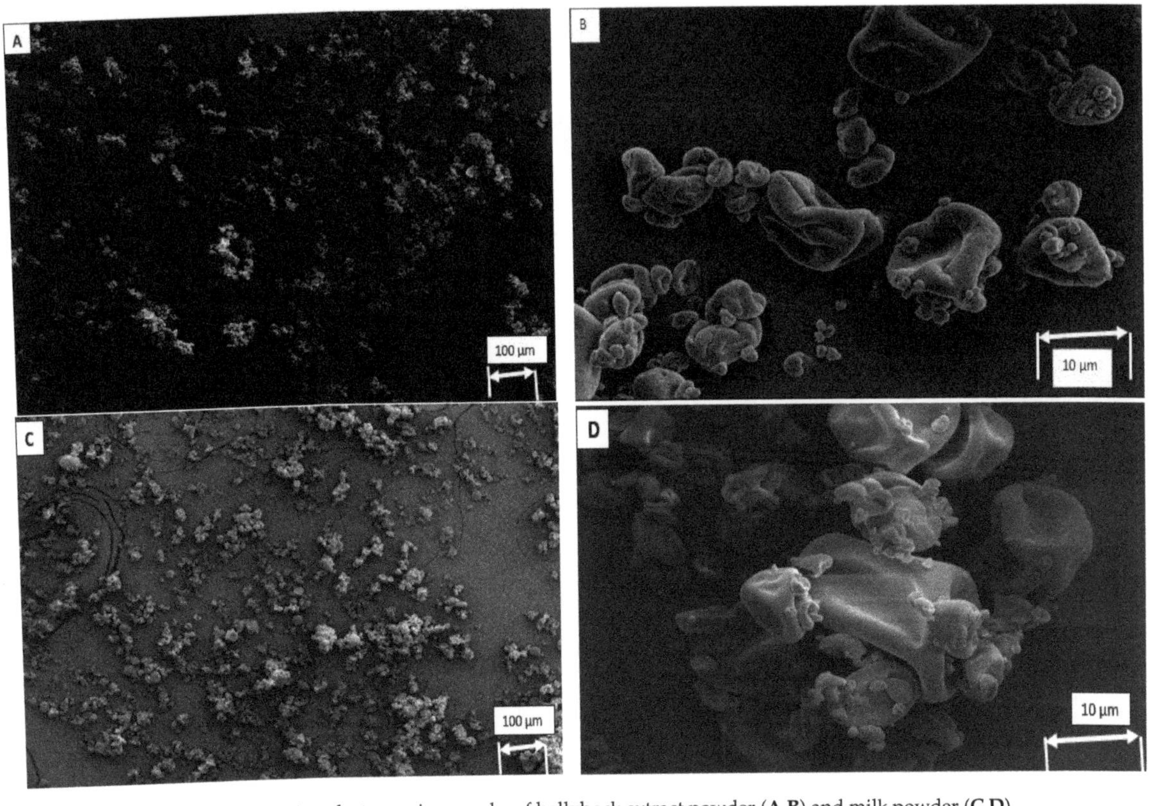

Figure 3. Scanning electron micrographs of hollyhock extract powder (**A**,**B**) and milk powder (**C**,**D**).

4. Conclusions

Although the hollyhock (*Alcea rosea*) plant and its various parts have been used in traditional medicine for hypoglycemic or hypolipidemic treatment, there is currently no processed industrial product available in the literature or markets. Based on the results of the aforementioned study, it is possible to develop a new powdered product from the traditional milky hollyhock drink. This new powdered product possesses distinct nutritional and morphological properties when compared to milk powder, including differences in terms of particle size, water activity, density, and flowability. Additionally, the inclusion of hollyhock can contribute to increased mineral content, such as sodium (Na) and potassium (K). Through sensory evaluation, it was determined that the optimal concentrations of hollyhock flower and the new powder were 5% for both the traditional milky drink and reconstituted drink.

This paper presents the application of spray-dried milk-based hollyhock flower extract for the production of a new ready-to-drink product derived from this medicinal plant. The resulting ready-to-drink powdered product can be used for general consumption due to its health benefits.

Funding: This research received no external funding.

Institutional Review Board Statement: Not applicable.

Informed Consent Statement: Not applicable.

Data Availability Statement: The data developed in this study will be made available upon request to the corresponding authors.

Conflicts of Interest: The author declares no conflict of interest.

References

1. Lim, T.K. *Edible Medicinal and Non-Medicinal Plants*; Springer: Berlin/Heidelberg, Germany, 2012.
2. Dudek, M.; Matławska, I.; Szkudlarek, M. Phenolic acids in the flowers of *Althaea rosea* var. nigra. *Acta Pol. Pharm.* **2006**, *63*, 207–211. [PubMed]
3. Hussain, L.; Akash, M.S.H.; Tahir, M.; Rehman, K.; Ahmed, K.Z. Hepatoprotective effects of methanolic extract of Alcea rosea against acetaminophen-induced hepatotoxicity in mice. *Bangladesh J. Pharmacol.* **2014**, *9*, 322–327. [CrossRef]
4. Ekpo, K.E.; Onigbinde, A.O.; Asia, I.O. Pharmaceutical potentials of the oils of some popular insects consumed in southern Nigeria. *Afr. J. Pharm. Pharmacol.* **2009**, *3*, 51–57.
5. Wang, D.F.; Shang, J.Y.; Yu, Q.H. Analgesic and anti-inflammatory effects of the flower of *Althaea rosea* (L.) Cav. *China J. Chin. Mater. Med.* **1989**, *4*, 46–48.
6. Lone, P.; Bhardwaj, A.; Bahar, F. Study of indigenous/traditional medicinal plant knowledge-An endeavour towards new drug discovery. *Afr. J. Tradit. Complement. Altern. Med.* **2015**, *12*, 73–95. [CrossRef]
7. Ammar, N.; El-Hakeem, R.A.; El-Kashoury, E.-S.; El-Kassem, L.A. Evaluation of the phenolic content and antioxidant potential of *Althaea rosea* cultivated in Egypt. *J. Arab. Soc. Med. Res.* **2013**, *8*, 48. [CrossRef]
8. Fahamiya, N.; Shiffa, M.; Mohd, A. A Comprehensive Review on *Althaea rosea* Linn. *Indo Am. J. Pharm. Res.* **2016**, *6*, 6888–6894.
9. Kim, M.S.; Chathuranga, K.; Kim, H.; Lee, J.-S.; Kim, C.-J. Anti-influenza properties of herbal extract of *Althaea rosea* in mice. *Korean J. Veter Res.* **2018**, *58*, 153–158. [CrossRef]
10. Shehzad, M.R.; Hanif, M.A.; Rehman, R.; Bhatti, I.A.; Hanif, A. Hollyhock. In *Medicinal Plants of South Asia*; Elsevier: Amsterdam, The Netherlands, 2020; pp. 381–391.
11. Yekta, R.; Mirmoghtadaie, L.; Hosseini, H.; Norouzbeigi, S.; Hosseini, S.M.; Shojaee-Aliabadi, S. Development and characterization of a novel edible film based on *Althaea rosea* flower gum: Investigating the reinforcing effects of bacterial nanocrystalline cellulose. *Int. J. Biol. Macromol.* **2020**, *158*, 327–337. [CrossRef]
12. AOAC. *Official Methods of Analysis of AOAC International*, 17th ed.; AOAC International: Rockville, MD, USA, 2020.
13. Li, Y.; Guo, C.; Yang, J.; Wei, J.; Xu, J.; Cheng, S. Evaluation of antioxidant properties of pomegranate peel extract in comparison with pomegranate pulp extract. *Food Chem.* **2020**, *96*, 254–260. [CrossRef]
14. Singh, R.P.; Murthy, K.N.C.; Jayaprakasha, G.K. Studies on the Antioxidant Activity of Pomegranate (*Punica granatum*) Peel and Seed Extracts Using In Vitro Models. *J. Agric. Food Chem.* **2020**, *50*, 81–86. [CrossRef] [PubMed]
15. Aidoo, H.; Sakyi-Dawson, E.; Tano-Debrah, K.; Saalia, F. Development and characterization of dehydrated peanut–cowpea milk powder for use as a dairy milk substitute in chocolate manufacture. *Food Res. Int.* **2010**, *43*, 79–85. [CrossRef]
16. Schokker, E.P.; Church, J.S.; Mata, J.P.; Gilbert, E.P.; Puvanenthiran, A.; Udabage, P. Reconstitution properties of micellar casein powder: Effects of composition and storage. *Int. Dairy J.* **2011**, *21*, 877–886. [CrossRef]
17. Mimouni, A.; Deeth, H.C.; Whittaker, A.K.; Gidley, M.J.; Bhandari, B.R. Rehydration process of milk protein concentrate powder monitored by static light scattering. *Food Hydrocoll.* **2009**, *23*, 1958–1965. [CrossRef]
18. Leite, T.S.; Augusto, P.E.; Cristianini, M. The use of high pressure homogenization (HPH) to reduce consistency of concentrated orange juice (COJ). *Innov. Food Sci. Emerg. Technol.* **2014**, *26*, 124–133. [CrossRef]
19. Gallo, L.; Ramírez-Rigo, M.V.; Piña, J.; Bucalá, V. A comparative study of spray-dried medicinal plant aqueous extracts. Drying performance and product quality. *Chem. Eng. Res. Des.* **2015**, *104*, 681–694. [CrossRef]
20. Walton, D.E. The morphology of spray-dried particles a qualitative view. *Dry. Technol.* **2000**, *18*, 1943–1986. [CrossRef]
21. Medina-Torres, L.; García-Cruz, E.; Calderas, F.; Laredo, R.G.; Sánchez-Olivares, G.; Gallegos-Infante, J.; Rocha-Guzmán, N.; Rodríguez-Ramírez, J. Microencapsulation by spray drying of gallic acid with nopal mucilage (Opuntia ficus indica). *LWT-Food Sci. Technol.* **2013**, *50*, 642–650. [CrossRef]
22. Pacheco, V.M.M.; Porras, A.O.O.; Velasco, E.; Morales-Valencia, E.M.; Navarro, A. Effect of the milk-whey relation over physicochemical and rheological properties on a fermented milky drink. *Ing. Compet.* **2017**, *19*, 83–91. [CrossRef]
23. Patil, A.T.; Meena, G.S.; Upadhyay, N.; Khetra, Y.; Borad, S.; Singh, A.K. Production and characterization of milk protein concentrates 60 (MPC60) from buffalo milk. *LWT* **2018**, *91*, 368–374. [CrossRef]
24. Fang, Z.; Bhandari, B. Effect of spray drying and storage on the stability of bayberry polyphenols. *Food Chem.* **2011**, *129*, 1139–1147. [CrossRef] [PubMed]
25. Song, F.-L.; Gan, R.-Y.; Zhang, Y.; Xiao, Q.; Kuang, L.; Li, H.-B. Total Phenolic Contents and Antioxidant Capacities of Selected Chinese Medicinal Plants. *Int. J. Mol. Sci.* **2010**, *11*, 2362–2372. [CrossRef]
26. Finkel, T.; Holbrook, N.J. Oxidants, oxidative stress and the biology of ageing. *Nature* **2000**, *408*, 239–247. [CrossRef] [PubMed]
27. Madhavi, D.L.; Deshpande, S.S.; Salunkhe, D.K. *Food Antioxidants: Technological, Toxicological, Health Perspective*, 1st ed.; Marcel Dekker: New York, NY, USA, 1996; pp. 1–32.
28. Valko, M.; Leibfritz, D.; Moncol, J.; Cronin, M.T.D.; Mazur, M.; Telser, J. Free radicals and antioxidants in normal physiological functions and human disease. *Int. J. Biochem. Cell Biol.* **2007**, *39*, 44–84. [CrossRef] [PubMed]
29. Cai, Y.M.; Ren, Y.R.; Wang, L. *Pharmacology and Clinical Application of Traditional Chinese Medicines*, 1st ed.; Huaxia Press: Beijing, China, 1999; pp. 348–387.

30. Wong, C.-C.; Li, H.-B.; Cheng, K.-W.; Chen, F. A systematic survey of antioxidant activity of Chinese medicinal plants using the ferric reducing antioxidant power assay. *Food Chem.* **2006**, *97*, 705–711. [CrossRef]
31. Kalušević, A.; Levic, S.; Čalija, B.; Pantić, M.; Belović, M.; Pavlović, V.; Bugarski, B.; Milić, J.; Žilić, S.; Nedović, V. Microencapsulation of anthocyanin-rich black soybean coat extract by spray drying using maltodextrin, gum Arabic and skimmed milk powder. *J. Microencapsul.* **2017**, *34*, 475–487. [CrossRef]
32. Quek, S.Y.; Chok, N.K.; Swedlund, P. The physicochemical properties of spray-dried watermelon powders. *Chem. Eng. Process. Process. Intensif.* **2007**, *46*, 386–392. [CrossRef]
33. Belscak-Cvitanovic, A.; Levic, S.; Kalusevic, A.; Špoljarić, I.; Đorđević, V.; Komes, D.; Mršić, G.; Nedovic, V. Efficiency Assessment of Natural Biopolymers as Encapsulants of Green Tea (Camellia sinensis L.) Bioactive Compounds by Spray Drying. *Food Bioprocess Technol.* **2015**, *8*, 2444–2460. [CrossRef]
34. Stuart, B.H. *Infrared Spectroscopy: Fundamentals and Applications*; John Wiley and Sons Ltd.: Chichester, UK, 2004; ISBN 0470854278.
35. Mudgil, D.; Barak, S.; Khatkar, B. X-ray diffraction, IR spectroscopy and thermal characterization of partially hydrolyzed guar gum. *Int. J. Biol. Macromol.* **2012**, *50*, 1035–1039. [CrossRef]
36. Bashir, M.; Haripriya, S. Assessment of physical and structural characteristics of almond gum. *Int. J. Biol. Macromol.* **2016**, *93*, 476–482. [CrossRef]
37. Ebrahiminezhad, A.; Barzegar, Y.; Ghasemi, Y.; Berenjian, A. Green synthesis and characterization of silver nanoparticles using Alcea rosea flower extract as a new generation of antimicrobials. *Chem. Ind. Chem. Eng. Q.* **2017**, *23*, 31–37. [CrossRef]
38. Khoei, S.; Azarian, M.; Khoee, S. Effect of Hyperthermia and Triblock Copolymeric Nanoparticles as Quercetin Carrier on DU145 Prostate Cancer Cells. *Curr. Nanosci.* **2012**, *8*, 690–696. [CrossRef]
39. Augusto, A.D.S.; Barsanelli, P.L.; Pereira, F.M.V.; Pereira-Filho, E.R. Calibration strategies for the direct determination of Ca, K, and Mg in commercial samples of powdered milk and solid dietary supplements using laser-induced breakdown spectroscopy (LIBS). *Food Res. Int.* **2017**, *94*, 72–78. [CrossRef] [PubMed]
40. Schuck, P.; Ouest, A. Milk powder: Physical and functional properties of milk powders. In *Encyclopedia of Dairy Sciences*; Fuquay, J.W., Fox, P.F., McSweeney, P.L.H., Eds.; Academic Press: San Diego, CA, USA, 2011; pp. 117–124. Available online: https://agris.fao.org/agris-search/search.do?recordID=FR20210181426 (accessed on 4 February 2023).
41. Fitzpatrick, J.; Iqbal, T.; Delaney, C.; Twomey, T.; Keogh, M. Effect of powder properties and storage conditions on the flowability of milk powders with different fat contents. *J. Food Eng.* **2004**, *64*, 435–444. [CrossRef]
42. Murphy, E.G. Infant Milk Formula Manufacture: Process and Compositional Interactions in High Dry Matter Wet-Mixes. Ph.D. Thesis, University College Cork, 2015. Available online: https://core.ac.uk/download/pdf/61579252.pdf (accessed on 27 January 2023).
43. Pugliese, A.; Cabassi, G.; Chiavaro, E.; Paciulli, M.; Carini, E.; Mucchetti, G. Physical characterization of whole and skim dried milk powders. *J. Food Sci. Technol.* **2017**, *54*, 3433–3442. [CrossRef]
44. Sharma, A.; Jana, A.; Chavan, R.S. Functionality of Milk Powders and Milk-Based Powders for End Use Applications-A Review. *Compr. Rev. Food Sci. Food Saf.* **2012**, *11*, 518–528. [CrossRef]
45. Bansal, V.; Sharma, H.K.; Nanda, V. Effect of honey addition on flowability and solubility of spray-dried low fat milk powder. *Int. Proc. Chem. Biol. Environ. Eng.* **2015**, *86*, 22–27.
46. Ilari, J.-L.; Mekkaoui, L. Physical properties of constitutive size classes of spray-dried skim milk powder and their mixtures. *Le Lait* **2010**, *85*, 279–294. [CrossRef]
47. Zouari, A.; Mtibaa, I.; Triki, M.; Jridi, M.; Zidi, D.; Attia, H.; Ayadi, M.A. Effect of spray-drying parameters on the solubility and the bulk density of camel milk powder: A response surface methodology approach. *Int. J. Dairy Technol.* **2020**, *73*, 616–624. [CrossRef]
48. Barbosa-Canovas, G.V.; Juliano, P. Physical and chemical properties of food powders. In *Encapsulated and Powdered Foods*; Onwulata, C., Ed.; Taylor and Francis Group, LLC: New York, NY, USA, 2005; pp. 39–71. [CrossRef]
49. Teferra, T.F. Engineering Properties of Food Materials. In *Handbook of Farm, Dairy and Food Machinery Engineering*; Academic Press: Cambridge, MA, USA, 2019; pp. 45–89. [CrossRef]
50. Ma, U.L.; Ziegler, G.; Floros, J. Effect of Sucrose on Physical Properties of Spray-Dried Whole Milk Powder. *J. Food Sci.* **2008**, *73*, E431–E438. [CrossRef]
51. da Silva, D.F.; Tziouri, D.; Ahrné, L.; Bovet, N.; Larsen, F.H.; Ipsen, R.; Hougaard, A. Reconstitution behavior of cheese powders: Effects of cheese age and dairy ingredients on wettability, dispersibility and total rehydration. *J. Food Eng.* **2020**, *270*, 109763. [CrossRef]
52. Lillford, P.; Fryer, P. Food Particles and the Problems of Hydration. *Chem. Eng. Res. Des.* **1998**, *76*, 797–802. [CrossRef]
53. Freudig, B.; Hogekamp, S.; Schubert, H. Dispersion of powders in liquids in a stirred vessel. *Chem. Eng. Process. Process. Intensif.* **1999**, *38*, 525–532. [CrossRef]
54. Alamilla-Beltrán, L.; Chanona-Pérez, J.; Jiménez-Aparicio, A.; Gutierrez, G. Description of morphological changes of particles along spray drying. *J. Food Eng.* **2005**, *67*, 179–184. [CrossRef]

sustainability

MDPI

Article

Sustainable Food in Teacher Training: Evaluation of a Proposal for Educational Intervention

Vanessa Ortega-Quevedo [1], Noelia Santamaría-Cárdaba [2,*] and Cristina Gil-Puente [3]

1 Department of Sciences Education, Social Sciences Education and Mathematics Education, Universidad Complutense de Madrid, 28040 Madrid, Spain; vanessao@ucm.es
2 Department of Pedagogy, Universidad de Valladolid, 40005 Segovia, Spain
3 Department of Sciences Education, Social Sciences Education and Mathematics Education, Universidad de Valladolid, 40005 Segovia, Spain; cristina.gil.puente@uva.es
* Correspondence: noelia.santamaria.cardaba@uva.es

Abstract: The sustainability of the food system is a particularly relevant issue today as it is a challenge to ensure environmental sustainability and the need to guarantee access to food in all parts of the world to promote social equity. Given this, the need to promote a sustainable food transition in consumption habits becomes imperative. It is crucial that teachers be sensitised to this issue so that they can try to develop a critical and globally aware student body that is committed to a social transformation towards sustainability. The main objective of this study is discovering the opinions of preservice teachers (PSTs) on sustainable food. It also has two specific objectives: (1) analyze the initial opinions of prospective teachers on sustainable food before and after carrying out an educational intervention on this topic, and (2) study the relationship between the opinions of PSTs and variables related to their personality or their relationship with nature, among others. A longitudinal pre-test-intervention-post-test design using quantitative methods was carried out to explore the opinions of 49 pre-service teachers studying for a degree in Primary Education before and after the educational intervention. The results show changes in the opinions of future teachers after the application of the educational intervention and differences towards food sustainability according to some socio-demographic variables such as gender.

Keywords: educational intervention; environmental sustainability; higher education; preservice teachers; sustainable development; sustainable food

1. Introduction

The sustainability of the current food system is a particularly relevant issue today as it is a challenge to ensure environmental sustainability and the need to guarantee access to food in all parts of the world while promoting social equity [1,2]. This raises the need to promote a sustainable food transition in food consumption habits, as it is of increasing concern that impoverished countries continue to suffer from famine and malnutrition [3,4]; while in developed countries, there is an increase in nutrition problems related to being overweight and obesity in populations of all ages [5,6].

The Food and Agriculture Organisation of the United Nations, better known as FAO, in its study published in 2020 on the state of food security and nutrition in the world, has confirmed that the Sustainable Development Goals established in the 2030 Agenda [2] on hunger and malnutrition are not going to be solved in ten years' time. Authors such as Sanahuja and Tezanos [7] and Lee [8] agree that a profound reform of the current economic system is necessary. Furthermore, they suggest that this reform should be approached from a cosmopolitan perspective, promoting a true "global governance" to ensure a fair distribution of opportunities and responsibilities in the development of humanity as a whole.

Therefore, the need to promote an urgent and crucial transformation in the food distribution and production system is clear, making this research's subject highly relevant to ensure social and environmental sustainability. Solving this challenge will be a fundamental issue in securing the future of the twentieth century [1]. Facing these questions requires training in this area for PST because if they are sensitised to the importance of food sovereignty and achieving sustainable transitions, they will be able to promote a transformative education. This is because education is key to forming behaviours and attitudes in favour of sustainability [9].

Despite the unquestionable urgency of addressing these issues to ensure a sustainable future, there is a paucity of research on food systems education and there is a need to promote critical studies on this topic [10,11]. Our study aims to make a novel contribution that seeks to help alleviate the lack of research on education for a sustainable food transition.

1.1. Food Consumption Habits of Young People

The food consumption habits of the population have been modified over time as they have been incorporating products into their diet and modifying their behaviour in a very different way than in the past [12,13]. In this way, a nutritional transition is taking place in which people are changing their eating habits, generally characterised by a shift from traditional meals to high-calorie, high-fat, and refined foods [14]. However, other authors such as Wang et al. [15], Cardona et al. [16], or Paul et al. [17] consider that there has also been a growing concern for health, which is reflected in the increased consumption of dietary products.

Despite this, the diet of university students is characterised by a high consumption of meat and dairy, resulting in an excess of saturated fats, cholesterol, and animal proteins, and an insufficient intake of fruits and vegetables. Therefore, the quality of their diet is poor and a food transition towards sustainable and balanced consumption should be made. [18]. In addition, Ruiz et al. [19] focused their analysis on the pattern of beverage consumption, which also allowed them to verify that the diet of young university students does not follow the recommendations of experts or those of the Mediterranean Diet.

This situation shows that the current population is suffering from a loss of food culture, displacing the Mediterranean Diet with a less healthy diet, as stated by authors such as Jacques and Jacques [20] and Bárbara and Ferreira-Pêgo [21]. In this way, globalisation has thus been gaining ground in markets, displacing local products, and transforming eating habits in many rural populations, leading to an imbalance between food intake and calorie consumption [22]. In this context, education plays an essential role in training people for food sovereignty and a healthier and more sustainable diet transition [10]. Indeed, education for food sovereignty must be understood as transformative ethical practice [23].

In summary, it is essential to train teachers who are aware of sustainable development, as they will be able to transmit these values to their students and stimulate a change towards sustainable food, thus improving their own eating habits. What do we mean, though, by responsible transition? Why is it important to include this issue in teacher training?

1.2. Responsible Transitions in Teacher Training: The Way to a Sustainable Food Future

The current food system is showing the effects of globalization, causing changes in diets, in the foods that are most consumed and sold, and even in the jobs linked to this sector, which are increasingly weakened. Because of this, as Garcés [24] explains:

> We need to rediscover and revalue the agricultural and food knowledge and practices of our countries and, at the same time, respond to the major contemporary challenges: job creation and the fight against poverty, sustainable management of natural resources and the fight against climate change, and the preservation of cultural heritage. (p. 261)

The need to promote responsible food transitions that ensure a more sustainable future for everyone, regardless of where they live, is readily apparent nowadays. Rastoin [25,26] highlights that there have been five food transitions: (1) the beginning of the use of fire modified the consumption of raw products; (2) the emergence of agriculture and the domestication of animals; (3) large cities began to emerge, leading to a division of labour; (4) the industrialisation of the production, processing, and distribution of food products; (5) during the twentieth century, consumer demand for certain products has led to changes in the production and distribution of food. Therefore, collective efforts towards a sustainable and responsible diet transition are key to shaping future food systems.

Based on the Sustainable Development Goals and taking as a basis the words of authors Mello-Théry [27] or Giunta and González [28], it is essential to think about a change in the system that promises a transition towards sustainability. This would be the way to ensure that all beings on the planet can survive while respecting the environment and avoiding increasing levels of degradation.

The future of the planet is in the hands of the new generations. Therefore, working on this issue in teacher training is essential for them to commit themselves to the social transformation towards sustainability and to transmit it to their future students [29]. In other words, PST will not be able to educate for sustainable development if they themselves do not have the necessary sustainability awareness and competences [30]. International organizations such as the World Commission on Environment and Development (WCED) have emphasized the importance of education to promote sustainable development and encourage responsible attitudes towards food consumption. Therefore, one of the main goals of education is to achieve an appropriate balance between current and future needs, so that current demand can be met without compromising the ability of future generations to meet their own needs [31].

2. Research Objectives

Based on the above, promoting education for sustainable development that promotes a food transition makes it necessary for trainee teachers to be aware of this issue so they can transmit it to their students; this is where the present study is framed, whose general objective is to analyze the opinions of PST on sustainable food. This objective is specified in the following specific objectives:

- To analyze the opinions of trainee teachers before and after taking a subject that deals with sustainable food.
- To study the relationship between PSTs' opinions and variables related to their personality, such as their relationship with nature, leisure activities in the natural environment, links with the cultivation of the land, and the size of their place of residence.

3. Materials and Methods

3.1. Design of the Study

This study is a quasi-experimental study, which is a study of a real situation (far from the control of variables typical of a laboratory) and with a non-random sample selection. The study design is longitudinal, as the evolution of the participants is analyzed over time. The methodology used in the design is quantitative. Consequently, a structure was established that initially evaluates the participants (pre-test), allows an intervention to be carried out in which content related to the objectives of the study is worked on and, finally, evaluates the evolution of the participants (post-test).

This design has been established, as it allows the general objective to be met—to analyse the opinions of PSTs on sustainable food. It also meets the specific objectives: (1) To analyse the opinions that PSTs have before and after taking a subject that deals with content on sustainable food; (2) to study the relationship between the PSTs' opinions and variables related to their personalities, such as their relationship with nature, leisure activities in the natural environment, links with the cultivation of the land, and the size of their place of residence.

The data were collected through an online questionnaire provided to the PSTs before the beginning of a class on the subject of Didactics of Experimental Sciences in a Spanish university.

3.2. Participants

In this study, 49 students participated in the Degree in Primary Education, 35 women and 14 men. This difference is based on the data published by the Ministry of Universities [32], which show that the percentage of female enrolments is 67.7% compared to 32.3% of male enrolments in Primary Education Degrees in Spain. Participation in the study was voluntary because the students had previously signed an informed consent form.

The participants were selected according to so-called convenience sampling. This type of sampling moves away from the randomness of probability sampling. It is carried out based on subjective criteria related to the research objectives and the possibilities of the research group. Specifically, the sampling was guided by the ease of access to the participants and by its simple execution. The only limitation may have been the number of participants, as not all those selected completed the questionnaire.

This study was carried out during one academic year on a group of students as a pilot to assess the impact of the proposal on PSTs. This, together with the fact that not all students agreed to publish their data, results in a very small sample.

3.3. Context

The subject in which the activity is framed belongs to the field of social sciences, more specifically educational sciences, and is based on specific didactic knowledge about the teaching–learning methods in experimental sciences with a dual purpose: to describe and explain these processes and to design, develop, and evaluate proposals for improving science education. One of the main challenges facing this discipline is facing the difficulties posed by the creation and dissemination of alternative proposals to the traditional way of teaching science and the school failure that this generates.

The didactic approach chosen not only in this activity, but also in the subject, in general, is based on activities that attempt to answer questions and problems of a professional and significant nature for the future teacher, as is the case with the study presented here.

The educational intervention was developed in three phases. The first was where students had to reflect on and analyze the content of documentary sources based on the use of vegetable gardens as a teaching resource (the documents were selected by the teachers of the subject). A second was where they visited vegetable gardens near the natural environment of the faculty to assess the importance of field activities for the learning of natural sciences and to learn about the contributions of this type of activity in the teaching–learning process. Finally, the students, organized in small groups, produced podcasts (one per group) with the aim of exploring, in depth, the use of vegetable gardens as a didactic resource in an informative way.

The evaluation procedure aligns with the didactic model and educational principles, as it aims to be comprehensive, formative, participatory, and continuous. This implies that the evaluation is integrated with the rest of the didactic process and the educational context, promoting continuous improvement of the teaching–learning process.

The work carried out by the students is assessed by the teachers (hetero-assessment) and in parallel by the students (self-assessment) based on a rubric containing the assessment criteria. These criteria are set in advance and are known to the students before the start of the activity.

3.4. Instrument and Data Collection

The data used in this research were collected via an online survey as part of a larger study that measured sustainable development. Although the questionnaire was conducted via an online platform, it was carried out in one of the class sessions. Therefore, the researchers were present in case any doubts arose during its development. This format was used because of the ease of data collection and analysis.

A higher number of participants was expected (the total number of students in the subject was 80). However, when the database was compiled, it was observed that many students had not taken the post-test, so the sample was reduced to those participants who had taken part in both assessment sessions.

The survey has three parts. The first part collects the questions that measure the opinion of the PSTs on sustainable development. This is a scale from one to six where one is identified as strongly disagree and six is identified as strongly agree (the response options ordered from lowest to highest value are: strongly disagree, do not agree at all, slightly agree, agree, quite agree, totally agree). The second gathered simple socio-demographic information. The third part consists of a scale from one to ten, which provides information on the personal background of the participants. Table 1 lists the questions from the three parts of the survey applied to this study. As can be seen, the questions in the first part of the survey selected for this study respond to the theme of sustainable food.

Table 1. Questions from the three parts of the survey applied to this study.

Part one. Opinion on...	It saddens me to see out-of-season products in the shops. For me it is important to choose food that pollutes the environment the least. I understand what food sovereignty means. Farmland is invaluable to me.
Part two. Socio-demographic questions	Gender Size of the municipality of residence
Part three. Questions on personal background	For one reason or another in my family we have been very close to the natural environment. I practice leisure activities related to nature (hiking, climbing, rafting, etc.). People close to me have cultivated the land. I am influenced by social networks when choosing my diet. The education I received at school has influenced the way I eat. I am concerned about achieving a sustainable society.

3.5. Data Analysis

Quantitative analysis was implemented using IBM SPSS Statistics software version 26.0. First, the Kolmogorov–Smirnov test was applied. The p-values < 0.01 (significant for all variables) were obtained, indicating a violation of the normality assumption. Consequently, the use of non-parametric statistics was determined. The Wilcoxon's test for two related samples was used to study the comparison between medians on the study variables. Hedges' G was applied to calculate the effect size of the intervention. Pearson's Chi-square was applied to contrast numerical and nominal variables. Spearman's test was applied to study the correlations between scale variables.

4. Results

The purpose of this study was to explore the opinions of PSTs on sustainable food. This was approached through the analysis of four variables: the consumption of seasonal products, the consumption of products that pollute the environment less during their production, processing and marketing, the concept of food sovereignty, and the importance of farmland.

4.1. Consumption of Products out of Season

As can be seen in Table 2, PSTs answer whether they are sad to consume seasonal products more often in the pre-test phase by marking the answers "slightly agree", "agree", or "quite agree". In terms of descriptive statistics, the mean value of 3.8, the median value of 4, and the standard deviation of 1.247 should be noted. In contrast, the same table shows a change in the frequency of responses in the measurement after the intervention. In this phase, the answers most frequently used by the participants are "agree", "quite agree", or "totally agree". Consequently, the value of the mean increases to 4.31, the value of the median remains at 4, and the value of the standard deviation is 1.194 (Figure 1).

Table 2. Results of the statement "I am sad to see out-of-season products in shops" in the pre- and post-test phases.

		It Saddens Me to See Out-of-Season Products in the Shops						
		Strongly Disagreed (1)	**Do Not Agree at All (2)**	**Slightly Agree (3)**	**Agree (4)**	**Quite Agree (5)**	**Totally Agree (6)**	**Total**
Pre-test	*f*	3	5	9	17	12	3	49
	%	6.1	10.2	18.4	34.7	24.5	6.1	100
Post-Test	*f*	0	5	5	18	12	9	49
	%	0	10.2	10.2	36.7	24.5	18.4	100

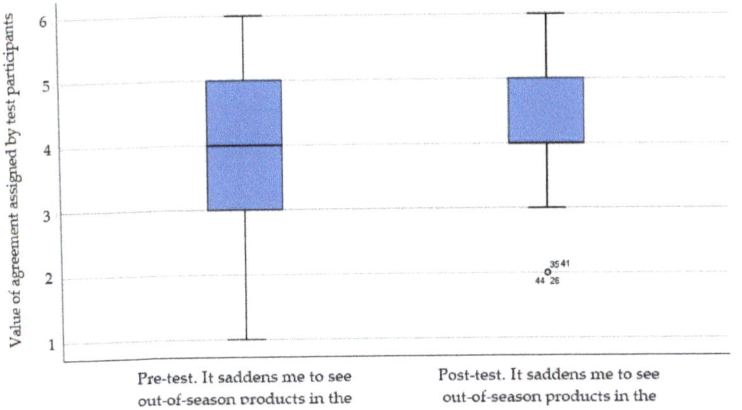

Figure 1. Boxplot of the statement "It saddens me to see out-of-season products in the shops".

After the descriptive study, the Wilcoxon's test was applied to determine whether the difference between the results of the measurement phases was significant. The *p*-value resulting from this test is 0.025. Therefore, significance is demonstrated. We then calculated the effect size of the intervention using Hedges' G, obtaining a value of 0.413, which we should interpret as small, although close to what is considered moderate (0.049) following Cohen's proposal [33].

There is a correlational study between item "It saddens me to see out-of-season products in the shops" and those items belonging to the scale that determined the participants' personal background. It should be noted that when applying Spearman's test between the results of the pre-test phase and the rest of the variables, correlation coefficients with significant values were obtained with the following items:

- For one reason or another in my family, we have been very close to the natural environment (correlation coefficient 0.378, *p*-value 0.007).
- I practice nature-related leisure activities (hiking, climbing, rafting, etc.) (correlation coefficient 0.454, *p*-value 0.001).

However, these significant correlations disappear when applying Spearman's test in the post-test measurement with the variables of the personal baggage scale.

When analyzing the study variable in the pre-test in contrast to the relationship with nature variable, it is observed that the mean (degree of agreement) with the statement increases as the relationship with nature increases, except in the maximum value of relationship with nature, where the value of the mean drops (for example, for a value of 6, the mean is 2.4, in value 9, the mean rises to 4.43, in value 10, the mean drops to 4.18). Finally, in the case of the variable consumption of seasonal products (pre-test) in contrast to the practice of leisure in nature, an increase in the mean of the degree of agreement is observed as the practice of leisure in nature increases (for example, at value 5 of the practice of leisure in nature, the mean takes a value of 2.25, increasing progressively until acquiring a value of 4.67 at the maximum value of the relationship with nature).

4.2. Consumption of Food That in Its Production and Marketing Has Polluted the Environment Less

The PSTs have shown their opinion on whether it is important for them to consume food that is less contaminated in its production, preservation, and marketing process (Table 3). From their answers in the pre-test phase, a majority slightly agree or agree. Specifically, a mean of 3.9, a median of 4, and a standard deviation with a value of 1.358 were obtained. In contrast, in the post-test phase, a greater environmental awareness was shown, as most of the responses were in the agree or quite agree range (Figure 2). In this second measurement, the median rises to 4.29, the median remains at 4, and the standard deviation takes a value of 1.19.

Table 3. Results of the statement "For me it is important to choose food that pollutes the environment the least".

		For Me It Is Important to Choose Food That Pollutes the Environment the Least						
		Strongly Disagreed (1)	**Do Not Agree at All (2)**	**Slightly Agree (3)**	**Agree (4)**	**Quite Agree (5)**	**Totally Agree (6)**	**Total**
Pre-test	*f*	2	6	10	15	9	7	49
	%	4.1	12.2	20.4	30.6	18.4	14.3	100
Post-Test	*f*	0	5	6	16	14	8	49
	%	0	10.2	12.2	32.7	28.6	16.3	100

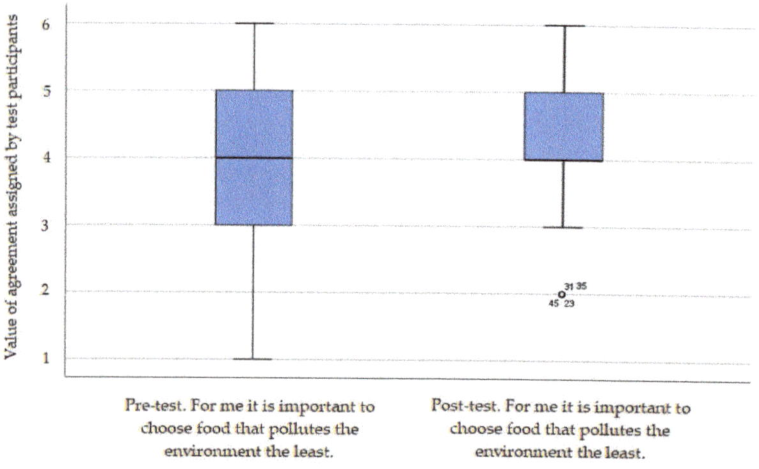

Figure 2. Boxplot of the statement "for me it is important to choose food that pollutes the environment the least".

Wilcoxson's test was applied to assess the significance of the improvement detected between the measurement phases. The *p*-value obtained in this test is < 0.05 (*p*-value 0.043), which demonstrates significance. Next, the effect size of the intervention is calculated using Hedges' G, which gives a value of 0.305, indicating a small effect.

Finally, it should be noted that in the correlational analysis (Spearman's test) between the study item and the personal baggage items, only one correlation was detected in the pre-test phase. This correlation is between the study item and the relationship with the natural environment (correlation coefficient 0.378, *p*-value 0.007). However, this relationship disappears after the intervention. When studying this correlation, it is observed that the mean of the agreement is higher when the relationship of the PSTs with nature is higher, except in the maximum value of relationship with nature where the mean decreases (e.g., for value 6, the mean is 3.67, for value 9, the mean is 4.64, and for value 10, the mean decreases to 4.09).

4.3. Understanding the Concept of "Food Sovereignty"

In this case, as seen in Table 4, the highest frequencies of answers given by PSTs are strongly disagree and agree. This indicates that 38.9% of the participating PSTs were unaware of the concept of food sovereignty at the time of the pre-test. However, the opinions of the PSTs show a change after the intervention, and the response with the highest frequency becomes agree, followed by slightly agree/do not agree at all. In the case of the pre-test values, 2.69 for the mean, 2 for the median, and 1.673 for the standard deviation were obtained. This is in contrast with the values obtained in the post-test, with a mean of 3.39, a median of 3, and a deviation of 1.565 (Figure 3). Applying the Wilcoxson's test to check the significance of these differences, a *p*-value of 0.017 was obtained, thus demonstrating significance. In the effect size study, the Hedges' G test 0.43 was applied, showing an effect size close to moderate.

Table 4. Results of the statement "I understand what food sovereignty means".

		I Understand What Food Sovereignty Means						
		Strongly Disagreed (1)	**Do Not Agree at All (2)**	**Slightly Agree (3)**	**Agree (4)**	**Quite Agree (5)**	**Totally Agree (6)**	**Total**
Pre-test	*f*	19	7	4	10	7	2	49
	%	38.9	14.3	8.2	20.4	14.3	4.1	100
Post-Test	*f*	7	9	9	11	8	5	49
	%	14.3	18.4	18.4	22.4	16.3	10.2	100

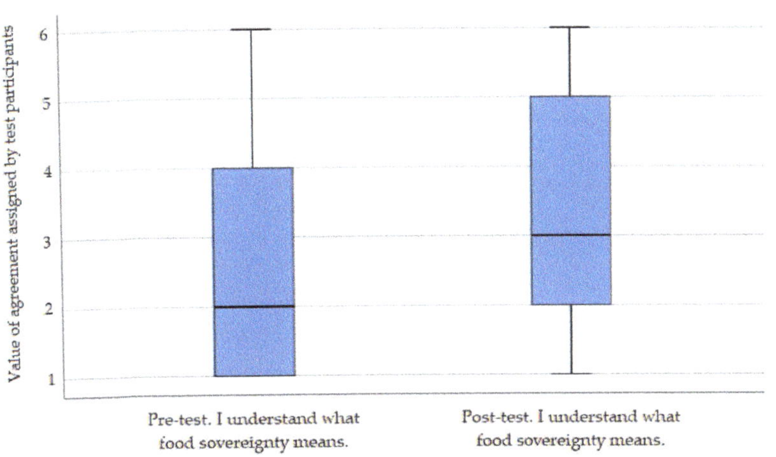

Figure 3. Boxplot of the statement I understand what food sovereignty means.

As for the correlation observed when applying Spearman's test between this item and those included in the personal baggage scale, it is worth noting that when applying the test to the pre-test results, significant correlations appear between the understanding of the concept of food sovereignty and the relationship with the environment (correlation coefficient 0.546, *p*-value 0.000); as well as between the understanding of the concept of food sovereignty and whether people close to the PSTs have cultivated the land (correlation coefficient 0.340, *p*-value 0.017). However, when applying the test in the post-test results, only the relationship between the understanding of the concept and the participating FSWs claiming to be connected to their natural environment is maintained (correlation coefficient 0.364, *p*-value 0.01).

When analysing the results in the pre-stage in contrast to the cultivation of the land, it is observed that the mean of the degree of agreement increases as the relationship between the social environment of the PSTs and the cultivation of the land increases (e.g., for value 5, the mean is 1.5, and for value 10, the mean rises to 3.29). However, there is an exception in those PSTs that value the relationship of their social environment with the cultivation of the land with 6 points. In this case, the average of the degree of agreement rises to 4.

4.4. The Value of Farmland

The PSTs have given their opinion on the importance of farmland for them (Table 5). The most frequent responses in the pre-test phase are "agree" and "totally agree". The mean obtained in this phase is 4.59, the median 5, and the standard deviation 1.258, i.e., the PSTs attach a certain value to farmland. However, in the post-test phase, the degree of agreement with the statement increases, with the most frequent responses being "totally agree" and "quite agree". In this case, the mean rises to 5.04, the median remains at 5, and the standard deviation is 1.117 (Figure 4).

Table 5. Results of the statement "Farmland is of inestimable value to me".

		Farmland Is of Inestimable Value to Me						
		Strongly Disagreed (1)	**Do Not Agree at All (2)**	**Slightly Agree (3)**	**Agree (4)**	**Quite Agree (5)**	**Totally Agree (6)**	**Total**
Pre-test	*f*	1	3	3	15	13	14	49
	%	2	6.1	6.1	30.6	26.5	28.6	100
Post-Test	*f*	0	2	3	8	14	22	49
	%	0	4.1	6.1	16.3	28.6	44.9	100

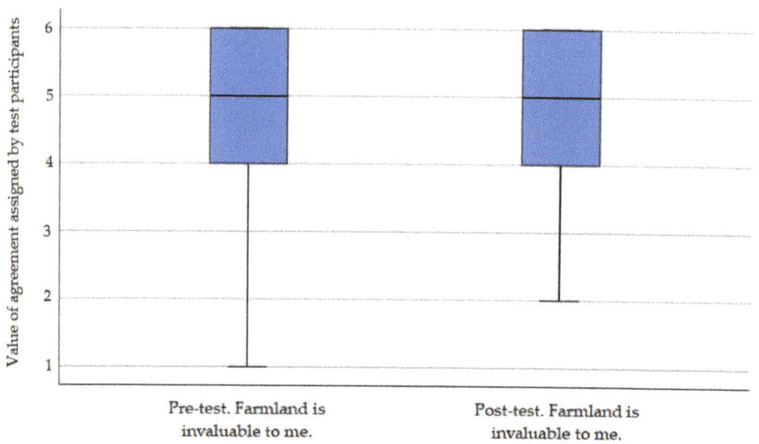

Figure 4. Boxplot of the statement Farmland is invaluable to me.

The study of the significance of the differences indicated using Wilcoxon's test determines that the differences are significant (p-value 0.01). The effect size is studied using the Hedges' G-test, which gives a value of 0.38, indicating a small effect size.

The study of this variable in contrast to the students' place of origin (whether they come from a small, medium, or large town or city) using Pearson's chi-square test shows no significant relationship between variables in the pre-test phase. However, there is a significant relationship between the variables in the post-test phase c2 21.5 with a p-value of 0.04. As seen in Table 6, PSTs, both in the pre-test and post-test phase, value farmland more highly if their origin is from a smaller locality.

Table 6. Descriptive statistics of "Farmland is of inestimable value to me" by habitat.

	Habitat	N	Media	Median	SD *
Pre-test	City	18	4.33	4	1.085
	Small locality (<10,000 inhabitants)	16	5.13	5.5	0.957
	Medium-sized locality (<10,000 inhabitants)	9	4.56	5	1.33
	Large locality (commercial head with small or medium-sized localities under its responsibility)	6	4	5	2
Post-test	City	18	4.67	4.5	1.188
	Small locality (<10,000 inhabitants)	16	5.5	6	0.632
	Medium-sized locality (<10,000 inhabitants)	9	5.11	6	1.269
	Large locality (commercial head with small or medium-sized localities under its responsibility)	6	4.83	5	1.472

* SD (Standard Deviation)

Regarding the study of the correlation between this variable in the pre-test phase with the personal baggage scale using Spearman's test, it should be noted that significant correlations are obtained between the value given to farmland and proximity to the natural environment (correlation coefficient 0.485, p-value 0.000) and whether people close to the PSTS have cultivated the land (correlation coefficient 0.333, p-value 0.02). In the comparison between the study variable in the post-test phase with the items of the personal baggage scale, the same correlations are detected as in the pre-test phase with proximity to the natural environment (correlation coefficient 0.675, p-value 0.000) and whether people close to the PST have cultivated the land (correlation coefficient 0.46, p-value 0.001).

However, at this stage, a further correlation is added with the item "I practice leisure in nature" (correlation coefficient 0.401, p-value 0.0004). When studying the results of the variable in the pre- and post-test phase by proximity to the natural environment, higher means are observed the more value is given to the relationship with nature (for example, in the pre-test phase, for a valuation of their relationship with nature of 6, a mean of 3.4 is obtained, in contrast to PSTs who value their relationship with nature with 10, who present a mean of 5.45).

5. Discussion

PST participants broadly agree, both in the pre-test and post-test, that seeing out-of-season products in the shops makes them feel sad. These feelings show a certain awareness of the issue of globalized agricultural trade, which, as several studies indicate, is a relevant issue from the perspective of sustainable food transitions due to its environmental and social consequences [34–36].

In the pre-test phase, PSTs showed significant relationships between sadness at seeing out-of-season products with family attachment to the natural environment and leisure in nature. However, in the post-test phase, these relationships disappear. In this sense, the significant relationship of family attachment to the environment is in line with the results of other studies that have already highlighted the importance of the family in forming citizens who understand the importance of sustainable development [37–39]. For example, the study by Molinario et al. [37] concludes that the family is of great importance in the development of children's sensitivity to nature. Something similar happens with people who practice leisure in nature, since, as already indicated by authors such as Horka and Hromádka [40], they tend to be more aware of the importance of caring for the environment. These significant relationships slow down in the post-test, which indicates that after the intervention was carried out with the PSTs, the awareness shown by these people does not show more affinity in terms of any of the contrast variables, not even with the aspects mentioned above.

PSTs broadly agree that it is important for them to choose foods that pollute the environment as little as possible. However, it should be noted that after the intervention, the percentage of PSTs who agreed or strongly agreed with this issue increased. Education on the importance of achieving sustainable food is key to choosing the products we buy. In their studies, Magnuson et al. [41] or Raptou and Manolas [42] also found that purchasing organic products is strongly related to awareness of both their benefits for human health and climate change. The increase in awareness of the need to change our attitudes to more sustainable ones after the intervention in our research does not correlate with any of the contrast variables. This allows us to conclude that the intervention boosts awareness regardless of the background and personal situation of the PSTs.

Regarding the concept of food sovereignty, 38.9% of the participating PSTs were unaware of the concept of food sovereignty at the time of the pre-test. However, the opinions of PSTs show a change after the intervention. This reveals that it is not a widespread concept and that without prior training, people do not understand what it is and why it is important [43,44]. The correlational analysis indicates that, initially, people who are closer to the natural environment and those who have people close to them who have cultivated the land perceive that they have a better understanding of the concept of food sovereignty. This again underlines the importance of the role of families in raising awareness of these environmental sustainability issues [37], as they are a key pillar in the formation of responsible citizenship [45,46]. Though, after the intervention, only the affinity between the understanding of the concept and the connection with the immediate environment remains, which indicates that training is an important element in raising awareness [47].

The value given to farmland by the PSTs is high and is especially remarkable in the post-test. This reveals that after appropriate training, the participants understood the importance of farmland and generally "totally agree" with its importance. Several studies such as the one conducted by Nous-Heen et al. [48] or Eugenio-Gozalbo et al. [49], conclude that PSTs who have received training on this topic value the garden as a resource that benefits them in their training and future teaching practice.

It is worth noting that the results revealed new positive correlations between the importance given to farmland, those who had a relationship with nature, and those who had people close to them who cultivated the land. In fact, in the post-test, this relationship was stronger and even showed a new positive correlation in terms of people who spend their leisure time in nature. These data are in line with those obtained in other studies, showing that when PSTs are trained in the importance of achieving environmental sustainability and a sustainable food transition, they show increasing environmental awareness and sensitivity [49,50]. Therefore, there is no doubt that teachers can significantly impact sustainability education because if they adopt sustainable attitudes in their daily behaviour, they can inspire their students and serve as role models on their way to sustainability [50].

6. Conclusions

To study the opinion of PSTs involved in the study on sustainable food, the opinion of participants on the following issues was investigated: "Consumption of products out of season", "Consumption of food that in its production and marketing has polluted the environment less", "Understanding the concept of food sovereignty", and "The value of farmland". It is concluded that:

PSTs involved in the study initially have an above-average opinion regarding negative feelings about the consumption of out-of-season products, the consumption of products that pollute more in the marketing process, and the importance of farmland. This average opinion improves after the intervention, which indicates a greater awareness of these issues after working on them. Likewise, the PSTs' knowledge of the concept of "food sovereignty" is below average before the intervention, indicating that they reach the third year of their degree in Education with little or no knowledge of this concept either in Higher Education or in Compulsory Education. However, after the intervention, their self-perception of their understanding of the concept increases.

As for the study of the relationship of the research variables with the scale of personal baggage of the participants, it is found that there are no differences in terms of gender in any of the variables and only in the case of the importance of farmland is a relationship with the locality of origin of the PSTs detected both in the pre- and post-test phase. In the variables consumption of products out of season and consumption of food that in its production and marketing has polluted the environment less, affinity is observed in the pre-test phase with the relationship of the PSTs involved in the study with nature and the practice of leisure in nature, and relationship with nature, respectively. However, this relationship is lost after the intervention, which serves to compensate for personal baggage when forming an opinion on these variables. In the variable on the understanding of the concept of "food sovereignty", affinity is found in the relationship with the natural environment and the connection of people close to the earth in the pre-test phase. This relationship is maintained in the case of connection with nature for the post-test phase. Regarding the last variable importance of farmland, affinities were detected in the pre-test phase with the connection with nature and the connection of close people with the cultivation of the land. In this case, the intervention highlights these connections, as they are more significant in the post-test phase. A further affinity variable is added, the practice of leisure in nature.

Consequently, it is concluded that the participants' views on sustainable food could initially be described as correct, but significantly improved after the intervention.

Finally, we would like to highlight as a strong point of the article the contribution it makes to the scientific literature on education for sustainable food. This is a subject that has been identified as needing study, although research is still limited. It also highlights the fact that this study not only analyses the opinion of PSTs involved in the study but also studies how this opinion changes after an intervention is carried out. It is considered that showing the evolution of the participants, after explicitly addressing the contents of the study in class, may motivate teachers to include these reflections on food sustainability in the classrooms of university degrees, especially in education degrees. As an example, based on the theoretical references cited in the study and our own results, it can be seen that activities such as these didactic proposals that connect students with local production, seasonal products in the area and the cultivation of the land favor the promotion of a reflection that contributes to sustainable awareness. On the other hand, it is pointed out that the small sample size does not allow for generalisations to be made and that the data are only representative of the participants In this study. That is, the study only allows us to get to know the reality studied and to understand the need to act in favour of sustainable awareness. As a result, it is proposed as a future line of research to extend the sample with students from different degrees and, if possible, institutions, to obtain generalisable results. In addition, once the sample and changes with the proposed intervention have been scaled up, other interventions will be explored that can further contribute to the development of responsible and sustainable attitudes among university students (belonging to degrees in Education and degrees in other areas of knowledge).

Author Contributions: Conceptualization, N.S.-C. and V.O.-Q.; methodology, V.O.-Q. and C.G.-P.; software, V.O.-Q. and N.S.-C.; formal analysis, V.O.-Q. and N.S.-C.; investigation, V.O.-Q., C.G.-P. and N.S.-C.; data curation, V.O.-Q., N.S.-C. and C.G.-P. writing—original draft preparation, V.O.-Q., N.S.-C. and C.G.-P.; writing—review and editing, V.O.-Q., N.S.-C. and C.G.-P. All authors have read and agreed to the published version of the manuscript.

Funding: This research received no external funding.

Institutional Review Board Statement: The institution did not consider the evaluation of the ethics committee necessary as long as the voluntary participation of students was ensured, and the intervention did not affect the development of the contents registered in the degree's verified report.

Informed Consent Statement: Informed consent was obtained from all subjects involved in the study.

Data Availability Statement: The data presented in this study are available upon reasonable request to the authors.

Acknowledgments: Thanks are due to the teaching innovation project "#PensaCiencia rethinking school science. University-school rapprochement" for its support for this study.

Conflicts of Interest: The authors declare no conflict of interest.

References

1. Friant, M.C. Comercio justo, seguridad alimentaria y globalización: Construyendo sistemas alimentarios alternativos [Fair trade, food security and globalization: Building alternative food systems]. *Íconos: Rev. Cienc. Soc.* **2016**, *55*, 215–240.
2. United Nations. Transforming Our World: The 2030 Sustainable Development Agenda. Available online: https://unctad.org/meetings/es/SessionalDocuments/ares70d1_es.pdf (accessed on 6 March 2023).
3. Wezel, A.; Herren, B.G.; Kerr, R.B.; Barrios, E.; Gonçalves, A.L.R.; Sinclair, F. Agroecological principles and elements and their implications for transitioning to sustainable food systems. A review. *Rev. Agron. Sustain. Dev.* **2020**, *40*, 40. [CrossRef]
4. Jha, C.K.; Ghosh, R.K.; Saxena, S.; Singh, V.; Mosnier, A.; Guzman, K.P.; Lotze-Campen, H. Pathway to achieve a sustainable food and land-use transition in India. *Sustain. Sci.* **2023**, *18*, 457–468. [CrossRef]
5. Saffari, M.; Chen, J.S.; Wu, H.C.; Fung, X.C.; Chang, C.C.; Chang, Y.L.; Lin, C.Y. Effects of weight-related self-stigma and smartphone addiction on female university students' physical activity levels. *Int. J. Environ. Res. Public Health* **2022**, *19*, 2631. [CrossRef] [PubMed]
6. Goday, A.; Castañer, O.; Benaiges, D. Tenemos un problema con la obesidad [We have a problem with obesity]. *Nutr. Hosp.* **2019**, *36*, 3–4.
7. Sanahuja, J.A.; Tezanos, S. Del milenio a la sostenibilidad: Retos y perspectivas de la Agenda 2030 para el desarrollo sostenible. [From millennium to sustainability: Challenges and prospects of the 2030 Agenda for sustainable development]. *Polít. Soc.* **2017**, *54*, 533–555.
8. Lee, M.I. Agenda 2030 de desarrollo sostenible: Comunidad epistémica de los límites planetarios y cambio climático. [Agenda 2030 for sustainable development: Epistemic community of planetary boundaries and climate change]. *Rev. Opera* **2019**, *24*, 69–93.
9. Esteve-Guirao, P.; Banos-González, I.; Pérez, M.V. The Relationship between Sustainability and Food Consumption in Teacher Training. *J. Teach. Educ. Sustain.* **2022**, *24*, 34–47. [CrossRef]
10. Meek, D.; Tarlau, R. Critical food systems education (CFSE): Educating for food sovereignty. *Agroecol. Sustain. Food Syst.* **2016**, *40*, 237–260. [CrossRef]
11. Berezowitz, C.K.; Yoder, A.B.B.; Schoeller, D.A. School Gardens Enhance Academic Performance and Dietary Outcomes in Children. *J. Sch. Health* **2015**, *85*, 508–518. [CrossRef]
12. Melchor, M.; Rodríguez, J.; Díaz, M. Comportamiento de compra y consumo de productos dietéticos en los jóvenes universitarios. [Purchasing behaviour and consumption of dietary products among young university students]. *Pensam. Gestión* **2016**, *41*, 174–193.
13. Yang, Y.; Shields, G.S.; Wu, Q.; Liu, Y.; Chen, H.; Guo, C. Cognitive training on eating behaviour and weight loss: A meta-analysis and systematic review. *Obes. Rev.* **2019**, *20*, 1628–1641. [CrossRef]
14. De Piero, A.; Bassett, N.; Rossi, A.; Sammán, N. Tendencia en el consumo de alimentos de estudiantes universitarios. [Trends in food consumption of university students]. *Nutr. Hosp.* **2015**, *31*, 1824–1831.
15. Wang, C.; Horby, P.W.; Hayden, F.G.; Gao, G.F. A novel coronavirus outbreak of global health concern. *Lancet* **2020**, *395*, 470–473. [CrossRef]
16. Aguilar-Zavala, H.; Tovar-Vega, A.R.; Quintero-Hernández, Y.; Calixto Olalde, M.G.; Negrete-Díaz, J.V. Consumo de alimentos ultraprocesados y obesidad en estudiantes universitarios [Consumption of ultra-processed foods and obesity in university students]. *Revista Nthe* **2019**, *26*, 33–39.
17. Paul, A.A.; Kumar, S.; Kumar, V.; Sharma, R. Milk Analog: Plant based alternatives to conventional milk, production, potential and health concerns. *Crit. Rev. Food Sci. Nutr.* **2020**, *60*, 3005–3023. [CrossRef]
18. Cervera, F.; Serrano, R.; Vico, C.; Milla, M.; García, M.J. Hábitos alimentarios y evaluación nutricional en una población universitaria [Eating habits and nutritional assessment in a university population]. *Nutr. Hosp.* **2013**, *28*, 438–446.
19. Ruiz, E.; Del Pozo, S.; Valero, T.; Ávila, J.M.; Varela-Moreiras, G. *Estudio de Hábitos Alimentarios y Estilos de Vida de los Universitarios Españoles [Study of Eating Habits and Lifestyles of Spanish University Students]*; Fundación Española de la Nutrición: Madrid, Spain, 2012.

20. Jacques, P.J.; Jacques, J.R. Monocropping cultures into ruin: The loss of food varieties and cultural diversity. *Sustainability* **2012**, *4*, 2970–2997. [CrossRef]
21. Bárbara, R.; Ferreira-Pêgo, C. Changes in eating habits among displaced and non-displaced university students. *Int. J. Environ. Res. Public Health* **2020**, *17*, 5369. [CrossRef]
22. Leyva, D.A.; Pérez, A. Pérdida de las raíces culinarias por la transformación en la cultura alimentaria [Loss of culinary roots due to the transformation of food culture]. *Rev. Mex. Cienc. Agrícolas* **2015**, *6*, 867–881. [CrossRef]
23. Mann, A. Education for food sovereignty as transformative ethical practice. *Policy Futures Educ.* **2019**, *17*, 862–877. [CrossRef]
24. Garcés, V. La transición agrícola y alimentaria en la región mediterránea [Agricultural and food transition in the Mediterranean region]. *Quad. Mediterr.* **2017**, *25*, 260–262.
25. Rastoin, J.L. Accélérer la transition vers une alimentation durable par un changement de paradigme scientifique et économique et des politiques publiques innovantes [Accelerate the transition to sustainable food through a scientific and economic paradigm shift and innovative public policies]. *Syst. Aliment./Food Syst. Class. Garnier* **2018**, *3*, 17–27.
26. Rastoin, J.L. Anclaje territorial, un activo importante para una transición alimentaria responsable y sostenible [Territorial anchoring, an important asset for a responsible and sustainable food transition]. *Agroalimentaria* **2020**, *26*, 227–234. [CrossRef]
27. Mello-Théry, N.A. A Segurança Alimentar Via Sistemas Alimentares Circulares e as Controvérsias da Transição para os ODS [Food Security via Circular Food Systems and the Controversies of the Transition to the SDGs]. *Rev. Gestão Polít. Públicas* **2018**, *8*, 215–225. [CrossRef]
28. Giunta, I.; González, J.D. Inclusive and sustainable economic growth in the 2030 Agenda: A critical analysis from a food sovereignty and rights of nature perspective. *Iberoam. J. Dev. Stud.* **2020**, *9*, 146–176.
29. Wakefield, W.; Weinberg, A.E.; Pretti, E.; Merritt, E.G.; Trott, C. "When I act consciously, I can see a brighter world around me": Preservice teacher readiness to support transformative sustainability learning. *Environ. Educ. Res.* **2022**, *28*, 1672–1690. [CrossRef]
30. Albareda-Tiana, S.; Vidal-Raméntol, S.; Pujol-Valls, M.; Fernández-Morilla, M. Holistic approaches to develop sustainability and research competencies in pre-service teacher training. *Sustainability* **2018**, *10*, 3698. [CrossRef]
31. World Commission on Environment and Development (WCED). *Our Common Future*; University Press: Oxford, UK, 1987.
32. Ministry of Universities. Education Figures in Spain. In *Statistics and Indicators for the 2019–2020 Academic Year*; Ministry of Universities: Madrid, Spain, 2021.
33. Cohen, J. *Statistical Power Analysis for the Behavioral Sciences*; Erlbaum: Hillsdale, NJ, USA, 1988.
34. Johnston, J.; Szabo, M. Reflexivity and the Whole Foods Market consumer: The lived experience of shopping for change. *Agric. Hum. Values* **2011**, *28*, 303–319. [CrossRef]
35. Robinson, G.M. Globalization of agriculture. *Annu. Rev. Resour. Econ.* **2018**, *10*, 133–160. [CrossRef]
36. Balogh, J.M.; Jámbor, A. The environmental impacts of agricultural trade: A systematic literature review. *Sustainability* **2020**, *12*, 1152. [CrossRef]
37. Molinario, E.; Lorenzi, C.; Bartoccioni, F.; Perucchini, P.; Bobeth, S.; Colléony, A.; Bonaiuto, M. From childhood nature experiences to adult pro-environmental behaviors: An explanatory model of sustainable food consumption. *Environ. Educ. Res.* **2020**, *26*, 1137–1163. [CrossRef]
38. Gillespie, A.H.; Smith, L.E. Food Decision-Making Framework: Connecting sustainable food systems to health and well-being. *J. Hunger. Environ. Nutr.* **2008**, *3*, 328–346. [CrossRef]
39. Moschis, G.P.; Mathur, A.; Shannon, R. Toward achieving sustainable food consumption: Insights from the life course paradigm. *Sustainability* **2020**, *12*, 5359. [CrossRef]
40. Horká, H.; Hromádka, Z. Active Leisure for health in context of environmental care. *Sch. Health* **2010**, *21*, 275–289.
41. Magnusson, M.K.; Arvola, A.; Hursti, U.K.K.; Åberg, L.; Sjödén, P.O. Choice of organic foods is related to perceived consequences for human health and to environmentally friendly behaviour. *Appetite* **2003**, *40*, 109–117. [CrossRef]
42. Raptou, E.; Manolas, E. Consumption Patterns and Public Attitudes toward Organic Foods: The Role of Climate Change Awareness. In *Climate Change in the Mediterranean and Middle Eastern Region*; Springer: Cham, Switzerland, 2022; pp. 365–393.
43. Wittman, H. Food sovereignty: A new rights framework for food and nature? *Environ. Soc.* **2011**, *2*, 87–105.
44. Portman, A. Food sovereignty and gender justice. *J. Agric. Environ. Ethics* **2018**, *31*, 455–466. [CrossRef]
45. Scott, W. Sustainable schools and the exercising of responsible citizenship—A review essay. *Environ. Educ. Res.* **2011**, *17*, 409–423. [CrossRef]
46. Anderson, A. Advancing global citizenship education through global competence and critical literacy: Innovative practices for inclusive childhood education. *Sage Open* **2019**, *9*, 2158244019826000. [CrossRef]
47. Warner, A.; Callaghan, E.; de Vreede, C. Promoting sustainable food and food citizenship through an adult education leisure experience. *Leisure/Loisir* **2013**, *37*, 337–360. [CrossRef]
48. Nousheen, A.; Zai, S.A.Y.; Waseem, M.; Khan, S.A. Education for sustainable development (ESD): Effects of sustainability education on pre-service teachers' attitude towards sustainable development (SD). *J. Clean. Prod.* **2020**, *250*, 119537. [CrossRef]

49. Eugenio-Gozalbo, M.; Ramos Truchero, G.; Valles Rapp, C. University gardens: Identifying learning dimensions as perceived by future teachers. *Ensen. Cienc.* **2019**, *37*, 111–127.
50. Afacan, O. Investigation of pre-service science teachers' behavior towards sustainable environmental education. *Int. Electron. J. Environ. Educ.* **2020**, *10*, 110–121.

MDPI AG
Grosspeteranlage 5
4052 Basel
Switzerland
Tel.: +41 61 683 77 34

Sustainability Editorial Office
E-mail: sustainability@mdpi.com
www.mdpi.com/journal/sustainability

www.ingramcontent.com/pod-product-compliance
Lightning Source LLC
Chambersburg PA
CBHW061509210926
51850CB00008B/1858
9783725853878